PETROLEUM REFINING PROCESSES

PETROLEUM REFINING PROCESSES

Edited by
Rakesh Rathi

2007

SBS Publishers & Distributors Pvt. Ltd.
New Delhi

ISBN 10 : 81-89741-57-8
ISBN 13 : 978-81-89741-57-0

Indian Price - Rs 1695
Foreign Price - USD 85

First Published in India in 2007

Published by:
SBS PUBLISHERS & DISTRIBUTORS PVT. LTD.
2/9, Ground Floor, Ansari Road, Darya Ganj,
New Delhi - 110002, INDIA
Tel: 23289119, 41563911
Email: mail@sbspublishers.com

Printed in India by:
Syndicate Binders, Noida

Preface

Petroleum Refining Processes is an invaluable reference for chemical, petroleum, and process engineers; chemists; chemical technologists; petroleum analysts; and students in these disciplines.

This book will be a welcome addition to the reference collection of large academic libraries at Universities with programs in petroleum engineering.

Apart from fundamentals and various Refinery Operations book also contains various UOP Processes, Technology Developments and description of production equipments.

By going through 62 FAQ's in Chapter-9 a layman can understand basic aspects of Petroleum Refinery.

Er. Rakesh Rathi

Contents

Various UOP Processes

Chapter 1

Introduction

Petroleum will remain an essential part of our energy mix for decades, if not centuries. At recent production rates, proven world petroleum reserves could last more than 40 years. With new discoveries and development of unconventional sources of petroleum such as the tar sands of western Canada, we could have enough for hundreds more years.

The petroleum industry is key to that progress. Refineries are making cleaner fuels and making them differently, emitting less and recycling more of the chemicals used in the refining process. Oil spills occur less often. Underground storage tanks and piping are checked regularly for leaks. Special equipment keeps gasoline vapors from entering the air when fuel is transferred from one tank to another. Service stations collect used oil from motorists for recycling and reuse.

While the latest technological advances could reduce the environmental impact of our mobile way of life still more, they have other drawbacks. Battery-powered cars have no emissions of their own, but are impractical for most people because their range is so limited. Now, automobile makers are introducing "hybrids" that use electric power for short hops and gasoline for highway driving. They can go as far as conventional cars before refueling and their cost is becoming competitive.

Fuel cells are also becoming more affordable. Like batteries, they rely on chemical reactions rather than combustion. In theory, they could produce no emissions except water vapor. But that would require using liquid hydrogen as fuel, which would in turn require cooling fuel tanks to hundreds of degrees below zero (at higher temperatures hydrogen turns into a gas and boils away).

Car makers are experimenting with fuel cells that take a more practical approach: extracting hydrogen from gasoline or methanol, an alcohol made from natural gas. Emissions are greater than they would be with pure hydrogen, but a big improvement over today.

Eventually, petroleum will become scarcer, providing further impetus to use it wisely. When that time comes, gasoline and other petroleum-based fuels could become unaffordable for everyday use. New petrochemical products could become more costly than recycled materials–and the plastic in today's landfills could become a valuable commodity.

As oil a half century ago ended the reign of King Coal–the driving force of the Industrial Revolution–something will someday replace petroleum as the driving force of progress. But for the foreseeable future, that distinction belongs to petroleum. And for good reason: no alternative is at once as cheap, abundant, and clean.

We may be living at the peak of the era historians will one day call "the age of petroleum." Certainly, the name is apt. Petroleum affects nearly everything people do the world over.

The petroleum industry is organized into four broad sectors: exploration and production of crude oil and natural gas; transport; refining; and marketing and distribution.

The petroleum industry began with the successful drilling of the first commercial oil well in 1859, and the opening of the first refinery two years later to process the crude into kerosene. The evolution of petroleum refining from simple distillation to today's sophisticated processes has created a need for health and safety management procedures and safe work practices. To those unfamiliar with the industry, petroleum refineries may appear to be complex and confusing places. Refining is the processing of one complex mixture of hydrocarbons into a number of other complex mixtures of hydrocarbons. The safe and orderly processing of crude oil into flammable gases and liquids at high temperatures and pressures using vessels, equipment, and piping subjected to stress and corrosion requires considerable knowledge, control, and expertise.

Petroleum refining has evolved continuously in response to changing consumer demand for better and different products. The original requirement was to produce kerosene as a cheaper and better source of light than whale oil. The development of the internal combustion engine led to the production of gasoline and diesel fuels. The evolution of the airplane created a need first for high-octane aviation gasoline and then for jet fuel, a sophisticated form of the original product, kerosene. Present-day refineries produce a variety of products including many required as feedstock's for the petrochemical industry.

Distillation Processes

The first refinery, opened in 1861, produced kerosene by simple atmospheric distillation. Its by-products included tar and naphtha. It was soon discovered that high-quality lubricating oils could be produced by distilling petroleum under vacuum. However, for the next 30 years kerosene was the product consumers wanted. Two significant events changed this situation:

1. invention of the electric light decreased the demand for kerosene, and
2. invention of the internal combustion engine created a demand for diesel fuel and gasoline (naphtha).

Thermal Cracking Processes

With the advent of mass production and World War I, the number of gasoline-powered vehicles increased dramatically and the demand for gasoline grew accordingly. However, distillation processes produced only a certain amount of gasoline from crude oil. In 1913, the thermal cracking process was developed, which subjected heavy fuels to both pressure and intense heat, physically breaking the large molecules into smaller ones to produce additional gasoline and distillate fuels. Visbreaking, another form of thermal cracking, was developed in the late 1930s to produce more desirable and valuable products.

Catalytic Processes

Higher-compression gasoline engines required higher-octane gasoline with better antiknock characteristics. The introduction of catalytic cracking and polymerization processes in the mid- to late 1930s met the demand by providing improved gasoline yields and higher octane numbers.

Alkylation, another catalytic process developed in the early 1940s, produced more high-octane aviation gasoline and petrochemical feedstock's for explosives and synthetic rubber. Subsequently, catalytic isomerization was developed to convert hydrocarbons to produce increased quantities of alkylation feedstock's. Improved catalysts and process methods such as hydrocracking and reforming were developed throughout the 1960s to increase gasoline yields and improve antiknock characteristics. These catalytic processes also produced hydrocarbon molecules with a double bond (alkenes) and formed the basis of the modern petrochemical industry.

Treatment Processes

Throughout the history of refining, various treatment methods have been used to remove nonhydrocarbons, impurities, and other constituents that adversely affect the properties of finished products or reduce the efficiency of the conversion processes. Treating can involve chemical reaction and/or physical separation. Typical examples of treating are chemical sweetening, acid treating, clay contacting, caustic washing, hydrotreating, drying, solvent extraction, and solvent dewaxing. Sweetening compounds and acids desulfurize crude oil before processing and treat products during and after processing.

Following the Second World War, various reforming processes improved gasoline quality and yield and produced higher-quality products. Some of these involved the use of catalysts and/or hydrogen to change molecules and remove sulfur. A number of the more commonly used treating and reforming processes are described in this chapter of the manual.

HISTORY OF REFINING

The oil and natural gas we use today are the remains of microscopic animals and plants–zooplankton and phytoplankton–that floated on the surface of lakes and seas millions of years ago. The plankton died and drifted downward, mixing with sediments of mud and silt.

As the centuries passed, layers of sediment covered the decaying animals and plants, burying them deeper and deeper. The pressure and temperature rose, transforming the plankton into hydrocarbons–the compoounds of hydrogen and carbon we call petroleum.

Petroleum naturally seeps to the Earth's surface along fault lines and cracks in rocks, where it gathers in pools as tar, asphalt or bitumen. From this property it gets its name: the Latin Petra, meaning "rock," and oleum, meaning "oil."

Natural gas emerges from the Earth as a colorless, odorless vapor (a chemical called mercaptan provides its characteristic scent). Crude oil ranges in color from almost clear to green, amber, brown or black. It may flow like water or creep like molasses. It is described as "sweet" or "sour depending on the presence or absence of sulfur and other impurities.

Mankind's use of petroleum is as old as recorded history. Ancient cultures found the stickiness of crude oil useful for binding things together, and its water repellent properties useful for keeping water from entering unwelcome places. Five thousand years ago, the Summerians used asphalt to inlay mosaics in walls and floors. The Mesopotamians used bitumen to line water canals, seal joints in wooden boats, and build roads. The Egyptians greased chariots with pitch and embalmed mummies with asphalt.

By 1500 B.C., state-of-the-art lighting consisted of a censer or firepan filled with liquid oil of just the right volatility, so it burned slowly without flaming dangerously or exploding. Eventually, the wick oil lamp replaced the firepan. It drew a flammale liquid up a wick by capillary action, like today's kerosine lanterns.

The Chinese were the first to discover underground deposits of oil, in salt wells. In 600 B.C., Confucius wrote of wells 100 feet deep yielding water and natural gas along the Tibetan border. To transport oil and natural gas from well sites to where they were needed, the Chinese built extensive pipelines of bamboo.

Year	*Process name*	*Purpose*	*By-products, etc.*
1862	Atmospheric distillation	Produce kerosene	Naphtha, tar, etc.
1870	Vacuum distillation	Lubricants (original) Cracking feedstocks (1930's)	Asphalt, residual coker feedstocks
1913	Thermal cracking	Increase gasoline	Residual, bunker fuel
1916	Sweetening	reduce sulfur & odor	Sulfur
1930	Thermal reforming	Improve octane number	Residual
1932	Hydrogenation	Remove sulfur	Sulfur
1932	Coking	Produce gasoline basestocks	Coke
1933	Solvent extraction	Improve lubricant viscosity index	Aromatics
1935	Solvent dewaxing	Improve pour point	Waxes
1935	Cat. polymerization	Improve gasoline yield & octane number	Petrochemical feedstocks
1937	Catalytic cracking	Higher octane gasoline	Petrochemical feedstocks
1939	Visbreaking	reduce viscosity	Increased distillate,tar
1940	Alkylation	Increase gasoline octane & yield	High-octane aviation gasoline
1940	Isomerization	Produce alkylation feedstock	Naphtha
1942	Fluid catalytic cracking	Increase gasoline yield & octane	Petrochemical feedstocks
1950	Deasphalting	Increase cracking feedstock	Asphalt
1952	Catalytic reforming	Convert low-quality naphtha	Aromatics
1954	Hydrodesulfurization	Remove sulfur	Sulfur
1956	Inhibitor sweetening	Remove mercaptan	Disulfides
1957	Catalytic isomerization	Convert to molecules with high octane number	Alkylation feedstocks
1960	Hydrocracking	Improve quality and reduce sulfur	Alkylation feedstocks
1974	Catalytic dewaxing	Improve pour point	Wax
1975	Residual hydrocracking	Increase gasoline yield from residual	Heavy residuals

The Romans used flaming containers of oil as war weapons. Cicero, the Roman orator, is said to have used a lamp that burned oil from surface deposits. By A.D. 1500—about the time Leonardo da Vinci drew an imaginary drilling machine—the Chinese were digging actual wells more than 2,000 feet deep.

Oil became an important commercial commodity after Peter the Great of Russia conquered the Khanate of Baku in Persia in 1723. Peter allowed private entrepreneurs to seek, refine, and sell oil from the region, on the condition that they give a per centage of their profits to the Crown. Naphtha, which occurred naturally in Baku, was among the products produced by these entrepreneurs. It became a key ingredient of the first waterproof raincoat: the British macintosh. (Invented in 1823, it was coated with rubber dissolved in naphtha.)

The modern petroleum industry was born in nineteenth century America. In the early 1800s, entrepreneurs built dams to make oil float to the water's surface in an area of western Pennsylvania called Oil Creek. They threw blankets in the water, let them soak up the oil, and retrieved the oil by wringing out the blankets. Then they sold the oil for $2 a gallon.

Chapter 2

Fundamentals of Petroleum Refinery

WHAT IS A REFINERY?

A refinery is a factory. Just as a paper mill turns lumber into paper, a refinery takes crude oil and turns it into gasoline and hundreds of other useful products. A typical refinery costs billions of dollars to build and millions more to maintain A refinery runs twenty-four hours a day, 365 days a year and requires a large number of employees to run. A refinery can occupy as much land as several hundred football fields. Workers ride bicycles to move from place to place inside the complex.

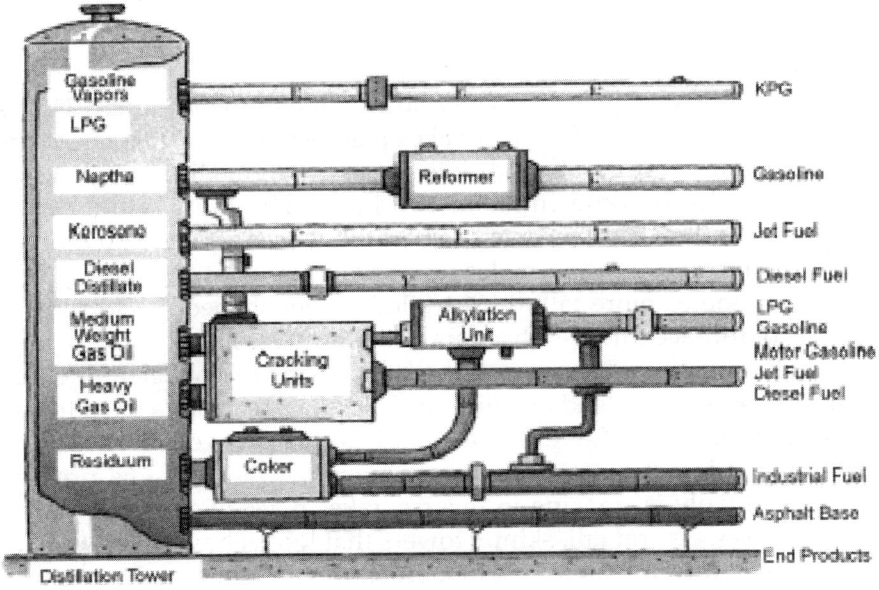

Figure 2.1: This simplified drawing shows many of a refinery's most important process.

The world needs gasoline and petroleum products to move merchandise and people; help make plastics; and do many other things. Today, some refineries turn more than half of every 42-gallon barrel of crude oil into gasoline. How does this transformation take place? Essentially, refining breaks crude oil down into its various components, which then are selectively reconfigured into new products. All refineries perform three basic steps: separation, conversion, and treatment.

Separation

Heavy petroleum fractions are on the bottom, light fractions are on the top. This allows the separation of the various petrochemicals. Modern separation involves piping oil through hot furnaces. The resulting liquids and vapors are discharged into distillation towers.

Inside the towers, the liquids and vapors separate into components or fractions according to weight and boiling point. The lightest fractions, including gasoline and liquid petroleum gas (LPG), vaporize and rise to the top of the tower, where they condense back to liquids. Medium weight liquids, including kerosene and diesel oil distillates, stay in the middle. (Heavier liquids, called gas oils, separate lower down, while the heaviest fractions with the highest boiling points settle at the bottom.)

Conversion

The finishing touches occur during the final treatment. To make gasoline, Cracking and rearranging molecules adds value to the products. This is where refining's fanciest footwork takes place—where fractions from the distillation towers are transformed into streams (intermediate components) that eventually become finished products. The most widely used conversion method is called cracking because it uses heat and pressure to "crack" heavy hydrocarbon molecules into lighter ones. A cracking unit consists of one or more tall, thick-walled, bullet-shaped reactors and a network of furnaces, heat exchangers and other vessels.

Cracking and coking are not the only forms of conversion. Other refinery processes, instead of splitting molecules, rearrange them to add value. Alkylation's, for example, makes gasoline components by combining some of the gaseous byproducts of cracking. The process, which essentially is cracking in reverse, takes place in a series of large, horizontal vessels and tall, skinny towers that loom above other refinery structures. Reforming uses heat, moderate pressure and catalysts to turn naphtha, a light, relatively low-value fraction, into high-octane gasoline components.

Treatment

The finishing touches occur during the final treatment. To make gasoline, refinery technicians carefully combine a variety of streams from the processing units. Among the variables that determine the blend are octane level, vapor pressure ratings and special considerations, such as whether the gasoline will be used at high altitudes.

Storage

Both the incoming crude oil and the outgoing final products need to be stored. These liquids are stored in large tanks on a tank farm. Pipelines carry the final products from the tank farm near the refinery to other tanks all across the country.

All of these activities are required to make the gasoline that powers our cars, the diesel fuel that brings our food to market, and the jet fuel that flies our planes. These provide us with the energy we need to get from place to place quickly and comfortably.

Imagine, for just a moment, how your lifestyle would be changed if you had no access to transportation via automobiles, trucks, motorcycles, planes, or any other equipment powered by internal combustion engines. How would you commute to work or to shop for consumer goods? Would you walk, ride a bicycle?or ride a horse? Imagine how the selection of goods available to you at local retailers would be negatively impacted by similar constraints. For example, you would have limited or no access to fresh fruits and vegetables except during the local growing season. The costs of goods produced other than locally would be appreciably higher. The lifestyle that we take for granted is made possible through the usage of the fuels (i.e., motor gasolines, diesel fuels, jet fuels, aviation gasolines, heating oils, heavy fuel oils, etc.) and specialty products (i.e., lubricants, waxes, asphalts and solvents), as well as the petrochemical industry feedstocks produced by the petroleum refining industry from crude oils. The modern petroleum refinery is a very sophisticated, capital-intensive industrial complex. To the casual observer, the typical petroleum refinery appears to be a maze of piping, with scattered, process units containing very tall equipment, and massive storage tanks too numerous to count taking up most of the refinery real estate.

Economy of scale has a very significant relevance to the economics of refinery operations. More than two-thirds of the 146 petroleum refineries in the United States have crude oil processing capacities exceeding 50,000 barrels/day (b/d, one barrel is 42 U.S. standard gallons), with the largest exceeding 500,000 b/d! In most cases, due to logistics economics, the primary consideration in deciding where to build a petroleum refinery is proximity to the product end-users or a product pipeline hub rather than the proximity to the area of crude oil production. This enables crude oils to be received by a few large pipelines and minimizes the length of the scores of product pipelines emanating from the refinery. The larger petroleum refineries are located near coastlines or navigable rivers, enabling crude oils to be received and product to be shipped via ships and barges.

No two major petroleum refineries are identical. This uniqueness stems predominantly from the geographical location of the refinery, which in conjunction with the refinery process configuration, determines how efficiently crude oils can be delivered to the refinery and products to the markets served by the refinery. An overview of the fuels processes typically incorporated into a modern petroleum refinery is provided in this short course. A simplified flow diagram of a typical fuels refinery is shown below.

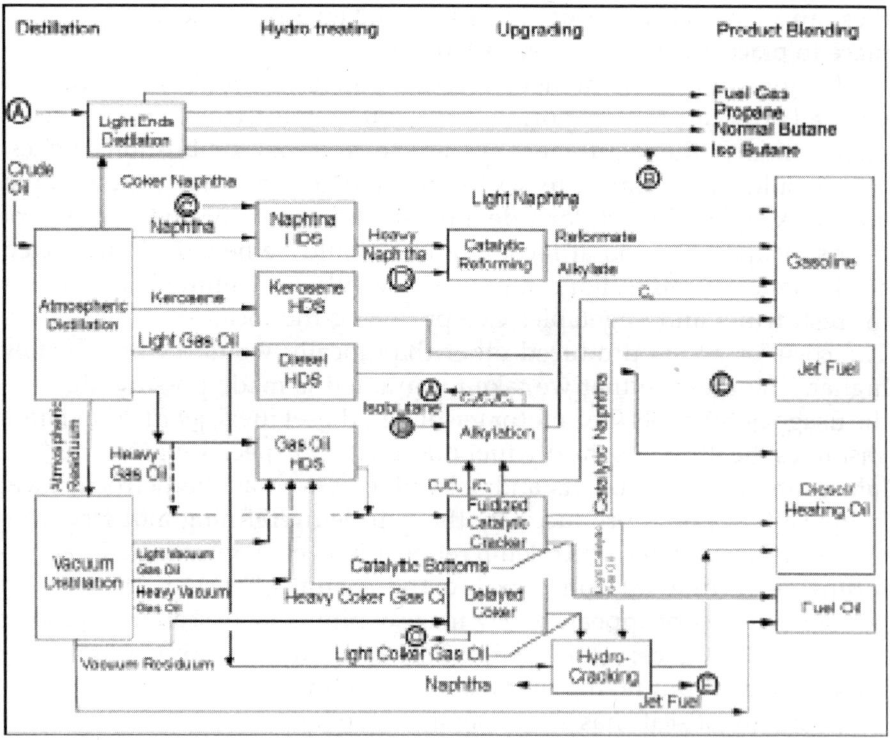

Figure 2.2: Simplified Refinery Process Flow Diagram

The key function of most refinery processes is to effect chemical reactions on the hydrocarbons being processed. Generally, the reactions are carried out at elevated temperatures in the 600-1,000°F range depending on the process, and in most cases at elevated pressures, from 200 pounds per square inch (psi) to as high as 3,000 psi. Those processes involving reactions will typically incorporate a fractionator to distill the reactor effluent into various product streams. The primary function of some refinery processes, such as Crude Distillation, for example, is fractionation only.

A general description of a refinery processing unit is that the unit feedstocks are pumped and/or compressed up to the required pressures, preheated via heat exchange with reactor effluent and/or product streams, and finally heated via heat exchange in a direct-fired furnace before entering the reactor(s) or distillation tower (if no reaction is intended). The reactor effluent is then cooled via heat exchange with unit feedstocks, fractionated into the desired product streams via distillation, which are then further cooled via heat exchange with unit feedstocks. As the individual refinery processes are described in the subsequent sections, simplified process flow diagrams will be provided to illustrate the specific process flow sequence for the applicable process.

The process units that constitute a refinery are continuous operations that run 24 hours per day, 365 days per year. Refinery process operations personnel typically work 12-hour shifts around-the-clock to operate the process equipment.

A process unit run-length is usually 1 to 5 years before it is shutdown for several weeks of turnaround maintenance. As a result, a refinery process unit's service factor (i.e., availability) is well above 90 per cent. The turnaround timing is scheduled based upon catalyst activity requiring a catalyst change out (if applicable) or based on predictive maintenance techniques applied to the critical process equipment.

Crude Oils

Crude oils are complex mixtures containing many different hydrocarbon compounds that vary in appearance and composition from one oil field to another. Crude oils range in consistency from water to tar-like solids, and in color from clear to black. An average crude oil contains about 84 per cent carbon, 14 per cent hydrogen, 1-3 per cent sulfur, and less than 1 per cent each of nitrogen, oxygen, metals, and salts. Crude oils are generally classified as paraffinic, naphthenic, or aromatic, based on the predominant proportion of similar hydrocarbon molecules. Mixed-base crudes have varying amounts of each type of hydrocarbon. Refinery crude base stocks usually consist of mixtures of two or more different crude oils.

Relatively simple crude-oil assays are used to classify crude oils as paraffinic, naphthenic, aromatic, or mixed. One assay method (United States Bureau of Mines) is based on distillation, and another method (UOP K factor) is based on gravity and boiling points. More comprehensive crude assays determine the value of the crude (i.e., its yield and quality of useful products) and processing parameters. Crude oils are usually grouped according to yield structure.

Table 2.1: Typical Approximate Characteristics and Properties and Gasoline Potential of Various Crudes (Representative Average Numbers)

Crude source	*Paraffins (% vol)*	*Aromatics (% vol)*	*Naphthenes (% vol)*	*Sulfur (% wt)*	*API gravity (approx.)*	*Napht. yield (% vol)*	*Octane no (typical)*
Nigerian -Light	37	9	54	0.2	36	28	60
Saudi -Light	63	19	18	2	34	22	40
Saudi -Heavy	60	15	25	2.1	28	23	35
Venezuela -Heavy	35	12	53	2.3	30	2	60
Venezuela -Light	52	14	34	1.5	24	18	50
USA -Midcont. Sweet	—	—	—	0.4	40	—	—
USA -W. Texas Sour	46	22	32	1.9	32	33	55

In almost all cases, crude oils have no inherent value without a petroleum refining industry to process them into usable specification products. The rare exceptions are very light crude oils or condensates that may be burned as low quality fuels in locations with very loose or non-existent environmental restrictions. However, as a result of the refining industry, crude oil is one of the most significant categories among the commodities in the world. Petroleum crude oils are composed of numerous hydrocarbons. Hydrocarbons are chemical compounds made up of predominantly carbon and hydrogen. Hydrocarbons found in crude oils generally also contain the elements sulfur and nitrogen. Many crude oils also contain absorbed levels of the toxic gas hydrogen sulfide (H_2S). Additionally, crude oils may contain trace amounts of metals such as nickel and vanadium, as well as salts. Most of the nonhydrogen, non-carbon elements found in crude oils are undesirable and are removed from the hydrocarbons in total or in part during refinery processing.

One of the key attributes for characterizing the hydrocarbons composing crude oils is by "boiling point" This attribute is determined through laboratory test methods by measuring the temperature at which the components of the crude oil will evaporate at a given pressure (typically atmospheric pressure unless stated to be a different pressure basis). A "True Boiling Point" (TBP) curve is developed as a part of the Crude Assay to plot or tabulate the liquid volume per cent of the crude oil that evaporates relative to temperature at atmospheric pressure. The numerous hydrocarbon components constituting crude oil will generally have individual boiling points ranging from less than 60°F to greater than 1200°F.

Crude oils are named and grouped into broad categories typically based on the geographic location of origin, along with the level of sulfur contained in the crude and/or density of the crude oil.

For example, West Texas Intermediate (WTI) and West Texas Sour (WTS) are two families of crude oils produced in the oilfields of West Texas—WTI is a light, sweet (i.e., low levels of sulfur relative to high sulfur "sour" crude oils) crude oil when compared to the heavier, higher sulfur content WTS. Higher sulfur crude oils are more corrosive than lower sulfur crude oils. To ensure a reasonable life expectancy for equipment processing the higher sulfur crude oils, refiners specify that such equipment be built from more expensive alloys with a higher corrosion resistance.

The American Petroleum Institute (API) has developed the term Degrees API Gravity (°API) which is widely used as another general characterization of the density of crude oils. The relationship is as follows:

°API = (141.5/Specific Gravity at 60 degrees Fahrenheit)—131.5

"Specific Gravity at 60 degrees Fahrenheit" is the density of the crude oil measured at 60°F divided by the density of water at 60°F.

Therefore, when comparing two crude oils, the higher density crude (i.e., the one with the highest specific gravity) will have a correspondingly lower °API. For example, the 34.5°API West African crude oil Bonny Light is heavier than the 40.4°API North Sea crude oil Forties. Crude oil assays are a compilation of the results of numerous laboratory analyses conducted on the whole crude oil or fractions of the crude oil. These tests characterize a crude oil and enable refiners to evaluate the feasibility and economics of processing a given crude in their refinery or competitor's refineries.

Crude oil assays vary widely in the degree of detail. However, qualities of interest with respect to the whole crude, as well as various fractions of the crude are presented in the assay. Refer to Appendix 2 for examples of crude oil assays.

Unfortunately, the boiling point curve for most crude oils does not match that of the products demanded. The United States demand for the lighter petroleum products boiling below 650°F (motor gasolines, diesel fuels, and jet fuels) represents roughly 85 per cent by volume of the crude oil processed in the U.S. However, an average crude mix has a much lower volumetric yield of hydrocarbons in the desired boiling range. Refinery process operations split the crude oils into fractions by boiling range and then convert some of the unwanted fractions (i.e., low demand and low value) into higher value, specification products.

Basics of Hydrocarbon Chemistry

Crude oil is a mixture of hydrocarbon molecules, which are organic compounds of carbon and hydrogen atoms that may include from one to 60 carbon atoms. The properties of hydrocarbons depend on the number and arrangement of the carbon and hydrogen atoms in the molecules. The simplest hydrocarbon molecule is one carbon atom linked with four hydrogen atoms: methane. All other variations of petroleum hydrocarbons evolve from this molecule.

Hydrocarbons containing up to four carbon atoms are usually gases; those with five to 19 carbon atoms are usually liquids; and those with 20 or more are solids. The refining process uses chemicals, catalysts, heat, and pressure to separate and combine the basic types of hydrocarbon molecules naturally found in crude oil into groups of similar molecules. The refining process also rearranges their structures and bonding patterns into different hydrocarbon molecules and compounds. Therefore it is the type of hydrocarbon, (paraffinic, naphthenic, or aromatic) rather than its specific chemical compounds that is significant in the refining process.

Three principal groups or series of hydrocarbon compounds that occur Naturally in crude oil:

Paraffins

The paraffinic series of hydrocarbon compounds found in crude oil have the general formula C_nH_{2n+2} and can be either straight chains (normal) or branched chains (isomers) of carbon atoms. The lighter, straight-chain paraffin molecules are found in gases and paraffin waxes. Examples of straight-chain molecules are methane, ethane, propane, and butane (gases containing from one to four carbon atoms), and pentane and hexane (liquids with five to six carbon atoms). The branched-chain (isomer) paraffins are usually found in heavier fractions of crude oil and have higher octane numbers than normal paraffins. These compounds are saturated hydrocarbons, with all carbon bonds satisfied, that is, the hydrocarbon chain carries the full complement of hydrogen atoms.

Aromatics

Aromatics are unsaturated ring-type (cyclic) compounds which react readily because they have carbon atoms that are deficient in hydrogen. All aromatics have at least one benzene ring (a single-ring compound characterized by three double bonds alternating with three single bonds between six carbon atoms) as part of their molecular structure. Naphthalenes are fused double-ring aromatic compounds. The most complex aromatics, polynuclears (three or more fused aromatic rings), are found in heavier fractions of crude oil.

Naphthenes

Naphthenes are saturated hydrocarbon groupings with the general formula C(n) H(2n), arranged in the form of closed rings (cyclic) and found in all fractions of crude oil except the very lightest. Single-ring naphthenes (monocycloparaffins) with five and six carbon atoms predominate, with two-ring naphthenes (dicycloparaffins) found in the heavier ends of naphtha.

Other Hydrocarbons

Alkenes

Alkenes are mono-olefins with the general formula C(n) H(2n) and contain only one carbon-carbon double bond in the chain. The simplest alkene is ethylene, with two carbon atoms joined by a double bond and four hydrogen atoms. Olefins are usually formed by thermal and catalytic cracking and rarely occur naturally in unprocessed crude oil.

Dienes and Alkynes

Dienes, also known as diolefins, have two carbon-carbon double bonds. The alkynes, another class of unsaturated hydrocarbons, have a carbon-carbon triple bond within the molecule. Both these series of hydrocarbons have the general formula C(n) H(2n-2). Diolefins such as 1,2-butadiene and 1,3-butadiene, and alkynes such as acetylene occur in C(5) and lighter fractions from cracking. The olefins, diolefins, and alkynes are said to be unsaturated because they contain less than the amount of hydrogen necessary to saturate all the valences of the carbon atoms. These compounds are more reactive than paraffins or naphthenes and readily combine with other elements such as hydrogen, chlorine, and bromine.

Nonhydrocarbons

Sulfur Compounds

Sulfur may be present in crude oil as hydrogen sulfide ($H(2) S$), as compounds (e.g., mercaptans, sulfides, disulfides, thiophenes, etc.), or as elemental sulfur. Each crude oil has different amounts and types of sulfur compounds, but as a rule the proportion, stability, and complexity of the compounds are greater in heavier crude-oil fractions. Hydrogen sulfide is a primary contributor to corrosion in refinery processing units. Other corrosive substances are elemental sulfur and mercaptans. Moreover, the corrosive sulfur compounds have an obnoxious odor.

Pyrophoric iron sulfide results from the corrosive action of sulfur compounds on the iron and steel used in refinery process equipment, piping, and tanks. The combustion of petroleum products containing sulfur compounds produces undesirables such as sulfuric acid and sulfur dioxide. Catalytic hydrotreating processes such as hydrodesulfurization remove sulfur compounds from refinery product streams. Sweetening processes either remove the obnoxious sulfur compounds or convert them to odorless disulfides, as in the case of mercaptans.

Oxygen Compounds

Oxygen compounds such as phenols, ketones, and carboxylic acids occur in crude oils in varying amounts.

Nitrogen Compounds

Nitrogen is found in lighter fractions of crude oil as basic compounds, and more often in heavier fractions of crude oil as nonbasic compounds that may also include trace metals such as copper, vanadium, and/or nickel. Nitrogen oxides can form in process furnaces. The decomposition of nitrogen compounds in catalytic cracking and hydrocracking processes forms ammonia and cyanides that can cause corrosion.

Trace Metals

Metals including nickel, iron, and vanadium are often found in crude oils in small quantities and are removed during the refining process. Burning heavy fuel oils in refinery furnaces and boilers can leave deposits of vanadium oxide and nickel oxide in furnace boxes, ducts, and tubes. It is also desirable to remove trace amounts of arsenic, vanadium, and nickel prior to processing as they can poison certain catalysts.

Salts

Crude oils often contain inorganic salts such as sodium chloride, magnesium chloride, and calcium chloride in suspension or dissolved in entrained water (brine). These salts must be removed or neutralized before processing to prevent catalyst poisoning, equipment corrosion, and fouling. Salt corrosion is caused by the hydrolysis of some metal chlorides to hydrogen chloride (HCl) and the subsequent formation of hydrochloric acid when crude is heated. Hydrogen chloride may also combine with ammonia to form ammonium chloride (NH_4Cl), which causes fouling and corrosion.

Carbon Dioxide

Carbon dioxide may result from the decomposition of bicarbonates present in or added to crude, or from steam used in the distillation process.

Naphthenic Acids

Some crude oils contain naphthenic (organic) acids, which may become corrosive at temperatures above 450 degrees F when the acid value of the crude is above a certain level.

Major Refinery Products

Gasoline

The most important refinery product is motor gasoline, a blend of hydrocarbons with boiling ranges from ambient temperatures to about 400° F. The important qualities for gasoline are octane number (antiknock), volatility (starting and vapor lock), and vapor pressure (environmental control). Additives are often used to enhance performance and provide protection against oxidation and rust formation.

Kerosene

Kerosene is a refined middle-distillate petroleum product that finds considerable use as a jet fuel and around the world in cooking and space heating. When used as a jet fuel, some of the critical qualities are freeze point, flash point, and smoke point. Commercial jet fuel has a boiling range of about 375-525 degrees F, and military jet fuel 130-550 degrees F.

Kerosene, with less-critical specifications, is used for lighting, heating, solvents, and blending into diesel fuel.

Liquefied Petroleum Gas (LPG)

LPG, which consists principally of propane and butane, is produced for use as fuel and is an intermediate material in the manufacture of petrochemicals. The important specifications for proper performance include vapor pressure and control of contaminants.

Distillate Fuels

Diesel fuels and domestic heating oils have boiling ranges of about 400-700 degrees F. The desirable qualities required for distillate fuels include controlled flash and pour points, clean burning, no deposit formation in storage tanks, and a proper diesel fuel cetane rating for good starting and combustion.

Residual Fuels

Many marine vessels, power plants, commercial buildings and industrial facilities use residual fuels or combinations of residual and distillate fuels for heating and processing. The two most critical specifications of residual fuels are viscosity and low sulfur content for environmental control.

Coke and Asphalt

Coke is almost pure carbon with a variety of uses from electrodes to charcoal briquets. Asphalt, used for roads and roofing materials, must be inert to most chemicals and weather conditions.

Solvents

A variety of products, whose boiling points and hydrocarbon composition are closely controlled, are produced for use as solvents. These include benzene, toluene, and xylene.

Petrochemicals

Many products derived from crude oil refining such as ethylene, propylene, butylene, and isobutylene are primarily intended for use as petrochemical feedstocks in the production of plastics, synthetic fibers, synthetic rubbers, and other products.

Lubricants

Special refining processes produce lubricating oil base stocks. Additives such as demulsifiers, antioxidants, and viscosity improvers are blended into the base stocks to provide the characteristics required for motor oils, industrial greases, lubricants, and cutting oils. The most critical quality for lubricating-oil base stock is a high viscosity index, which provides for greater consistency under varying temperatures.

Common Refinery Chemicals

Leaded Gasoline Additives

Tetraethyl lead (TEL) and tetramethyl lead (TML) are additives formerly used to improve gasoline octane ratings but are no longer in common use except in aviation gasoline.

Oxygenates

Ethyl tertiary butyl ether (ETBE), methyl tertiary butyl ether (MTBE), tertiary amyl methyl ether (TAME), and other oxygenates improve gasoline octane ratings and reduce carbon monoxide emissions.

Caustics

Caustics are added to desalting water to neutralize acids and reduce corrosion. They are also added to desalted crude in order to reduce the amount of corrosive chlorides in the tower overheads. They are used in some refinery treating processes to remove contaminants from hydrocarbon streams.

Sulfuric Acid and Hydrofluoric Acid

Sulfuric acid and hydrofluoric acid are used primarily as catalysts in alkylation processes. Sulfuric acid is also used in some treatment processes.

Drilling Methods

Oil naturally flows toward the Earth's surface, forced upward by the pressure of natural gas and water. But much of the oil gets trapped along the way and comes to rest in the pores of subterranean rocks. To extract it, we drill wells.

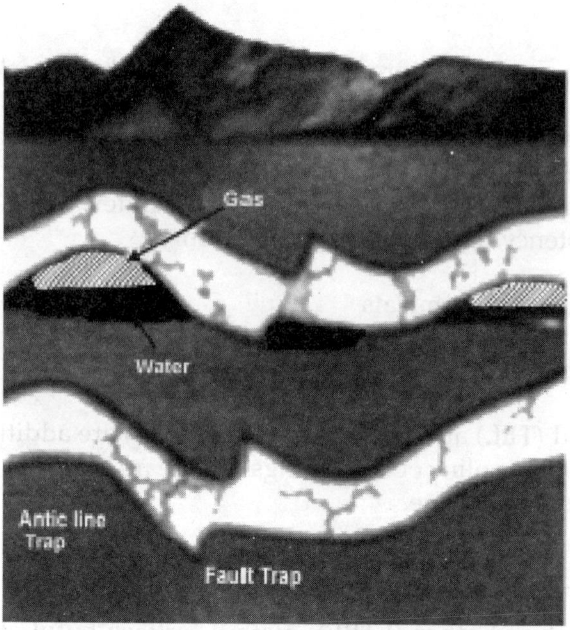

Figure 2.3:

(Oil and gas, produced by decaying organic matter, are trapped in the Earth in rocks riddled with holes like sponges. Geologists classify the traps by the forces of nature that shape layers of rock)

At first, people drilled wells near spots where oil seeped naturally to the surface. Or they made haphazard guesses about where to drill, often with disappointing results. Even with the modern technology, the search for oil is fraught with uncertainty. The odds are against striking oil in a new location. And when oil is found, rarely is there enough to make production commercially viable.

To locate underground structures that may contain oil, geologists survey the Earth with technologically sophisticated gravity meters and magnetometers. With seismographs similar to those used to measure earthquakes, they explore what they cannot see by sending sound waves underground. (How long they take to return helps determine the depth and characteristics of the rock layers through which they travel.)

Using computer models, geologists create three-dimensional pictures of sedimentary basins. With CAT scans and magnetic resonance imaging–the same technology doctors use to peer inside the human body–they visualize the flow of oil through the rocks. Exploration teams get more information from sound waves generated by trucks equipped with vibrators, dynamite explosions, and air guns specially designed for underwater use; the patterns of the waves reflected back reveal subterranean structures.

Often discovered in dome-shaped structures called anticlines, oil does not reside in underground lakes as is commonly supposed. Instead, it is trapped in rocks with holes like sponges. Natural gas is dissolved in the oil or separates and rests atop the oil as a separate layer.

The porous rocks holding oil and natural gas may be far below the surface, so extracting them requires drilling a well. A rotary tool with a tough diamond bit drills through the rock. Drilling mud–a fluid composed of clay, water, and chemicals–keeps the bit cool. It is pumped down with the bit and returns through the space between the drilling string and the bore-hole, carrying rock fragments with it. Drilling proceeds at a rate of 30 to 60 feet per hour. In the United States, the average well is more than a mile deep; the deepest is nearly seven miles.

In the days of the "gushers," the pressure of natural gas released when the drill reached its target sometimes forced petroleum to explode to the surface, polluting the environment and wasting precious oil. To prevent this, drilling rigs now have instruments that measure well pressure and special valves called blow-out preventers. The mud that lubricates the bit also helps control pressure

Still, most of the oil trapped underground needs to be pumped out. To extract as much as possible, carbon dioxide, other gases, water or chemicals are injected to maintain well pressure. An assembly of pipes and valves called a Christmas tree controls the flow at the wellhead.

Roughly a third of the world's oil comes from offshore wells, mainly in the North Sea, the Persian Gulf, and the Gulf of Mexico. Offshore wells are usually drilled from fixed platforms, which may be as tall as skyscrapers. Wells can also be drilled from semi-submersible structures or ships. The distance between the water's surface and the ocean floor, where the well starts, can be a mile or more.

The Heating Value

The heating value or calorific value of a substance, usually a fuel or food, is the amount of heat released during the combustion of a specified amount of it. The calorific value is a characteristic for each substance. It is measured in units of energy per unit weight: Btu/m^3, kcal/kg, kJ/kg, J/mol, or other units. Heating value is commonly determined by use of a bomb calorimeter.

The quantity known as higher heating value (or gross calorific value or gross energy) is determined by bringing all the products of combustion back to the original pre-combustion temperature, and in particular condensing any vapour produced. The quantity known as lower heating value (or net calorific value) is determined by subtracting the heat of vaporization of the water produced by combustion from the higher heating value.

Most applications which burn fuel produce water vapour which is not used, wasting its heat content; the net heating value is the applicable measure. This is particularly relevant for natural gas, whose high hydrogen content produces much water. The gross calorific value is relevant for gas burnt in condensing boilers which condense the water vapour produced by combustion, recovering heat which would otherwise be wasted.

Comparison of Fuels

	Energy per kilogram		
Fuel	*MJ/kg*	*BTU/lb*	*kJ/mol*
Hydrogen	141.79	61,000	286
Gasoline	47.3	20,400	—
Diesel	44.8	19,300	—
Ethanol	29.7	12,800	1,300
Propane	50.35	—	2,219
Butane	49.51	20,900	2,800
Wood	15	6,500	—
Coal	15-27	8,000-14,000	—
Natural uranium	500,000	—	—

Fuel needed to run a lightbulb (100 W) for a year (= 876kWh)

- 400 kg (876 lb) of coal
- 230 kg (508 lb) of oil
- 170 kg (377 lb) which equals 255 m3 of natural gas
- 147 kg (324 lb) equalling 220 m3 of natural gas (using cct)

Table 2.1: Gross Calorific Value of Natural Gas by Source of Production

Source	*kJ/m³*
Russia	38231
United States	38416
Canada	38200
Netherlands	33320
United Kingdom	39710
Indonesia	40600
Algeria	42000
Uzbekistan	37889
Saudi Arabia	38000
Norway	39877
Bangladesh	36000

The Wobbe Index (WI) is the main indicator of the interchangeability of fuel gases such as natural gas LPG and Town Gas and is frequently defined in the specifications of gas supply and transport utilities. If VC is the higher heating value, or calorific value, and GS is the specific gravity, the Wobbe Index, IW, is defined as: $I_w] \frac{V_c}{\sqrt{G_s}}$

In details, the above equation can be translated into: Wobbe Index = higher heating value/(square root of gas specific gravity)

The Wobbe Index is used to compare the combustion energy output of different composition fuel gases in an appliance (fire, cooker etc.). If two fuels have identical Wobbe Indices then for a given valve setting the energy output will also be identical.

Table 2.2: Wobbe Index (kcal per standard cubic meter) of Common Fuel Gases

Fuel Gas	*Upper Index*	*Lower Index*
Hydrogen	11,528	9,715
Methane	12,735	11,452
Ethane	16,298	14,931
Ethylene	15,253	14,344
Natural gas	12,837	11,597
Propane	19,376	17,817
Propylene	18,413	17,180
n-butane	22,066	20,336
Iso-butane	21,980	20,247
Butylene-1	21,142	19,728
LPG	20,755	19,106
Acetylene	14,655	14,141
Carbon monoxide	3,060	3,060

Typically variations of up to 5 per cent are allowed as these would not be noticeable to the consumer. The Wobbe Index is a critical factor to minimise the impact of the changeover when analyzing the use of SNG fuels such as propane-air mixtures.

Usage

The Wobbe index is commonly expressed in Btu per standard cubic foot or megajoules per standard cubic meter (1000 BTU/scf = 37.3 Mj/Sm^3). In the case of natural gas (Mol. weight 17), the typical heating value is around 1,050 Btu per cubic foot and the specific gravity is approximately 0.59, giving a typical Wobbe index of 1,367.

There are three ranges or 'families' of fuel gasses that have been internationally agreed based on Wobbe Index. Family 1 covers manufactured gases, family 2 covers natural gases (with high and low ranges) and family 3 covers liquefied petroleum gas (LPG). Combustion equipment is typically designed to burn a fuel gas within a a particular family—hydrogen-rich Town Gas, Natural Gas or LPG.

Family	*Wobbe Index range (MJ/Sm^3)*	*Type of gas*
1	22.5 – 30	Town gas/Syn Gas
2 L	39.0 – 45	Natural
2 H	45.5 – 55	
3	73.5 – 87.5	LPG

Other flame characteristics and composition limits may determine the acceptability of the replacement gas, e.g. flame speed, 'yellow tipping' due to incomplete combustion, sulphur content, oxygen content etc.

Heat Exchangers, Coolers, and Process Heaters

Heating Operations

Process heaters and heat exchangers preheat feedstock in distillation towers and in refinery processes to reaction temperatures. Heat exchangers use either steam or hot hydrocarbon transferred from some other section of the process for heat input. The heaters are usually designed for specific process operations, and most are of cylindrical vertical or box-type designs. The major portion of heat provided to process units comes from fired heaters fueled by refinery or natural gas, distillate, and residual oils. Fired heaters are found on crude and reformer preheaters, coker heaters, and large-column reboilers.

Cooling Operations

Heat also may be removed from some processes by air and water exchangers, fin fans, gas and liquid coolers, and overhead condensers, or by transferring heat to other systems. The basic mechanical vapor-compression refrigeration system, which may serve one or more process units, includes an evaporator, compressor, condenser, controls, and piping. Common coolants are water, alcohol/water mixtures, or various glycol solutions.

Health and Safety Considerations

(a) *Fire protection and prevention*: A means of providing adequate draft or steam purging is required to reduce the chance of explosions when lighting fires in heater furnaces. Specific start-up and emergency procedures are required for each type of unit. If fire impinges on fin fans, failure could occur due to overheating. If flammable product escapes from a heat exchanger or cooler due to a leak, fire could occur.

(b) *Safety*: Care must be taken to ensure that all pressure is removed from heater tubes before removing header or fitting plugs. Consideration should be given to providing for pressure relief in heat-exchanger piping systems in the event they are blocked off while full of liquid. If controls fail, variations of temperature and pressure could occur on either side of the heat exchanger. If heat exchanger tubes fail and process pressure is greater than heater pressure, product could enter the heater with downstream consequences. If the process pressure is less than heater pressure, the heater stream could enter into the process fluid. If loss of circulation occurs in liquid or gas coolers, increased product temperature could affect downstream operations and require pressure relief.

(c) *Health*: Because these are closed systems, exposures under normal operating conditions are expected to be minimal. Depending on the fuel, process operation, and unit design, there is a potential for exposure to hydrogen sulfide, carbon monoxide, hydrocarbons, steam boiler feed-water sludge, and water-treatment chemicals. Skin contact should be avoided with boiler blowdown, which may contain phenolic compounds. Safe work practices and/or appropriate personal protective equipment against hazards may be needed during process maintenance, inspection, and turnaround activities and for protection from radiant heat, superheated steam, hot hydrocarbon, and noise exposures.

Steam Generation

Heater and Boiler Operations

Steam is generated in main generation plants, and/or at various process units using heat from flue gas or other sources. Heaters (furnaces) include burners and a combustion air system, the boiler enclosure in which heat transfer takes place, a draft or pressure system to remove flue gas from the furnace, soot blowers, and compressed-air systems that seal openings to prevent the escape of flue gas. Boilers consist of a number of tubes that carry the water-steam mixture through the furnace for maximum heat transfer. These tubes run between steam-distribution drums at the top of the boiler and water-collecting drums at the bottom of the boiler. Steam flows from the steam drum to the superheater before entering the steam distribution system.

Heater Fuel

(a) Heaters may use any one or combination of fuels including refinery gas, natural gas, fuel oil, and powdered coal. Refinery off-gas is collected from process units and combined with natural gas and LPG in a fuel-gas balance drum. The balance drum provides constant system pressure, fairly stable Btu-content fuel, and automatic separation of suspended liquids in gas vapors, and it prevents carryover of large slugs of condensate into the distribution system. Fuel oil is typically a mix of refinery crude oil with straight-run and cracked residues and other products. The fuel-oil system delivers fuel to process-unit heaters and steam generators at required temperatures and pressures. The fuel oil is heated to pumping temperature, sucked through a coarse suction strainer, pumped to a temperature-control heater, and then pumped through a fine-mesh strainer before being burned.

(b) In one example of process-unit heat generation, carbon monoxide boilers recover heat in catalytic cracking units as carbon monoxide in flue gas is burned to complete combustion. In other processes, waste-heat recovery units use heat from the flue gas to make steam.

Steam Distribution

The distribution system consists of valves, fittings, piping, and connections suitable for the pressure of the steam transported. Steam leaves the boilers at the highest pressure required by the process units or electrical generation. The steam pressure is then reduced in turbines that drive process pumps and compressors. Most steam used in the refinery is condensed to water in various types of heat exchangers. The condensate is reused as boiler feedwater or discharged to wastewater treatment. When refinery steam is also used to drive steam turbine generators to produce electricity, the steam must be produced at much higher pressure than required for process steam. Steam typically is generated by heaters (furnaces) and boilers combined in one unit.

Feedwater

(a) Feedwater supply is an important part of steam generation. There must always be as many pounds of water entering the system as there are pounds of steam leaving it. Water used in steam generation must be free of contaminants including minerals and dissolved impurities that can damage the system or affect its operation. Suspended materials such as silt, sewage, and oil, which form scale and sludge, must be coagulated or filtered out of the water. Dissolved gases, particularly carbon dioxide and oxygen, cause boiler corrosion and are removed by deaeration and treatment. Dissolved minerals including metallic salts, calcium, carbonates, etc., that cause scale, corrosion, and turbine blade deposits are treated with lime or soda ash to precipitate them from the water. Recirculated cooling water must also be treated for hydrocarbons and other contaminants.

(b) Depending on the characteristics of raw boiler feedwater, some or all of the following six stages of treatment will be applicable:

- Clarification
- Sedimentation
- Filtration
- Ion exchange
- Deaeration
- Internal treatment

Health and Safety Considerations

(a) *Fire protection and prevention*: The most potentially hazardous operation in steam generation is heater startup. A flammable mixture of gas and air can build up as a result of loss of flame at one or more burners during light-off. Each type of unit requires specific startup and emergency procedures including purging before lightoff and in the event of misfire or loss of burner flame.

(b) *Safety*: If feedwater runs low and boilers are dry, the tubes will overheat and fail. Conversely, excess water will be carried over into the steam distribution system and damage the turbines. Feedwater must be free of contaminants that could affect operations. Boilers should have continuous or intermittent blowdown systems to remove water from steam drums and limit buildup of scale on turbine blades and superheater tubes. Care must be taken not to overheat the superheater during startup and shut-down. Alternate fuel sources should be provided in the event of loss of gas due to refinery unit shutdown or emergency. Knockout pots provided at process units remove liquids from fuel gas before burning.

(c) *Health*: Safe work practices and/or appropriate personal protective equipment may be needed for potential exposures to feedwater chemicals, steam, hot water, radiant heat, and noise, and during process sampling, inspection, maintenance, and turnaround activities.

Pressure-Relief and Flare Systems

Pressure-Relief Systems

Pressure-relief systems control vapors and liquids that are released by pressure-relieving devices and blow-downs. Pressure relief is an automatic, planned release when operating pressure reaches a predetermined level. Blowdown normally refers to the intentional release of material, such as blowdowns from process unit startups, furnace blowdowns, shutdowns, and emergencies. Vapor depressuring is the rapid removal of vapors from pressure vessels in case of fire. This may be accomplished by the use of a rupture disc, usually set at a higher pressure than the relief valve.

Safety Relief Valve Operations

Safety relief valves, used for air, steam, and gas as well as for vapor and liquid, allow the valve to open in proportion to the increase in pressure over the normal operating pressure. Safety valves designed primarily to release high volumes of steam usually pop open to full capacity. The overpressure needed to open liquid-relief valves where large-volume discharge is not required increases as the valve lifts due to increased spring resistance. Pilot-operated safety relief valves, with up to six times the capacity of normal relief valves, are used where tighter sealing and larger volume discharges are required. Nonvolatile liquids are usually pumped to oil-water separation and recovery systems, and volatile liquids are sent to units operating at a lower pressure.

Flare Systems

A typical closed pressure release and flare system includes relief valves and lines from process units for collection of discharges, knockout drums to separate vapors and liquids, seals, and/or purge gas for flashback protection, and a flare and igniter system which combusts vapors when discharging directly to the atmosphere is not permitted. Steam may be injected into the flare tip to reduce visible smoke.

Pressure Relief Health and Safety Considerations

(a) *Fire protection and prevention*: Vapors and gases must not discharge where sources of ignition could be present.

(b) *Safety*: Liquids should not be discharged directly to a vapor disposal system. Flare knockout drums and flares need to be large enough to handle emergency blowdowns. Drums should be provided with relief in the event of overpressure.
Pressure relief valves must be provided where the potential exists for overpressure in refinery processes due to the following causes:

- Loss of cooling water, which may greatly reduce pressure in condensers and increase the pressure in the process unit.
- Loss of reflux volume, which may cause a pressure drop in condensers and a pressure rise in distillation towers because the quantity of reflux affects the volume of vapors leaving the distillation tower.
- Rapid vaporization and pressure increase from injection of a lower boiling-point liquid including water into a process vessel soperating at higher temperatures.
- Expansion of vapor and resultant over-pressure due to overheated process steam, malfunctioning heaters, or fire.
- Failure of automatic controls, closed outlets, heat exchanger failure, etc.
- Internal explosion, chemical reaction, thermal expansion, or accumulated gases.

Maintenance is important because valves are required to function properly. The most common operating problems are listed below:

- Failure to open at set pressure, because of plugging of the valve inlet or outlet, or because corrosion prevents proper operation of the disc holder and guides.
- Failure to reseat after popping open due to fouling, corrosion, or deposits on the seat or moving parts, or because solids in the gas stream have cut the valve disc.
- Chattering and premature opening, because operating pressure is too close to the set point.

(c) *Health*: Safe work practices and/or appropriate personal protective equipment may be needed to protect against hazards during inspection, maintenance, and turnaround activities.

Wastewater Treatment

Wastewater treatment is used for process, runoff, and sewerage water prior to discharge or recycling. Wastewater typically contains hydrocarbons, dissolved materials, suspended solids, phenols, ammonia, sulfides, and other compounds. Wastewater includes condensed steam, stripping water, spent caustic solutions, cooling tower and boiler blowdown, wash water, alkaline and acid waste neutralization water, and other process-associated water.

Pretreatment Operations

Pretreatment is the separation of hydrocarbons and solids from wastewater. API separators, interceptor plates, and settling ponds remove suspended hydrocarbons, oily sludge, and solids by gravity separation, skimming, and filtration. Some oil-in-water emulsions must be heated to assist in separating the oil and water. Gravity separation depends on the specific gravity differences between water and immiscible oil globules and allows free oil to be skimmed off the surface of the wastewater. Acidic wastewater is neutralized using ammonia, lime, or soda ash. Alkaline wastewater is treated with sulfuric acid, hydrochloric acid, carbon dioxide-rich flue gas, or sulfur.

Secondary Treatment Operations

After pretreatment, suspended solids are removed by sedimentation or air flotation. Wastewater with low levels of solids may be screened or filtered. Flocculation agents are sometimes added to help separation. Secondary treatment processes biologically degrade and oxidize soluble organic matter by the use of activated sludge, unaerated or aerated lagoons, trickling filter methods, or anaerobic treatments. Materials with high adsorption characteristics are used in fixed-bed filters or added to the wastewater to form a slurry which is removed by sedimentation or filtration. Additional treatment methods are used to remove oils and chemicals from wastewater. Stripping is used on wastewater containing sulfides and/or ammonia, and solvent extraction is used to remove phenols.

Tertiary Treatment Operations

Tertiary treatments remove specific pollutants to meet regulatory discharge requirements. These treatments include chlorination, ozonation, ion exchange, reverse osmosis, activated carbon adsorption, etc. Compressed oxygen is diffused into wastewater streams to oxidize certain chemicals or to satisfy regulatory oxygen-content requirements. Wastewater that is to be recycled may require cooling to remove heat and/or oxidation by spraying or air stripping to remove any remaining phenols, nitrates, and ammonia.

Health and Safety Considerations

(a) *Fire protection and prevention*: The potential for fire exists if vapors from wastewater containing hydrocarbons reach a source of ignition during treatment.
(b) *Health*: Safe work practices and/or appropriate personal protective equipment may be needed for exposures to chemicals and waste products during process sampling, inspection, maintenance, and turnaround activities as well as to noise, gases, and heat.

Cooling Towers

Cooling towers remove heat from process water by evaporation and latent heat transfer between hot water and air. The two types of towers are crossflow and counterflow. Crossflow towers introduce the airflow at right angles to the water flow throughout the structure. In counterflow cooling towers, hot process water is pumped to the uppermost plenum and allowed to fall through the tower. Numerous slats or spray nozzles located throughout the length of the tower disperse the water and help in cooling. Air enters at the tower bottom and flows upward against the water. When the fans or blowers are at the air inlet, the air is considered to be forced draft. Induced draft is when the fans are at the air outlet.

Cooling Water

Recirculated cooling water must be treated to remove impurities and dissolved hydrocarbons. Because the water is saturated with oxygen from being cooled with air, the chances for corrosion are increased. One means of corrosion prevention is the addition of a material to the cooling water that forms a protective film on pipes and other metal surfaces.

Health and Safety Considerations

(a) *Fire prevention and protection*: When cooling water is contaminated by hydrocarbons, flammable vapors can be evaporated into the discharge air. If a source of ignition is present, or if lightning occurs, a fire may start. A potential fire hazard also exists where there are relatively dry areas in induced-draft cooling towers of combustible construction.

(b) *Safety*: Loss of power to cooling tower fans or water pumps could have serious consequences in the operation of the refinery. Impurities in cooling water can corrode and foul pipes and heat exchangers, scale from dissolved salts can deposit on pipes, and wooden cooling towers can be damaged by microorganisms.

(c) *Health*: Cooling-tower water can be contaminated by process materials and by-products including sulfur dioxide, hydrogen sulfide, and carbon dioxide, with resultant exposures. Safe work practices and/or appropriate personal protective equipment may be needed during process sampling, inspection, maintenance, and turnaround activities; and for exposure to hazards such as those related to noise, water-treatment chemicals, and hydrogen sulfide when wastewater is treated in conjunction with cooling towers.

Electric Power

Refineries may receive electricity from outside sources or produce their own power with generators driven by steam turbines or gas engines. Electrical substations receive power from the utility or power plant for distribution throughout the facility. They are usually located in nonclassified areas, away from sources of vapor or cooling-tower water spray. Transformers, circuit breakers, and feed-circuit switches are usually located in substations. Substations feed power to distribution stations within the process unit areas. Distribution stations can be located in classified areas, providing that classification requirements are met. Distribution stations usually have a liquid-filled transformer and an oil-filled or air-break disconnect device.

Health and Safety Considerations

(a) *Fire Protection and Prevention*: Generators that are not properly classified and are located too close to process units may be a source of ignition should a spill or release occur.

(b) *Safety*: Normal electrical safety precautions including dry footing, high-voltage warning signs, and guarding must be taken to protect against electrocution. Lockout/tagout and other appropriate safe work practices must be established to prevent energization while work is being performed on high-voltage electrical equipment.

(c) *Health*: Safe work practices and/or the use of appropriate personal protective equipment may be needed for exposures to noise, for exposure to hazards during inspection and maintenance activities, and when working around transformers and switches that may contain a dielectric fluid which requires special handling precautions.

Gas and Air Compressors

Both reciprocating and centrifugal compressors are used throughout the refinery for gas and compressed air. Air compressor systems include compressors, coolers, air receivers, air dryers, controls, and distribution piping. Blowers are used to provide air to certain processes. Plant air is provided for the operation of air-powered tools, catalyst regeneration, process heaters, steam-air decoking, sour-water oxidation, gasoline sweetening, asphalt blowing, and other uses. Instrument air is provided for use in pneumatic instruments and controls, air motors and purge connections.

Health and Safety Considerations

(a) *Fire protection and prevention*: Air compressors should be located so that the suction does not take in flammable vapors or corrosive gases. There is a potential for fire should a leak occur in gas compressors.

(b) *Safety*: Knockout drums are needed to prevent liquid surges from entering gas compressors. If gases are contaminated with solid materials, strainers are needed. Failure of automatic compressor controls will affect processes. If maximum pressure could potentially be greater than compressor or process-equipment design pressure, pressure relief should be provided. Guarding is needed for exposed moving parts on compressors. Compressor buildings should be properly electrically classified, and provisions should be made for proper ventilation.
Where plant air is used to back up instrument air, interconnections must be upstream of the instrument air drying system to prevent contamination of instruments with moisture. Alternate sources of instrument air supply, such as use of nitrogen, may be needed in the event of power outages or compressor failure.

(c) *Health*: Safe work practices and/or appropriate personal protective equipment may be needed for exposure to hazards such as noise and during inspection and maintenance activities. The use of appropriate safeguards must be considered so that plant and instrument air is not used for breathing or pressuring potable water systems.

Marine, Tank Car, and Tank Truck Loading and Unloading

Facilities for loading liquid hydrocarbons into tank cars, tank trucks, and marine vessels and barges are usually part of the refinery operations. Product characteristics, distribution needs, shipping requirements, and operating criteria are important when designing loading facilities. Tank trucks and rail tank cars are either top- or bottom-loaded, and vapor-recovery systems may be provided where required. Loading and unloading liquefied petroleum gas (LPG) require special considerations in addition to those for liquid hydrocarbons.

Health and Safety Considerations

(a) *Fire Protection and Prevention*: The potential for fire exists where flammable vapors from spills or releases can reach a source of ignition. Where switch-loading is permitted, safe practices need to be established and followed. Bonding is used to equalize the electrical charge between the loading rack and the tank truck or tank car. Grounding is used at truck and rail loading facilities to prevent flow of stray currents. Insulating flanges are used on marine dock piping connections to prevent static electricity buildup and discharge. Flame arrestors should be installed in loading rack and marine vapor-recovery lines to prevent flashback.

(b) *Safety*: Automatic or manual shutoff systems at supply headers are needed for top and bottom loading in the event of leaks or overfills. Fall protection such as railings are needed for top-loading racks where employees are exposed to falls. Drainage and recovery systems may be provided for storm drainage and to handle spills and leaks. Precautions must be taken at LPG loading facilities not to overload or overpressurize tank cars and trucks.

(c) *Health*: The nature of the health hazards at loading and unloading facilities depends upon the products being loaded and the products previously transported in the tank cars, tank trucks, or marine vessels. Safe work practices and/or appropriate personal protective equipment may be needed to protect against hazardous exposures when loading or unloading, cleaning up spills or leaks, or when gauging, inspecting, sampling, or performing maintenance activities on loading facilities or vapor-recovery systems.

Turbines

Turbines are usually gas- or steam-powered and are typically used to drive pumps, compressors, blowers, and other refinery process equipment. Steam enters turbines at high temperatures and pressures, expands across and drives rotating blades while directed by fixed blades.

Health and Safety Considerations

(a) *Safety*: Steam turbines used for exhaust operating under vacuum should have safety relief valves on the discharge side, both for protection and to maintain steam in the event of vacuum failure. Where maximum operating pressure could be greater than design pressure, steam turbines should be provided with relief devices. Consideration should be given to providing governors and overspeed control devices on turbines.

(b) *Health*: Safe work practices and/or appropriate personal protective equipment may be needed for noise, steam and heat exposures, and during inspection and maintenance activities.

Pumps, Piping and Valves

(a) Centrifugal and positive-displacement (i.e., reciprocating) pumps are used to move hydrocarbons, process water, fire water, and wastewater through piping within the refinery. Pumps are driven by electric motors, steam turbines, or internal combustion engines. The pump type, capacity, and construction materials depend on the service for which it is used.

(b) Process and utility piping distribute hydrocarbons, steam, water, and other products throughout the facility. Their size and construction depend on the type of service, pressure, temperature, and nature of the products. Vent, drain, and sample connections are provided on piping, as well as provisions for blanking.

(c) Different types of valves are used depending on their operating purpose. These include gate valves, bypass valves, globe and ball valves, plug valves, block and bleed valves, and check valves. Valves can be manually or automatically operated.

Health and Safety Considerations

(a) *Fire protection and prevention*: The potential for fire exists should hydrocarbon pumps, valves, or lines develop leaks that could allow vapors to reach sources of ignition. Remote sensors, control valves, fire valves, and isolation valves should be used to limit the release of hydrocarbons at pump suction lines in the event of leakage and/or fire.

(b) *Safety*: Depending on the product and service, backflow prevention from the discharge line may be needed. The failure of automatic pump controls could cause a deviation in process pressure and temperature. Pumps operated with reduced or no flow can overheat and rupture. Pressure relief in the discharge piping should be provided where pumps can be overpressured. Provisions may be made for pipeline expansion, movement, and temperature changes to avoid rupture. Valves and instruments that require servicing or other work should be accessible at grade level or from an operating platform. Operating vent and drain connections should be provided with double-block valves, a block valve and plug, or blind flange for protection against releases.

(c) *Health*: Safe work practices and/or appropriate personal protective equipment may be needed for exposure to hazards such as those related to liquids and vapors when opening or draining pumps, valves, and/or lines, and during product sampling, inspection, and maintenance activities.

Tank Storage

Atmospheric storage tanks and pressure storage tanks are used throughout the refinery for storage of crudes, intermediate hydrocarbons (during the process), and finished products. Tanks are also provided for fire water, process and treatment water, acids, additives, and other chemicals. The type, construction, capacity and location of tanks depends on their use and materials stored.

Health and Safety Considerations

(a) *Fire prevention and protection*: The potential for fire exists should hydrocarbon storage tanks be overfilled or develop leaks that allow vapors to escape and reach sources of ignition. Remote sensors, control valves, isolation valves, and fire valves may be provided at tanks for pump-out or closure in the event of a fire in the tank, or in the tank dike or storage area.

(b) *Safety*: Tanks may be provided with automatic overflow control and alarm systems, or manual gauging and checking procedures may be established to control overfills.

(c) *Health*: Safe work practices and/or appropriate personal protective equipment may be needed for exposure to hazards related to product sampling, manual gauging, inspection, and maintenance activities including confined space entry where applicable.

Chapter 3

Petroleum Refining Operation

The petroleum industry is organized into four broad sectors: exploration and production of crude oil and natural gas; transport; refining; and marketing and distribution.

Crude oil is fractionated into liquefied petroleum gas, naphtha (used to produce gasoline by blending with octane boosters), kerosene/aviation turbine fuel, diesel oil, and residual fuel oil.

Catalytic cracking and reforming, thermal cracking, and other secondary processes are used to achieve the desired product specifications. Certain refineries also produce feedstocks for the manufacture of lubricating oils and bitumens. Some refineries also manufacture coke.

Petroleum refining begins with the distillation, or fractionation, of crude oils into separate hydrocarbon groups. The resultant products are directly related to the characteristics of the crude processed. Most distillation products are further converted into more usable products by changing the size and structure of the hydrocarbon molecules through cracking, reforming, and other conversion processes as discussed in this chapter. These converted products are then subjected to various treatment and separation processes such as extraction, hydrotreating, and sweetening to remove undesirable constituents and improve product quality. Integrated refineries incorporate fractionation, conversion, treatment, and blending operations and may also include petrochemical processing.

Fractionation

Fractionation utilizes a mass separation technique called distillation in which the feedstock is distilled into various cuts of target boiling ranges or even separated into individual hydrocarbon compounds. Distillation is accomplished by imposing a temperature profile across the tower enabling differences in the equilibrium compositions of the vapor and liquid phases to change the compositions throughout the distillation tower.

Heat is added to the hydrocarbons at the bottom of the tower through heat exchange in a "reboiler" which vaporizes a portion of the tower bottoms liquid for recirculation to the bottom of the tower. Heat is removed from the top of the tower through heat exchange in an overhead condenser and then returning a portion of the condensed hydrocarbons back to the tower as "reflux". This heat addition at the bottom and heat removal from the top of the tower establishes the temperature profile across the tower. In some applications, additional heat is removed by heat exchange with circulating liquid "pumparound" streams which are withdrawn and returned at intermediate levels of the tower. Perforated tray decks or packed bed sections allow intimate contacting of the liquid and vapor phases followed by separation.

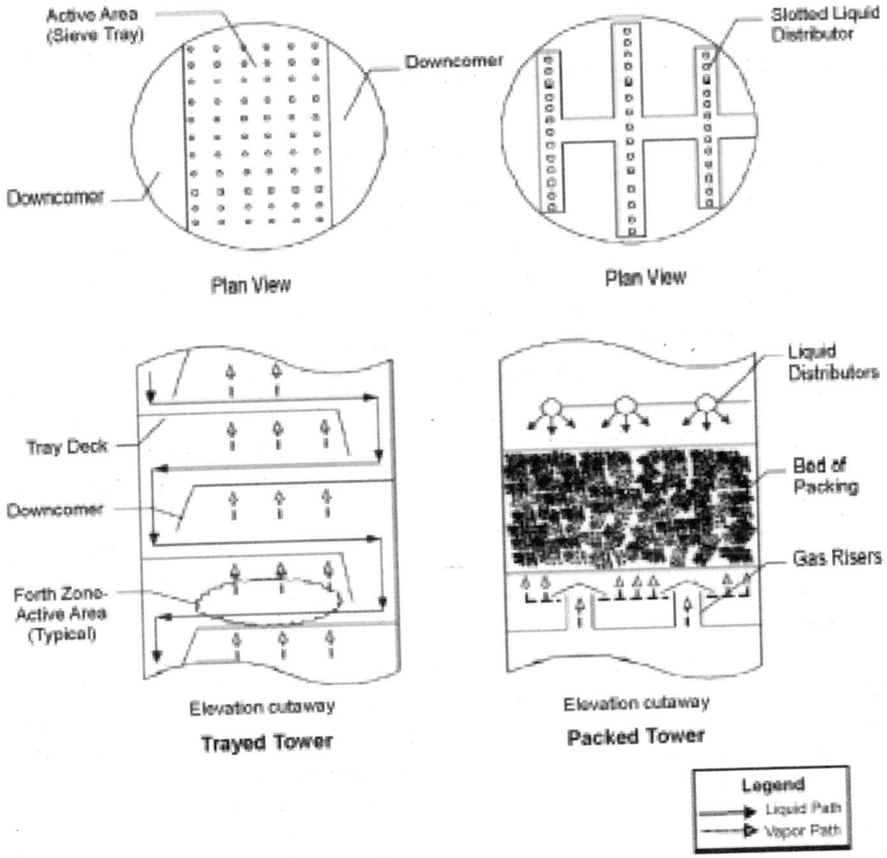

Figure 3.1: Distillation Tower Internals

Distillation concentrates the lower boiling point material towards the top of the tower. The lowest boiling point product is the tower overhead vapors which are condensed as distillate. Progressively higher boiling point material is present further down the tower. Gravity forces the liquid phase to flow down the tower. Additional intermediate boiling range streams may be withdrawn at various levels from the tower as side-stream products. The highest boiling range material is the liquid product withdrawn from the bottom of the tower.

Without going into a complete course on distillation theory, an oversimplification is that the degree of fractionation or sharpness of the distillation cut between the fractions is a function of physically how tall the tower's fractionation section(s) are and how much energy is applied via heating and cooling to promote the intimate contacting between the vapor and liquid phases.

Crude Oil Distillation (Primary Fractionation)

The first refinery process unit that a crude oil or mixture of crude oils is charged to is a Crude Distillation unit (commonly referred to as a pipe still), which consists of an Atmospheric Distillation tower and typically also includes a Vacuum Distillation tower. The crude oil or mixture of crude oils that represents the feed to a Crude Distillation unit is pumped from crude oil storage tanks into banks of heat exchangers which pre-heat the crude oil to approximately 250°F by cross exchanging heat with various product streams. Most crude oils contain appreciable levels of inorganic salts which would cause downstream corrosion and equipment fouling. The crude oil is then desalted by emulsifying the crude oil with water. The salts are dissolved in the water and the brine phase is then separated from the oil phase and withdrawn. The crude oil is further preheated to the maximum possible temperature (typically about 500-550°F) through heat exchange with distillation product streams. Finally, the crude oil is heated to approximately 750°F in a direct-fired furnace and fed to the Atmospheric Distillation tower.

The most common fractions withdrawn from an Atmospheric Distillation unit (progressing from lightest to heaviest) are naphthas, kerosene, gas oils, and residuum (the liquid bottoms stream). An Atmospheric Distillation unit operates at a pressure of just a few psi above atmospheric pressure in the tower overhead accumulator.

The maximum process temperature in the Atmospheric Distillation unit is approximately 750°F. At temperatures above 750°F, thermal cracking of the petroleum into light gases and coke occurs. Coke is essentially pure carbon and is a solid. The presence of coke is undesirable in refinery process units (with the exception of Coking units which produce coke purposefully and will be discussed in a later section) because solid coke formation fouls refinery process equipment and severely reduces equipment performance.

The Atmospheric Residuum stream is usually further fractionated in a Vacuum Distillation tower. Hydrocarbons existing as a liquid at a given temperature at atmospheric pressure will boil at a lower temperature when the pressure is sufficiently reduced. By pulling a vacuum with steam ejectors, the absolute operating pressure in a Vacuum Distillation tower can be 20 mm of mercury or less. In addition, steam is injected with the vacuum tower feed and at the bottom of the vacuum tower to reduce the effective (i.e., hydrocarbon partial sea level is an absolute pressure of approximately 760 mm of mercury. Vacuum distillation thereby enables hydrocarbons with boiling points well above 750°F to be further fractionated into light and heavy vacuum gas oils and vacuum residuum streams.

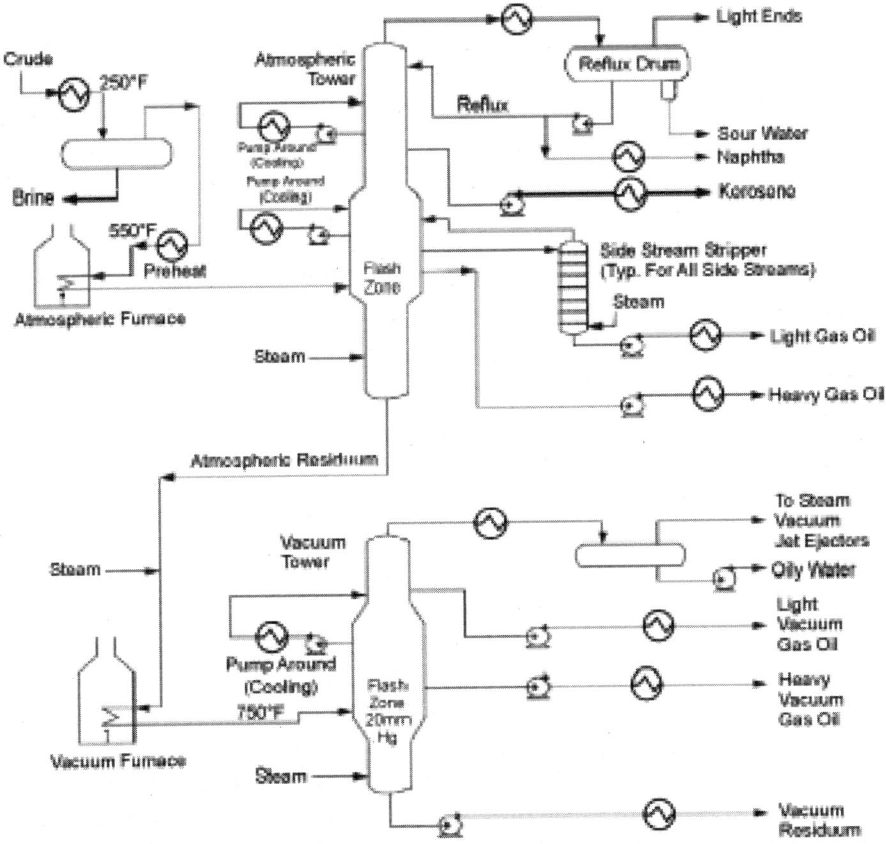

Figure 3.2: Crude Distillation Simplified Process Flow

Typical primary fractions and TBP cut-point ranges from Crude Oil Distillation (i.e., both Atmospheric and Vacuum tower products) are presented in the table below. The actual cut-point target within the range for individual fractions is predicated upon downstream processing objectives and product specifications.

	Initial TBP range (°F)	*Final TBP range (°F)*
Light naphtha	60-90	180-220
Heavy naphtha	180-220	30-430
Kerosene	330-380	80-550
Light gas oil	420-520	610-650
Heavy gas oil	610-650	750-850
Vacuum gas oil	750-800	950-1,050
Vacuum residuum	950-1,050	

Light Ends Fractionation

Light ends are the hydrocarbons boiling at the lowest temperatures—methane, ethane, propane, butanes, and pentanes, which contain one to five carbon atoms in their molecular structure, respectively. Butane and pentanes are mentioned in the plural because different straight and branched chain variants (called isomers) with the same carbon number are present. These hydrocarbons accumulate in the gas streams produced from the various refinery process units. Light ends streams are fractionated via distillation and treated with amine contacting to remove any H_2S present.

Unsaturated light ends containing the olefins ethylene, propylene, butylenes, and pentylenes (the two to five carbon olefins, respectively, from the Fluidized Catalytic Cracking and Delayed Coking processes) are fractionated separately from the saturated light ends (from the Crude Oil Distillation, Hydrotreating, Hydrocracking, and Catalytic Reforming processes). This separation enables the olefins to be removed from the saturated components for alternative processing (i.e., ethylene and propylene to petrochemical operations, if applicable, and propylene, butylenes, and pentylenes to Alkylation, as economics dictate). The saturated components are then further distilled to meet the finished product specifications for propane, normal butane (i.e., the four carbon number straight chain molecule), and isobutane (i.e, the four carbon number branched chain molecule). The methane and ethane are either fed to the Hydrogen Generation process, if applicable, or burned as fuel gas in the refinery process direct-fired furnaces. The bottoms stream from Light

Ends Fractionation?consisting of the pentanes and any heavier components are utilized as a blendstock for motor gasolines.

Treating

Hydrotreating

With rare exceptions, the intermediate hydrocarbon product streams from the Crude Distillation unit contain levels of sulfur that exceed the specifications for the finished product stream and/or the catalyst specifications for downstream processing units. Hydrotreating is the most common process configuration utilized to remove the sulfur from the intermediate stream. Hydrotreating may also reduce the levels of nitrogen contained in the stream. In addition, some of the metals (such as nickel and vanadium) may be removed from the hydrocarbon stream during hydrotreating. Hydrotreaters may be designated to continuously process one particular hydrocarbon feedstock, or may alternate processing of different feed streams (i.e., a batch-continuous operating mode).

Hydrotreating is a refinery process in which hydrogen gas is mixed with the hydrocarbon stream and contacted with a fixed-bed of catalyst in a reactor vessel at a sufficiently high enough temperature and pressure to effect the hydrodesulfurization (HDS) reactions. The catalyst is a solid consisting of a base of alumina impregnated with metal oxides that promote (i.e., catalyze) the desired reactions. These catalysts are usually formed into small pellets (approximately 1/8 inch diameter by less than an inch in length), typically shaped as cylinders or trilobes, to maximize the surface area available for contacting the reactants in the reactors. For HDS reactions, the most common metal oxides impregnated in the catalyst are those of cobalt and molybdenum. For reactions where hydrodenitrogenation (HDN) reactions in addition to HDS reactions are desired, the metal oxides impregnated in the catalyst are those of nickel and molybdenum.

In the HDS reaction, the bond between the carbon and sulfur atoms is broken, and the sulfur atom is replaced with a hydrogen atom. The sulfur atom combines with additional hydrogen to form the toxic gas hydrogen sulfide (H_2S).

The general chemical formula for the HDS reaction occurring is:

HDS reaction: $2(\ldots C\text{-}S) + 3H2 \longrightarrow 2(\ldots C\text{-}H) + 2H_2S$

Similarly, in the HDN reaction, the bond between the carbon and nitrogen atoms is broken, and the nitrogen atom is replaced with a hydrogen atom. The nitrogen combines with additional hydrogen to form ammonia (NH_3).

HDN reaction: ...... C-N + $2H_2 \longrightarrow$ ...... C-H + NH_3

The HDS and HDN reactions occur faster (i.e., a higher reactor severity) the higher the reactor temperature, the higher the reactor pressure (which results in an increased partial pressure of hydrogen) and the higher the volume of catalyst in the reactor relative to the volume of oil being processed. For a given crude, when comparing two different boiling point fractions, the fraction with the higher boiling point range generally has the highest concentrations of sulfur and nitrogen in each fraction.

In addition, the sulfur and nitrogen are more easily removed from lower boiling compounds. As a result, the reactor severity must be increased, the higher the boiling range of the fraction. HDN reactions generally required a much higher degree of reactor severity than HDS reactions. Hydrotreating process conditions range from the relatively mild reactor conditions of as low as 400 psi and 500°F for naphthas to very severe conditions of up to 2,000 psi and 800°F for heavy gas oils and vacuum residuum. The amount of hydrogen consumed per barrel of feedstock, and correspondingly the amount of hydrogen required in the reactor (called "treat gas") increases significantly as the feedstocks become heavier. At the higher reactor temperatures and hydrogen partial pressures, in addition to the HDS and HDN reactions, some cracking of heavy molecules into lighter molecules followed by hydrogenation occurs. As a result, very high severity Hydrotreating of heavy gas oils or vacuum residuum is often referred to as Hydrorefining since an appreciable yield of naphtha and distillate hydrocarbons occurs.

Generally, the process flow for a Hydrotreater process unit is that the hydrocarbon feedstock and hydrogen streams are both preheated through heat exchange with reactor effluent, then combined either before or after the final heating from a direct-fired furnace and then the mixed hydrocarbon and hydrogen stream is passed through the reactor, flowing from top to bottom. The reactor effluent (hydrogen, light hydrocarbons, H_2S and NH_3) is cooled through heat exchange with unit feed followed by separation of the vapor and liquid phases.

The liquid stream is sent to a stripper tower in which steam (or nitrogen in some cases) is employed to "strip" the hydrogen sulfide and any naphtha and lighter boiling components generated in the reactor from any higher boiling range product streams. Since the resulting naphtha stream contains light ends components, it is referred to as unstabilized naphtha or "wild naphtha". The stripped liquid product stream is then further cooled prior to disposition to storage tanks for additional refinery processing or finished product blending.

The separated reactor effluent vapor stream, which is predominantly hydrogen gas, may be compressed and recycled back to the reactor. A hydrogen "makeup" gas stream (with a hydrogen purity of 75-100 per cent hydrogen, depending on the source of hydrogen) is combined with any recycled hydrogen above. The purge stream of the effluent gas is scrubbed in an amine contactor to absorb the H_2S prior to the light ends disposition to fuel gas, light products or petrochemical operations.

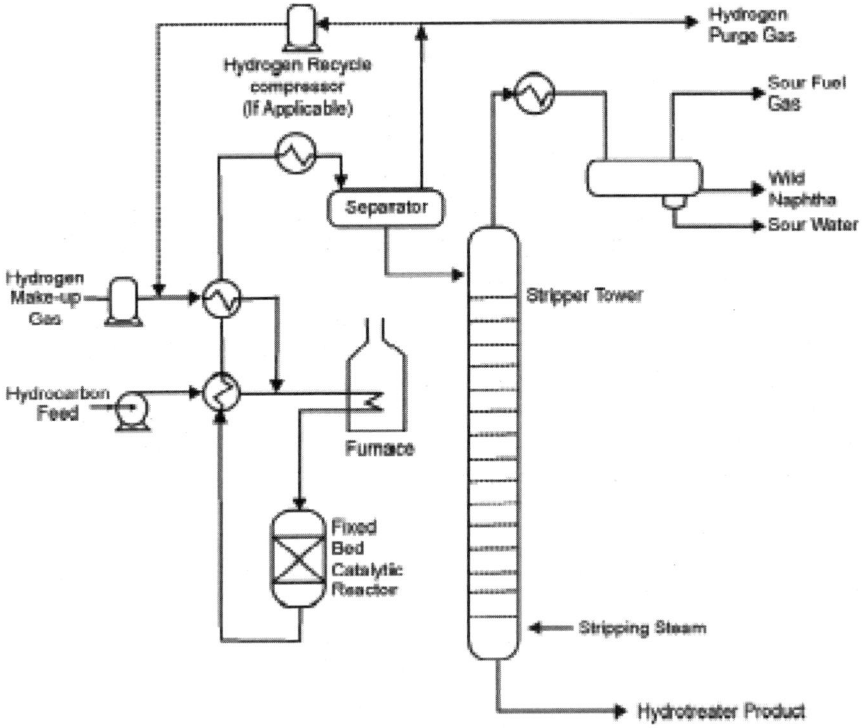

Figure 3.3: Hydrotreater Simplified Process Flow

The activity of the reactor catalyst bed is reduced over time as coke (i.e., essentially pure carbon) and metals attach to the individual catalyst surface and block active catalyst surface area from the oil being processed. To compensate for this catalyst deactivation, the reactor temperature is gradually increased as required to meet product specifications over the life of the catalyst bed. Upon reaching the maximum operating reactor temperature (typically 750°F) the unit will be scheduled for a shutdown to allow the catalyst bed to be replaced with either fresh or regenerated catalyst. Depending on the catalyst, feedstock and processing severity requirements, typical run lengths for catalyst beds are anywhere from one to four years.

Miscellaneous Product Treating

Many fuels products are treated as a finishing step prior to being shipped as "finished" products. Treating removes impurities which cause objectionable odors, unwanted colors or corrosivity of the product. Hydrogen sulfide (H_2S) and other sulfur compounds such as mercaptans are examples of such impurities.

Amine contacting using aqueous amine solutions such as monoethanolamine (MEA), diethanolamine (DEA), or methyldiethanol amine (MDEA) are commonly used to remove H2S from light ends streams prior to disposition as fuel gas or propane (i.e., Liquefied Petroleum Gas or "LPG") product. The amine solution is then regenerated in a still in which the application of heat drives off the H_2S. The H_2S-rich stream produced from the still (called "acid gas") is then routed to the Sulfur Conversion process. In gasolines, the presence of mercaptans causes objectionable odors.

Sweetening processes utilize caustic and in some cases air to either convert the mercaptans to disulfides or remove them from the gasoline products. Jet fuels which have not been hydrotreated are typically caustic washed to neutralize naphthenic acids. Jet fuels may also be clay treated by passing the fuel through a fixed bed of clay to remove any surfactants as well as color bodies.

Upgrading

Upgrading is the broad term applied to refinery processing which significantly increases the market value of the hydrocarbons processed. This is accomplished through chemical reactions to yield more desirable hydrocarbon compounds. The upgrading reactions result in either improving product specification qualities or rearranging the molecular structure (i.e., converting) so that the hydrocarbons boil in a more desirable boiling range.

Catalytic Reforming

Although motor gasolines have numerous specifications that must be satisfied to provide the performance demanded by our high-performance motor vehicles, the most widely recognized gasoline specification is the octane number.

Gasolines are typically retailed in grades of regular, mid-grade and premium, which are differentiated by the posted octane number. The Octane Number of a test fuel refers to the per centage by volume of isooctane in a mixture of isooctane and heptane in a reference fuel that when tested in a laboratory engine, matches the antiknock quality, as measured by a knockmeter, of the fuel being tested under the same conditions. The octane number posted at the gasoline pump is actually the average of the Research Octane Number (RON) and Motor Octane Number (MON), commonly referred to as (R+M)/2. RON and MON are two different test methods that quantify the antiknock qualities of a fuel. Since the MON is a test under more severe conditions than the RON test, for any given fuel, the RON is always higher than the MON.

Unfortunately, the desulfurized light and heavy naphtha fractions of crude oils have very low octane numbers?the heavy naphtha fraction is roughly 50 (R+M)/2. Catalytic Reforming is the refinery process that reforms the molecular structure of the heavy naphtha to increase the per centage of high-octane components while reducing the per centage of low-octane components.

The hydrocarbon compounds that constitute heavy naphtha are classified into four different categories: paraffins, olefins (a very low per centage of olefins occur in the heavy naphthas from crude), naphthenes and aromatics. In lieu of a complete course in organic chemistry, simplistically the paraffins and olefins are compounds with straight or branched carbon chains, whereas the naphthenes and aromatics are carbon rings. The paraffins and naphthenes are saturated hydrocarbons. Saturated means that they have the maximum number of hydrogen atoms attached to the carbon atoms. The olefins and aromatics, however, are unsaturated hydrocarbons because the compounds contain carbon atoms that are double bonded to other carbon atoms. The straight chain, compounds exhibit progressively higher octane numbers, while the unsaturated compounds exhibit very high octane numbers. Catalytic Reforming uses a precious metal catalyst (platinum supported by an alumina base) in conjunction with very high temperatures to reform the paraffins and napthenes into high-octane components. Sulfur is a poison to the Catalytic Reforming catalyst, which requires that virtually all the sulfur must be removed from the heavy naphtha through Hydrotreating prior to Catalytic Reforming. Several different types of chemical reactions occur in the Catalytic Reforming reactors?olefins are converted to paraffins, paraffins are isomerized to branched chains and to a lesser extent to naphthenes, and naphthenes are converted to aromatics. Aromatic compounds are essentially unchanged. The resulting reformate product stream from Catalytic Reforming has a RON from 96-102 depending on the reactor severity and feedstock quality. The dehydrogenation reactions which convert the saturated naphthenes into unsaturated aromatics produce hydrogen. This hydrogen is available for distribution to other refinery processes which consume hydrogen.

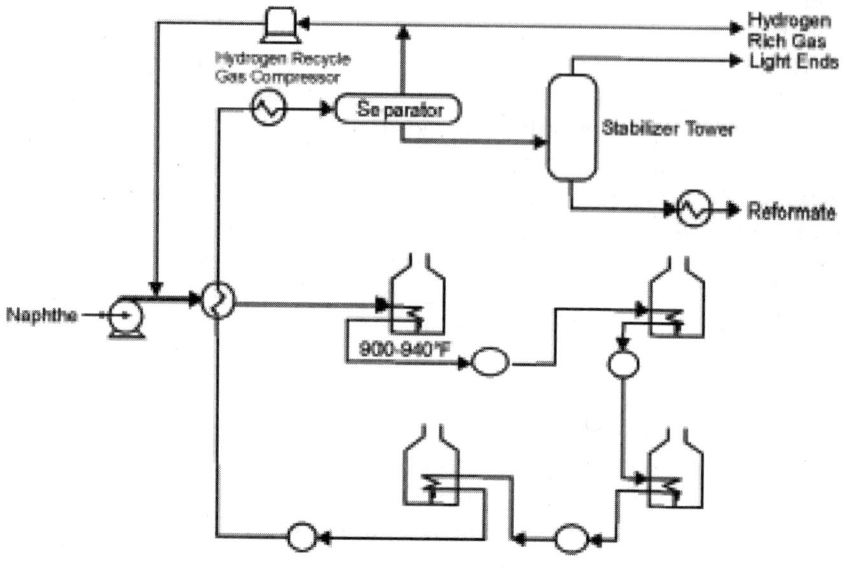

Figure 3.4: Catalytic Reforming Simplified Process Flow

The Catalytic Reforming process consists of a series of several spherical reactors which operate at temperatures of approximately 900°F. The reforming reactions are ?endothermic? meaning that the reactions cool the hydrocarbons. The hydrocarbons are re-heated by direct-fired furnaces in between the subsequent reforming reactors. As a result of the very high temperatures, the catalyst becomes deactivated by the formation of ?coke? (i.e., essentially pure carbon) on the catalyst which reduces the surface area available to contact with the hydrocarbons. A simplified process flow for the Catalytic Reforming process is presented above.

There are several types of Catalytic Reforming process configurations which differ in the manner that they accommodate the regeneration of the reforming catalyst. Catalyst regeneration involves burning off the coke with oxygen. Semiregenerative is the simplest configuration, which requires that the unit be shutdown for catalyst regeneration in which all reactors (typically four) are regenerated. The cyclic configuration utilizes an additional "swing" reactor which enables one reactor at a time to be taken off-line for regeneration while the other four remain in service. The continuous catalyst regeneration (CCR) configuration, which is the most complex, enables catalyst to be continuously removed for regeneration and replaced after regeneration. The benefits to the more complex configurations are that operating severity may be increased as a result of higher catalyst activity, which come at an increased capital cost for the process.

Fluidized Catalytic Cracking

The Fluidized Catalytic Cracking (FCC) process unit is considered by many refiners to be the heart of the petroleum refinery. This derives from the fact that the FCC is a key tool to correct the imbalance reflected by the markets demand for predominantly lighter, lower boiling petroleum products, whereas fractionated crude oils typically provide an excess of heavy, high boiling range oils. The FCC process converts heavy gas oils into lighter products which are then used as blendstocks for gasoline and diesel fuels. The olefinic FCC catalytic naphtha product exhibits a very high-octane value for gasoline blending.

The FCC process cracks the heavy gas oils by breaking carbon-to-carbon bonds in the large molecules comprising the gas oils and splitting them into multiple smaller molecules which boil at a much lower temperatures. The FCC may achieve conversions of 70-80 per cent of the feed hydrocarbons boiling above the gasoline range (i.e., 430°F) to products boiling below 430°F. The lower density of the FCC products relative to the gas oil feedstocks has the added benefit of producing a volume gain in which the combined volume of the liquid product streams is greater than the volume of the unit feed stream. Since most petroleum products are bought and sold on a volume basis, the volume gain aspect of the FCC process is a key aspect in how it enhances refinery profitability. The resulting FCC product hydrocarbons are highly olefinic (i.e., unsaturated). Virgin is a term used to distinguish straight-run (i.e., crude distillation and possibly hydrotreated only) hydrocarbons stocks from those that are products of conversion units such as the FCC.

The FCC cracking reactions are catalytically promoted at very high temperatures of 950-1,020°F. At these temperatures, coke (i.e., essentially pure carbon) formation deactivates the catalyst by blocking catalyst surface area which prevents intimate contact between the catalyst and the hydrocarbons. To retain catalyst activity, the FCC utilizes a very fine powdery, zeolite catalyst that behaves like a fluid (i.e., is able to flow). The fluidized catalyst is continuously circulated in the FCC from the reactor to a regenerator vessel and then returned to the reactor. Coke is removed from the catalyst in the regenerator vessel through the controlled incomplete combustion of the carbon with oxygen to form carbon monoxide and carbon dioxide.

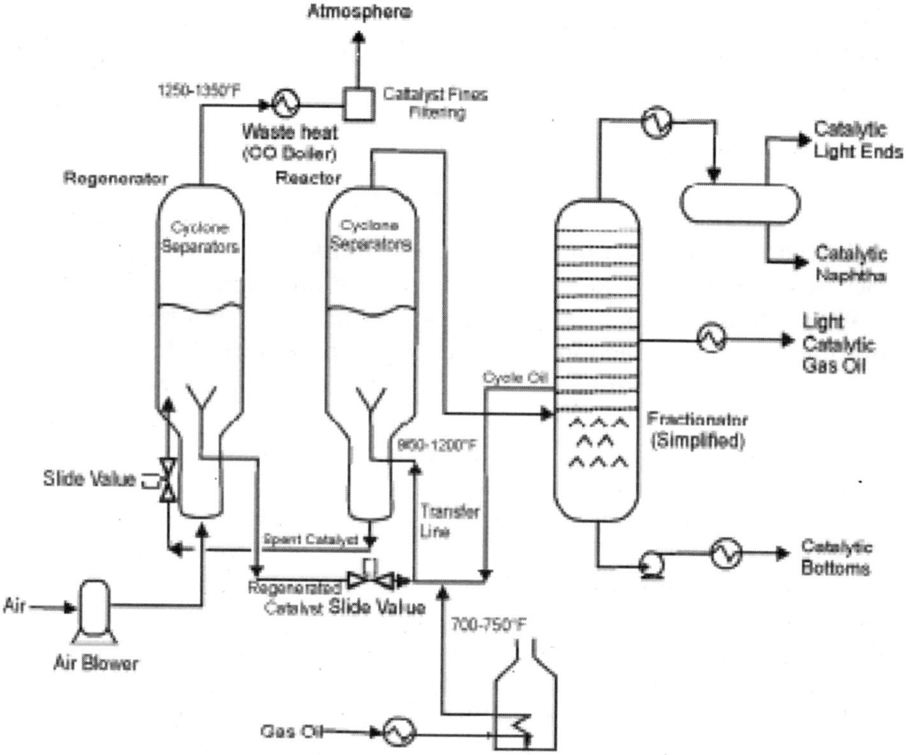

Figure 3.5: Fluidized Catalytic Cracker Simplified Process Flow

The gas oil feed to the FCC is preheated via heat exchange with reactor products and then a direct-fired furnace before being mixed with the hot (1,200-1,350°F) regenerated catalyst. The hot catalyst vaporizes the gas oil and heats the oil to the reactor temperature. To prevent ?overcracking? which produces excessive light ends at the expense of gasoline yield, the contacting time between the oil and catalyst is minimized. The bulk of the cracking reactions occur in the transfer line from the initial point of contact between the catalyst and gas oil feed and the "reactor" vessel. The primary purpose of the FCC reactor vessel is to disengage the catalyst/oil mixture. Primary and secondary cyclone separators are utilized to effect the separation of the vaporized hydrocarbons from the powdery solid catalyst. The hydrocarbons are condensed and distilled in the fractionator into the product streams: catalytic naphtha, light catalytic gas oil, heavy catalytic gas oil and catalytic bottoms. The heavy catalytic gas oil (also referred to as cycle oil) is typically recycled back to the reactor with the unit feedstock to increase the overall conversion of gas oil to lower boiling components. The catalyst then flows to the regenerator vessel where the coke is burned off by the introduction of air. The flow of air to the regenerator vessel is controlled to provide only enough oxygen to partially combust the carbon into carbon monoxide which prevents overheating the catalyst. The carbon monoxide is completely combusted to carbon dioxide after the catalyst is no longer present to provide waste heat for steam production.

Hydrocracking

Hydrocracking is similar to FCC to the extent that this process catalytically cracks the heavy molecules that comprise gas oils by splitting them into smaller molecules which boil in the gasoline, jet fuel, and diesel fuel boiling ranges. The fundamental difference is that Hydrocracking reactions are carried out in an extremely hydrogen-rich environment. In Hydrocracking, two different reactions are occurring in the reactor(s). First, a carbon-to-carbon bond is broken (endothermic cracking reaction), followed by the attachment of hydrogen to the carbon atom (exothermic hydrogenation reaction). The resulting Hydrocracking reactor products are saturated. The net effect of the endothermic (consumes heat) and exothermic (creates heat) reactions is a temperature increase across each Hydrocracking reactor bed because more heat is created in the hydrogenation reactions than is consumed in the cracking reactions.

Typical Hydrocracking feedstocks are the light catalytic gas oil from the FCC (which is very unsaturated), the light coker gas oil and virgin, light gas oils. The heavy naphtha produced from Hydrocracking makes an excellent Catalytic Reformer feedstock due to the significant presence of naphthenes created by saturating aromatics in the gas oil feedstock with hydrogen. Hydrocracking also produces an excellent blendstock for jet fuels. The yields across a Hydrocracking unit may exhibit liquid volume gains of as much as 20-25 per cent! Typically, Hydrocracking reactors contain fixed, multiple catalyst beds.

The catalyst pellets are shaped similarly to Hydrotreating catalysts, however, the active metals impregnated in the silica-alumina catalyst base are typically palladium, platinum, or nickel, depending on the catalyst licensor. Hydrocracking catalysts are poisoned by sulfur and organic nitrogen. This requires the feedstock to be processed in a very high-severity HDS/HDN reactor (containing catalyst beds comprised of alumina-based pellets impregnated with nickel/molybdenum oxides) prior to entering the Hydrocracking reactor(s).

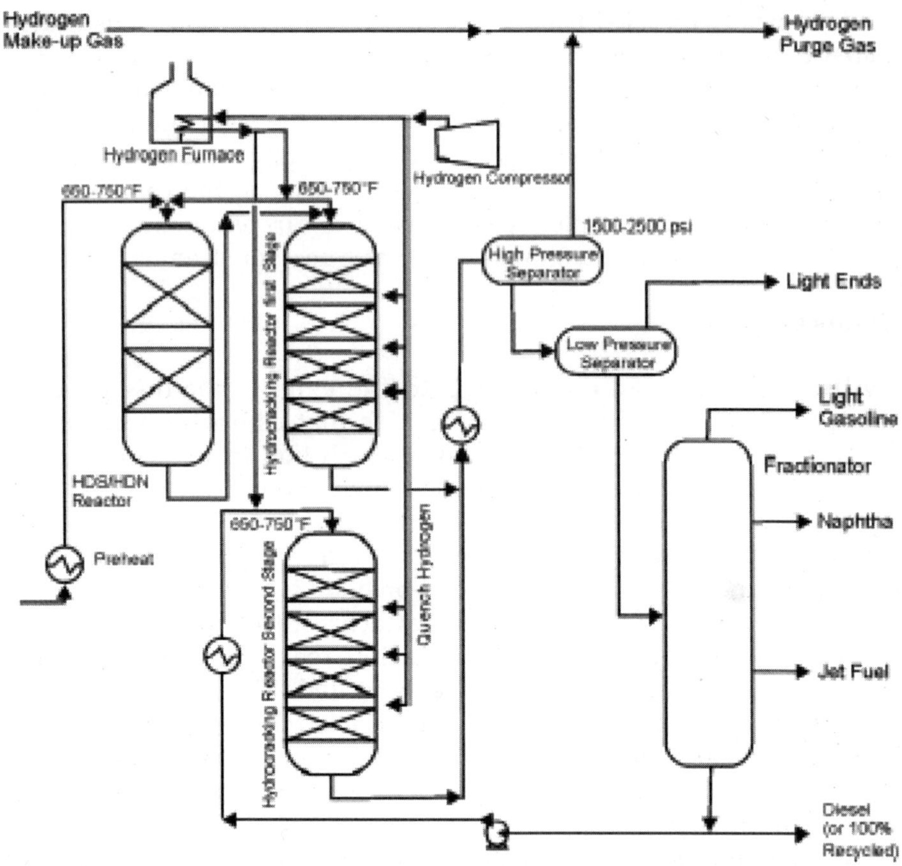

Figure 3.6: Hydrocracking (Two-Stage) Simplified Process Flow

HDS/HDN and Hydrocracking reactors operate at high pressures of 1,500 to 3,000 psi to provide the hydrogen-rich (i.e., hydrogen partial pressure) environment required for the high severity HDS/HDN and Hydrocracking reactions. The reactor temperatures will typically vary from ?start-of-run? levels of approximately 650°F to end-of-run at approximately 800°F. The actual temperature requirements are dependent on type of unit feeds, catalyst activity and target yields based on economics. Hydrocracking reactors contain multiple beds to enable quenching with cool hydrogen between the beds to prevent uncontrolled ?temperature runaways? which could result in catastrophic equipment failure. When the catalyst bed reaches end of run temperatures, the Hydrocracking unit will be scheduled for a shutdown to allow the catalyst beds to be replaced with either fresh or regenerated catalyst.

Hydrocracking units are configured for either a single stage or two stages of Hydrocracking reactors. The second stage reactor enables the overall unit conversion of light gas oils into lower boiling range components to be increased.

This is accomplished by recycling (depending on economics, "recycling to extinction") the diesel oil boiling range stream from the fractionator tower bottoms back to the second stage Hydrocracking reactor where it is converted into lighter boiling range material.

Alkylation

Alkylation is the refinery process that provides an economically feasible outlet for several of the very light olefins produced from the FCC. Propylenes, butylenes, and pentylenes (also known as amylenes) are the names for the olefinic (i.e., having one carbon to carbon double bond instead of being saturated with hydrogen) hydrocarbon groups with three, four, and five carbon atoms, respectively, in their molecular structures. Propylene alkylation is the normal disposition for FCC propylenes unless a petrochemical disposition is readily available. Although some FCC butylenes could be blended into gasolines, the very high vapor pressure of butylenes would prevent a lot of low cost butane blendstock from being blended into gasoline, thereby carrying a very high opportunity cost for this option. In the Alkylation process, the FCC propylenes, butylenes, and when economical, pentylenes are combined with isobutane (a branched, saturated four carbon molecule) in the catalyzed alkylation reaction to produce branched, saturated seven, eight, or nine carbon molecules, respectively. The resulting alkylate consisting of isoheptanes, isooctanes, and isononanes (from the propylenes, butylenes, and pentylenes, respectively) is a low vapor pressure (relative to the feedstocks), very high-octane gasoline blendstock. The high-octane value makes alkylate an excellent blendstock for premium grades of gasolines. Since alkylate contains no olefins, no aromatics, and no sulfur, it is also an excellent blendstock for use in reformulated gasolines.

The Alkylation reaction is catalyzed by the presence of very strong acids either sulfuric acid or hydrofluoric acid. Both sulfuric acid and hydrofluoric acid are extremely corrosive. The fact that hydrofluoric acid exists in a vapor state at typical ambient conditions, dictates that extreme measures must be taken to ensure that this toxic substance is contained inside the process equipment.

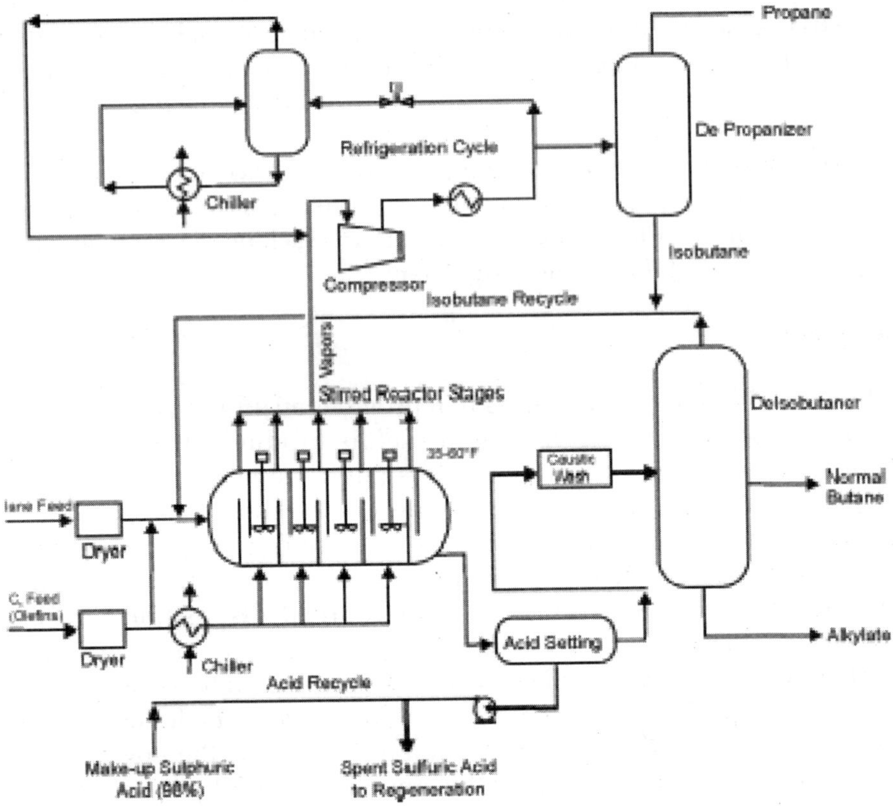

Figure 3.7: Alkylation (Sulfuric Acid) Simplified Process Flow

The Alkylation reactors typically operate at temperatures 35-60°F (70°F maximum to minimize polymerization of the olefins to form undesirable hydrocarbons) for the sulfuric acid process. The hydrofluoric acid process, which is less sensitive to polymerization at warmer temperatures, typically operates at reactor temperatures of 70-100°F. Isobutane concentrations are maintained very high (i.e., at ratios of 4:1 or more above the reaction requirements) in the reactor vessels to ensure that all of the olefins are reacted. The reactor effluent is distilled to separate the propane, isobutane, and alkylate boiling fractions. The propane is routed to propane product treating, the isobutane is recycled back to the alkylation reactors and the alkylate is routed to gasoline blending, or in some cases to additional solvents refinery processing.

Coking

With the exception of the Coking process, the formation of coke (i.e., essentially pure carbon) is an undesirable result of refinery processing operations because coke formation fouls equipment and reduces catalyst activity. However, in the Coking process, coke is intentionally formed to maximize the conversion of the bottom of the barrel vacuum residuum from low value fuel oils to higher value products.

Although there are other Coking process configurations in use today, essentially all of the Coking units in modern petroleum refineries are of the Delayed Coking variety. The Delayed Coking process thermally cracks vacuum residuum (the bottom of the barrel crude oil fraction) into lighter boiling range products such as gas, naphtha, gas oils, and coke. Although the liquid volume yield varies based on feedstock properties, a 75 per cent liquid products yield is approximate.

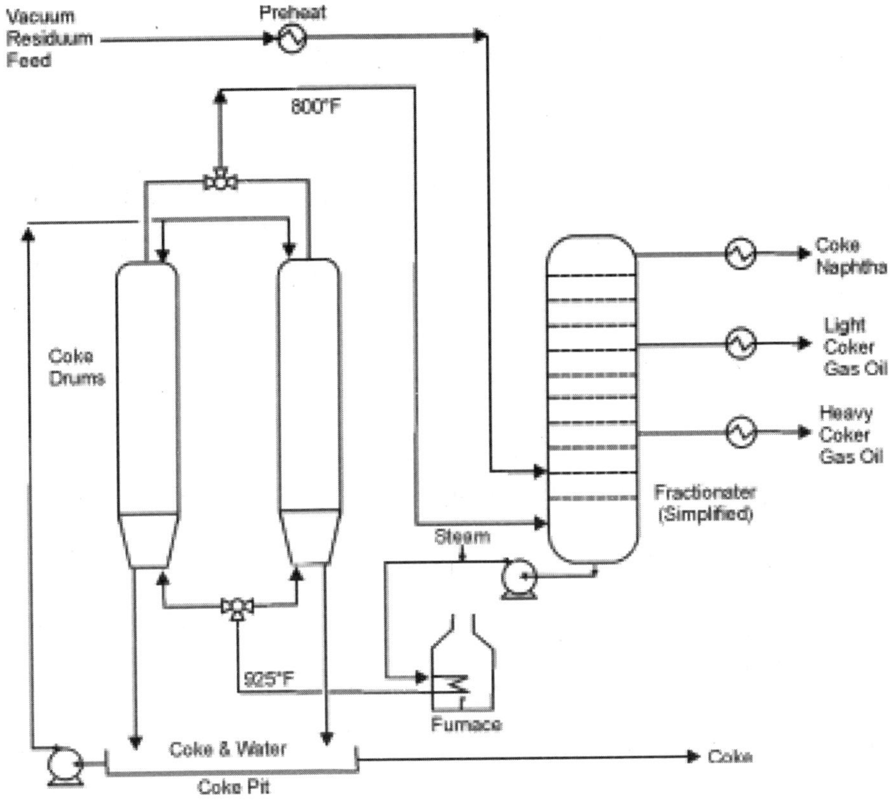

Figure 3.8: Delayed Coking Simplified Process Flow

As previously noted, petroleum coke tends to form whenever hydrocarbons are exposed to temperatures in excess of 750°F. In the Delayed Coking process, the vacuum residuum is first fed to the coker fractionator to remove as many lighter boiling components as possible. The fractionator bottoms is then heated to 900+°F in a direct-fired heater by utilizing high velocities in the heater tubes in conjunction with the addition of steam to minimize the deposition of coke in the heater tubes. The 900°+F residuum stream is then introduced into a coke drum where adequate hold-up time enables coke to form in that specific location. The vaporized, non-coke products from the coke drum are routed to a coker fractionator enabling the distillation of the coker products. Delayed cokers employ two, three, four or more coke drums in parallel which are controllably switched between on-line and off-line status. As one coke drum is being filled, the others are off-line for mechanical or hydraulic drilling to remove of the solid coke.

The coker fractionator distills the unit feed and coking reaction product stream into gas, coker naphtha, light and heavy coker gas oils, and bottoms. The bottoms stream is recycled to extinction via the heater and coke drums. The Delayed Coker fractionated products are very highly olefinic (i.e., unsaturated with carbon to carbon double bonds), even containing diolefins (molecules with two carbon to carbon double bonds). Coker naphtha may be routed to Catalytic Reforming (after Hydrotreating to remove sulfur and saturate the diolefins) or Hydrocracking, depending on current economics. Light coker gas oil may be routed to Hydrocracking or to FCC. Heavy coker gas oil is typically routed to the FCC via Hydrotreating. Coker gas oils may alternatively be routed to fuel oil product blending—possibly after Hydrotreating for sulfur removal.

Ancillary Processes

Hydrogen Production Refineries contain processes that consume hydrogen (Hydrotreaters and Hydrocracking) and those that produce hydrogen (Catalytic Reforming). In some cases, nearby petrochemical operations such as ethylene crackers may provide an additional source of hydrogen to the refinery. In many cases, however, the refinery demand for high-purity hydrogen exceeds the supply. The shortfall is produced in a Hydrogen Generation process unit. A Hydrogen Generation process unit produces a 90+ per cent hydrogen gas product stream from light ends feedstocks through a process called Steam Reforming. The Steam Reforming process employs several different chemical reactions including steam reforming, followed by shift conversion, and then methanation. Sulfur represents a severe poison to the steam reforming catalyst, which requires that essentially all the sulfur in the feed gas must be removed, usually by passing the gas through a fixed bed of zinc oxide which reacts with the sulfur.

Steam Reforming reactions occur at temperatures of approximately 1,500°F in the presence of a nickel based catalyst typically shaped in hollow rings located inside the tubes of a direct-fired furnace. The reactions are:

$$CH_4 + H_2O \longrightarrow CO + 3\,H_2 \text{ (most)}$$

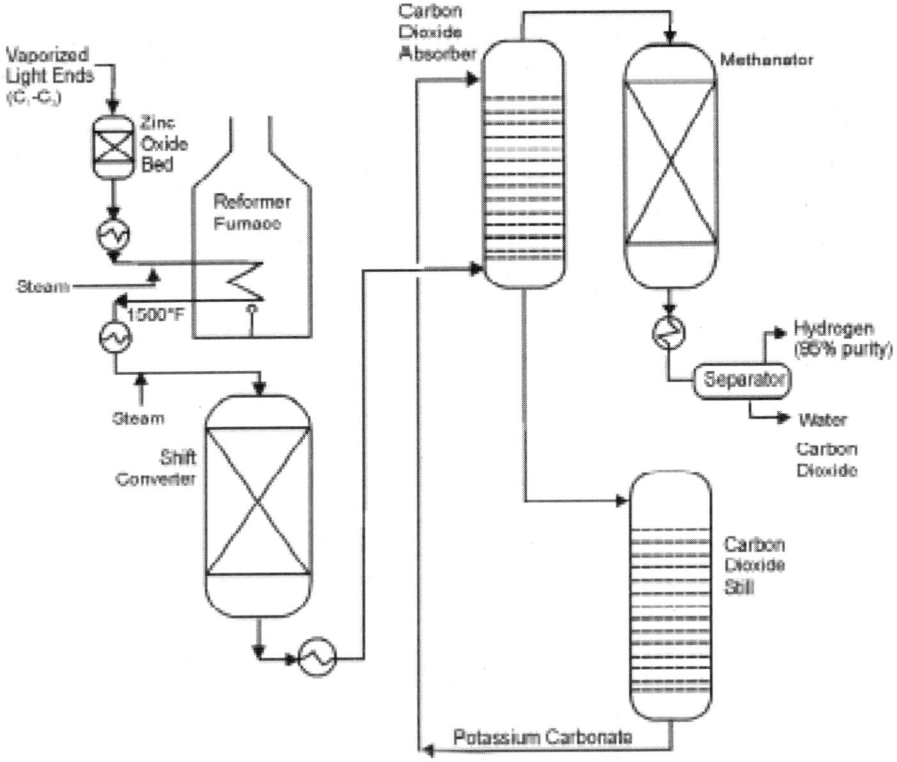

Figure 3.9: Hydrogen Generation Simplified Process Flow

$$CH_4 + 2\,H_2O \longrightarrow CO_2 + 4\,H_2 \text{ (some)}$$

The Steam Reforming reactor products are then cooled by heat exchange to generate waste heat and more steam is added prior to entering the Shift Conversion reactor. The Shift Conversion reaction takes place in a fixed-bed reactor consisting of an iron or iron/chromium catalyst. The Shift Conversion reaction converts the carbon monoxide to carbon dioxide and produces additional hydrogen according to the following reaction:

$$CO + H_2O \longrightarrow CO_2 + H_2$$

The Shift Converter product stream is then scrubbed, usually through absorption with a potassium carbonate solution to remove the carbon dioxide. The potassium carbonate solution is regenerated in a Carbon Dioxide Still by applying reboiler heat to the tower bottoms. This heat drives off the carbon dioxide from the solution which is then re-circulated.

Since carbon monoxide (CO) and carbon dioxide (CO_2) are poisons to the catalysts of some of the hydrogen consuming refinery processes, Methanation is employed as the final step to remove any remaining CO and CO2 in the hydrogen stream. The Methanation reaction takes place in a fixed-bed reactor consisting of a nickel-based catalyst. The resulting hydrogen product stream is typically approximately 95 per cent hydrogen and the balance methane with only trace amounts of CO and CO_2. The Methanation reactions are:

$$CO + 3H_2 \longrightarrow CH_4 + H_2O \text{ and } CO_2 + 4H_2 \longrightarrow CH_4 + 2H_2O$$

Depending on the refiners' overall hydrogen balance and requirements for high-purity hydrogen, the Hydrogen Plant may be used in effect as a hydrogen purification unit by processing low-purity hydrogen feed streams. The hydrogen contained in the feed passes through the Hydrogen Plant unchanged, whereas the hydrocarbon impurities in the feed stream are converted into hydrogen.

Sulfur Conversion Hydrogen sulfide (H_2S) is a highly toxic gas that originates in crude oils and is also produced in the Hydrotreating, Hydrocracking, FCC, and Delayed Coking process reactions. Simply burning H2S as a fuel gas component is precluded by safety and environmental considerations since one of the combustion products is sulfur dioxide (SO_2), which is also toxic. Special operating and maintenance procedures are in place for any refinery process equipment containing H_2S in concentrations above 50 parts per million. H_2S is typically removed from the refinery light ends gas streams through amine absorption which entails contacting the gas stream with an amine (usually an aqueous solution of monoethanolamine or diethanolamine) in a contactor and then regenerating the amine by applying heat to drive off the H_2S into a concentrated gas stream referred to as acid gas.

The Sulfur Conversion process takes the acid gas stream and converts the toxic H_2S into the benign products of elemental sulfur and water. The Sulfur Conversion process utilized in most modern refineries is a variant of the process first invented by Claus in the late 19th century—hence the name "Claus process".

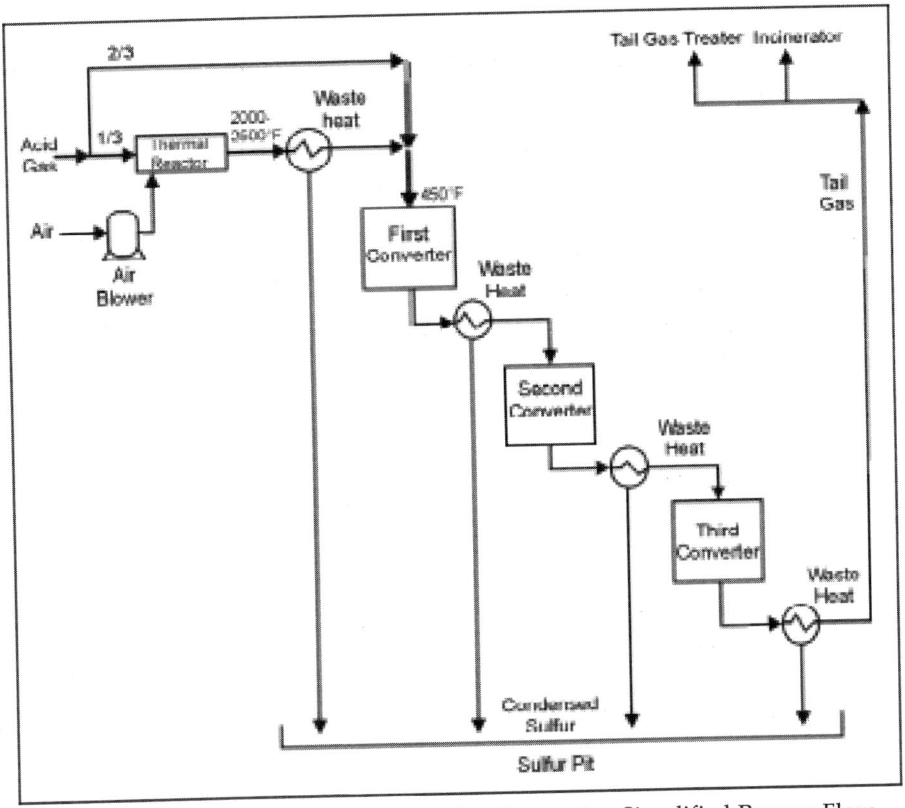

Figure 3.10: Split-Flow Claus-Type Sulfur Conversion Simplified Process Flow

In general, the Claus process entails combusting one-third of the H_2S to sulfur dioxide (SO2) and then reacting the SO_2 with the remaining H_2S in the presence of a fixed bed of activated alumina, cobalt molybdenum catalyst to form elemental sulfur. The combustion and conversion chemical reactions occurring are:

$$2H_2S \quad 3O_2 \quad 2So_2 \quad 2H_2O \qquad \text{(Combustion)}$$

$$2H_2S \quad SO_2 \quad 3S \quad 2H_2O \qquad \text{(Conversion)}$$

Different process flow configurations are in use to achieve the proper "Claus" 2:1 hydrogen sulfide to sulfur dioxide ratio in the conversion reactors. In a "Split-Flow" configuration, a one-third split of the acid gas stream is completely combusted and the combustion products are then combined with the noncombusted acid gas upstream of the conversion reactors. In a "Once-Through" configuration, the acid gas stream is partially combusted by only providing sufficient oxygen in the combustion chamber to combust one-third of the acid gas. Depending on the total level of H_2S

conversion required, two or in most cases 3 stages of conversion reactors are required. Each additional stage provides incrementally less conversion than the previous stage. Conversion of 96-97 per cent of the H_2S to elemental sulfur is achievable in a Claus-type process.

Finally, if required to meet air quality standards, a Tail Gas Treater is utilized to remove essentially all of the remaining H_2S in the tail gas from the Sulfur Conversion process. A Tail Gas Treater employs a proprietary solution to absorb the H_2S and convert it to elemental sulfur.

Product Blending

Refinery products are typically the result of blending several component streams or ?blendstocks?. In most cases, product blending is accomplished by controlling the volumes of blendstocks from individual component storage tanks that are mixed in the finished product storage tank. Samples of the finished blend are then analyzed by laboratory testing for all product specifications prior to shipping. Alternatively, ?in line? blending refers to pipeline shipments in which the finished product is actually blended directly into the product pipeline (as opposed to a "standing" product storage tank).

The applicable specifications vary by product but typically include density, distillation, and sulfur content specifications. Examples of other product specifications include Octane number, Cetane number, Reid Vapor Pressure, Flash, Pour Point, Smoke Point, etc. Many of these product specifications do not blend linearly by component volumes. In these circumstances, the finished blend properties are predicted using experience-based algorithms for the applicable blendstock components. These algorithms are very sophisticated and their precision is absolutely critical for in line blending?the costs associated with offsite re-processing or re-blending of any off-specification product are prohibitive.

Refiners blend hundreds of different specification products. For example, gasoline is not a single product. We are all familiar with the regular, mid-grade, and premium grades of gasoline segregated at the retail pump by octane number. However, gasolines are subject to different specifications based on the geographic location specific temperature and humidity zones, as well as, the season of the year (i.e., summer or winter) to optimize performance in modern automobile engines. Environmental regulations have even further complicated the picture by introducing additional classifications such as conventional, oxygenated, and reformulated gasolines.

Lubricant, Wax, and Grease Manufacturing Processes

Description

Lubricating oils and waxes are refined from the residual fractions of atmospheric and vacuum distillation. The primary objective of the various lubricating oil refinery processes is to remove asphalts, sulfonated aromatics, and paraffinic and isoparaffinic waxes from residual fractions. reduced crude from the vacuum unit is deasphalted and combined with straight-run lubricating oil feedstock, preheated, and solvent-extracted (usually with phenol or furfural) to produce raffinate.

Wax Manufacturing Process

Raffinate from the extraction unit contains a considerable amount of wax that must be removed by solvent extraction and crystallization. The raffinate is mixed with a solvent (propane) and precooled in heat exchangers. The crystallization temperature is attained by the evaporation of propane in the chiller and filter feed tanks. The wax is continuously removed by filters and cold solvent-washed to recover retained oil. The solvent is recovered from the oil by flashing and steam stripping. The wax is then heated with hot solvent, chilled, filtered, and given a final wash to remove all oil.

Lubricating Oil Process

The dewaxed raffinate is blended with other distillate fractions and further treated for viscosity index, color, stability, carbon residue, sulfur, additive response, and oxidation stability in extremely selective extraction processes using solvents (furfural, phenol, etc.). In a typical phenol unit, the raffinate is mixed with phenol in the treating section at temperatures below 400° F. Phenol is then separated from the treated oil and recycled. The treated lube-oil base stocks are then mixed and/or compounded with additives to meet the required physical and chemical characteristics of motor oils, industrial lubricants, and metal working oils.

Table 3.1: Lubricating Oil and Wax Manufacturing Processes

Feedstock	*From*	*Process*	*Typicalproducts*	*To*
Lube feedstock and additives	Vacuum tower, solvent dewaxing, hydrotreating solvent extraction, etc.	Treatment	Dewaxed raffinate	Lube blend or compound, grease compounding
			Wax	Storage or shipping

Grease Compounding

Grease is made by blending metallic soaps (salts of long-chained fatty acids) and additives into a lubricating oil medium at temperatures of 400°-600° F. Grease may be either batch-produced or continuously compounded. The characteristics of the grease depend to a great extent on the metallic element (calcium, sodium, aluminum, lithium, etc.) in the soap and the additives used.

Safety and Health Considerations

(a) *Fire protection and prevention*: The potential for fire exists if a product or vapor leak or release in the lube blending and wax processing areas reaches a source of ignition. Storage of finished products, both bulk and packaged, should be in accordance with recognized practices.
While the potential for fire is reduced in lube oil blending, care must be taken when making metal-working oils and compounding greases due to the use of higher blending and compounding temperatures and lower flash point products.

(b) *Safety*: Control of treater temperature is important as phenol can cause corrosion above 400° F. Batch and in-line blending operations require strict controls to maintain desired product quality. Spills should be cleaned and leaks repaired to avoid slips and falls. Additives in drums and bags need to be handled properly to avoid strain. Wax can clog sewer or oil drainage systems and interfere with wastewater treatment.

(c) *Health*: When blending, sampling, and compounding, personal protection from steam, dusts, mists, vapors, metallic salts, and other additives is appropriate. Skin contact with any formulated grease or lubricant should be avoided. Safe work practices and/or appropriate personal protection may be needed for exposures to chemicals and other hazards such as noise and heat; during inspection, maintenance, and turnaround activities; and while sampling and handling hydrocarbons and chemicals during the production of lubricating oil and wax.

Heat Exchangers, Coolers, and Process Heaters

Heating Operations

Process heaters and heat exchangers preheat feedstock in distillation towers and in refinery processes to reaction temperatures. Heat exchangers use either steam or hot hydrocarbon transferred from some other section of the process for heat input. The heaters are usually designed for specific process operations, and most are of cylindrical vertical or box-type designs. The major portion of heat provided to process units comes from fired heaters fueled by refinery or natural gas, distillate, and residual oils. Fired heaters are found on crude and reformer preheaters, coker heaters, and large-column reboilers.

Cooling Operations

Heat also may be removed from some processes by air and water exchangers, fin fans, gas and liquid coolers, and overhead condensers, or by transferring heat to other systems. The basic mechanical vapor-compression refrigeration system, which may serve one or more process units, includes an evaporator, compressor, condenser, controls, and piping. Common coolants are water, alcohol/water mixtures, or various glycol solutions.

Health and Safety Considerations

(a) *Fire Protection and Prevention*: A means of providing adequate draft or steam purging is required to reduce the chance of explosions when lighting fires in heater furnaces. Specific start-up and emergency procedures are required for each type of unit. If fire impinges on fin fans, failure could occur due to overheating. If flammable product escapes from a heat exchanger or cooler due to a leak, fire could occur.

(b) *Safety*: Care must be taken to ensure that all pressure is removed from heater tubes before removing header or fitting plugs. Consideration should be given to providing for pressure relief in heat-exchanger piping systems in the event they are blocked off while full of liquid. If controls fail, variations of temperature and pressure could occur on either side of the heat exchanger. If heat exchanger tubes fail and process pressure is greater than heater pressure, product could enter the heater with downstream consequences. If the process pressure is less than heater pressure, the heater stream could enter into the process fluid. If loss of circulation occurs in liquid or gas coolers, increased product temperature could affect downstream operations and require pressure relief.

(c) *Health*: Because these are closed systems, exposures under normal operating conditions are expected to be minimal. Depending on the fuel, process operation, and unit design, there is a potential for exposure to hydrogen sulfide, carbon monoxide, hydrocarbons, steam boiler feed-water sludge, and water-treatment chemicals. Skin contact should be avoided with boiler blowdown, which may contain phenolic compounds. Safe work practices and/or appropriate personal protective equipment against hazards may be needed during process maintenance, inspection, and turnaround activities and for protection from radiant heat, superheated steam, hot hydrocarbon, and noise exposures.

Steam Generation

Heater and Boiler Operations

Steam is generated in main generation plants, and/or at various process units using heat from flue gas or other sources. Heaters (furnaces) include burners and a combustion air system, the boiler enclosure in which heat transfer takes place, a draft or pressure system to remove flue gas from the furnace, soot blowers, and compressed-air systems that seal openings to prevent the escape of flue gas. Boilers consist of a number of tubes that carry the water-steam mixture through the furnace for maximum heat transfer. These tubes run between steam-distribution drums at the top of the boiler and water-collecting drums at the bottom of the boiler. Steam flows from the steam drum to the superheater before entering the steam distribution system.

Heater Fuel

(a) Heaters may use any one or combination of fuels including refinery gas, natural gas, fuel oil, and powdered coal. Refinery off-gas is collected from process units and combined with natural gas and LPG in a fuel-gas balance drum. The balance drum provides constant system pressure, fairly stable Btu-content fuel, and automatic separation of suspended liquids in gas vapors, and it prevents carryover of large slugs of condensate into the distribution system. Fuel oil is typically a mix of refinery crude oil with straight-run and cracked residues and other products. The fuel-oil system delivers fuel to process-unit heaters and steam generators at required temperatures and pressures. The fuel oil is heated to pumping temperature, sucked through a coarse suction strainer, pumped to a temperature-control heater, and then pumped through a fine-mesh strainer before being burned.

(b) In one example of process-unit heat generation, carbon monoxide boilers recover heat in catalytic cracking units as carbon monoxide in flue gas is burned to complete combustion. In other processes, waste-heat recovery units use heat from the flue gas to make steam.

Steam Distribution

The distribution system consists of valves, fittings, piping, and connections suitable for the pressure of the steam transported. Steam leaves the boilers at the highest pressure required by the process units or electrical generation. The steam pressure is then reduced in turbines that drive process pumps and compressors. Most steam used in the refinery is condensed to water in various types of heat exchangers. The condensate is reused as boiler feedwater or discharged to wastewater treatment. When refinery steam is also used to drive steam turbine generators to produce electricity, the steam must be produced at much higher pressure than required for process steam. Steam typically is generated by heaters (furnaces) and boilers combined in one unit.

Health and Safety Considerations

Crude Oil Pretreatment (desalting)

(a) *Fire Prevention and Protection*: The potential exists for a fire due to a leak or release of crude from heaters in the crude desalting unit. Low boiling point components of crude may also be released if a leak occurs.

(b) *Safety*: Inadequate desalting can cause fouling of heater tubes and heat exchangers throughout the refinery. Fouling restricts product flow and heat transfer and leads to failures due to increased pressures and temperatures. Corrosion, which occurs due to the presence of hydrogen sulfide, hydrogen chloride, naphthenic (organic) acids, and other contaminants in the crude oil, also causes equipment failure. Neutralized salts (ammonium chlorides and sulfides), when moistened by condensed water, can cause corrosion. Overpressuring the unit is another potential hazard that causes failures.

(c) *Health*: Because this is a closed process, there is little potential for exposure to crude oil unless a leak or release occurs.
Where elevated operating temperatures are used when desalting sour crudes, hydrogen sulfide will be present. There is the

possibility of exposure to ammonia, dry chemical demulsifiers, caustics, and/or acids during this operation. Safe work practices and/or the use of appropriate personal protective equipment may be needed for exposures to chemicals and other hazards such as heat, and during process sampling, inspection, maintenance, and turnaround activities.

Depending on the crude feedstock and the treatment chemicals used, the wastewater will contain varying amounts of chlorides, sulfides, bicarbonates, ammonia, hydrocarbons, phenol, and suspended solids. If diatomaceous earth is used in filtration, exposures should be minimized or controlled. Diatomaceous earth can contain silica in very fine particle size, making this a potential respiratory hazard.

Crude Oil Distillation (Fractionation)

(a) *Fire Prevention and Protection*: Even though these are closed processes, heaters and exchangers in the atmospheric and vacuum distillation units could provide a source of ignition, and the potential for a fire exists should a leak or release occur.

(b) *Safety*: An excursion in pressure, temperature, or liquid levels may occur if automatic control devices fail. Control of temperature, pressure, and reflux within operating parameters is needed to prevent thermal cracking within the distillation towers. Relief systems should be provided for overpressure and operations monitored to prevent crude from entering the reformer charge. The sections of the process susceptible to corrosion include (but may not be limited to) preheat exchanger (HCl and H_2S), preheat furnace and bottoms exchanger (H_2S and sulfur compounds), atmospheric tower and vacuum furnace (H_2S, sulfur compounds, and organic acids), vacuum tower (H_2S and organic acids), and overhead (H_2S, HCl, and water). Where sour crudes are processed, severe corrosion can occur in furnace tubing and in both atmospheric and vacuum towers where metal temperatures exceed 450° F. Wet H_2S also will cause cracks in steel. When processing high-nitrogen crudes, nitrogen oxides can form in the flue gases of furnaces. Nitrogen oxides are corrosive to steel when cooled to low temperatures in the presence of water.

Chemicals are used to control corrosion by hydrochloric acid produced in distillation units. Ammonia may be injected into the overhead stream prior to initial condensation and/or an alkaline solution may be carefully injected into the hot crude-oil feed.

If sufficient wash-water is not injected, deposits of ammonium chloride can form and cause serious corrosion. Crude feedstock may contain appreciable amounts of water in suspension which can separate during startup and, along with water remaining in the tower from steam purging, settle in the bottom of the tower. This water can be heated to the boiling point and create an instantaneous vaporization explosion upon contact with the oil in the unit.

(c) *Health*: Atmospheric and vacuum distillation are closed processes and exposures are expected to be minimal. When sour (high-sulfur) crudes are processed, there is potential for exposure to hydrogen sulfide in the preheat exchanger and furnace, tower flash zone and overhead system, vacuum furnace and tower, and bottoms exchanger. Hydrogen chloride may be present in the preheat exchanger, tower top zones, and overheads. Wastewater may contain water-soluble sulfides in high concentrations and other water-soluble compounds such as ammonia, chlorides, phenol, mercaptans, etc., depending upon the crude feedstock and the treatment chemicals. Safe work practices and/or the use of appropriate personal protective equipment may be needed for exposures to chemicals and other hazards such as heat and noise, and during sampling, inspection, maintenance, and turnaround activities.

Solvent Extraction and Dewaxing

(a) *Fire prevention and protection*: Solvent treatment is essentially a closed process and, although operating pressures are relatively low, the potential exists for fire from a leak or spill contacting a source of ignition such as the drier or extraction heater. In solvent dewaxing, disruption of the vacuum will create a potential fire hazard by allowing air to enter the unit.

(b) *Health*: Because solvent extraction is a closed process, exposures are expected to be minimal under normal operating conditions. However, there is a potential for exposure to extraction solvents such as phenol, furfural, glycols, methyl ethyl ketone, amines, and other process chemicals. Safe work practices and/or the use of appropriate personal protective equipment may be needed for exposures to chemicals and other hazards such as noise and heat, and during repair, inspection, maintenance, and turnaround activities.

Thermal Cracking

(a) *Fire protection and prevention*: Because thermal cracking is a closed process, the primary potential for fire is from leaks or releases of liquids, gases, or vapors reaching an ignition source such as a heater. The potential for fire is present in coking operations due to vapor or product leaks. Should coking temperatures get out of control, an exothermic reaction could occur within the coker.

(b) *Safety*: In thermal cracking when sour crudes are processed, corrosion can occur where metal temperatures are between 450° and 900° F. above 900° F coke forms a protective layer on the metal. The furnace, soaking drums, lower part of the tower, and high-temperature exchangers are usually subject to corrosion. Hydrogen sulfide corrosion in coking can also occur when temperatures are not properly controlled above 900° F.
Continuous thermal changes can lead to bulging and cracking of coke drum shells. In coking, temperature control must often be held within a 10°-20° F range, as high temperatures will produce coke that is too hard to cut out of the drum. Conversely, temperatures that are too low will result in a high asphaltic-content slurry. Water or steam injection may be used to prevent buildup of coke in delayed coker furnace tubes. Water must be completely drained from the coker, so as not to cause an explosion upon recharging with hot coke. Provisions for alternate means of egress from the working platform on top of coke drums are important in the event of an emergency.

(c) *Health*: The potential exists for exposure to hazardous gases such as hydrogen sulfide and carbon monoxide, and trace polynuclear aromatics (PNA's) associated with coking operations. When coke is moved as a slurry, oxygen depletion may occur within confined spaces such as storage silos, since wet carbon will adsorb oxygen. Wastewater may be highly alkaline and contain oil, sulfides, ammonia, and/or phenol. The potential exists in the coking process for exposure to burns when handling hot coke or in the event of a steam-line leak, or from steam, hot water, hot coke, or hot slurry that may be expelled when opening cokers. Safe work practices and/or the use of appropriate personal protective equipment may be needed for exposures to chemicals and other hazards such as heat and noise, and during process sampling, inspection, maintenance, and turnaround activities. (Note: coke produced from petroleum is a different product from that generated in the steel-industry coking process.)

Fluid Catalytic Cracking

(a) *Fire prevention and protection*: Liquid hydrocarbons in the catalyst or entering the heated combustion air stream should be controlled to avoid exothermic reactions. Because of the presence of heaters in catalytic cracking units, the possibility exists for fire due to a leak or vapor release. Fire protection including concrete or other insulation on columns and supports, or fixed water spray or fog systems where insulation is not feasible and in areas where firewater hose streams cannot reach, should be considered.
In some processes, caution must be taken to prevent explosive concentrations of catalyst dust during recharge or disposal. When unloading any coked catalyst, the possibility exists for iron sulfide fires. Iron sulfide will ignite spontaneously when exposed to air and therefore must be wetted with water to prevent it from igniting vapors. Coked catalyst may be either cooled below 120° F before it is dumped from the reactor, or dumped into containers that have been purged and inerted with nitrogen and then cooled before further handling.

(b) *Safety*: Regular sampling and testing of the feedstock, product, and recycle streams should be performed to assure that the cracking process is working as intended and that no contaminants have entered the process stream. Corrosives or deposits in the feedstock can foul gas compressors. Inspections of critical equipment including pumps, compressors, furnaces, and heat exchangers should be conducted as needed. When processing sour crude, corrosion may be expected where temperatures are below 900° F. Corrosion takes place where both liquid and vapor phases exist, and at areas subject to local cooling such as nozzles and platform supports.
When processing high-nitrogen feedstock, exposure to ammonia and cyanide may occur, subjecting carbon steel equipment in the FCC overhead system to corrosion, cracking, or hydrogen blistering. These effects may be minimized by water wash or corrosion inhibitors. Water wash may also be used to protect overhead condensers in the main column subjected to fouling from ammonium hydrosulfide. Inspections should include checking for leaks due to erosion or other malfunctions such as catalyst buildup on the expanders, coking in the overhead feeder lines from feedstock residues, and other unusual operating conditions.

(c) *Health*: Because the catalytic cracker is a closed system, there is normally little opportunity for exposure to hazardous substances during normal operations. The possibility exists of exposure to extremely hot (700° F) hydrocarbon liquids or vapors during process sampling or if a leak or release occurs. In addition, exposure to hydrogen sulfide and/or carbon monoxide gas may occur during a release of product or vapor.

Catalyst regeneration involves steam stripping and decoking, and produces fluid waste streams that may contain varying amounts of hydrocarbon, phenol, ammonia, hydrogen sulfide, mercaptan, and other materials depending upon the feedstock, crudes, and processes. Inadvertent formation of nickel carbonyl may occur in cracking processes using nickel catalysts, with resultant potential for hazardous exposures. Safe work practices and/or the use of appropriate personal protective equipment may be needed for exposures to chemicals and other hazards such as noise and heat; during process sampling, inspection, maintenance and turnaround activities; and when handling spent catalyst, recharging catalyst, or if leaks or releases occur.

Hydrocracking

(a) *Fire prevention and protection*: Because this unit operates at very high pressures and temperatures, control of both hydrocarbon leaks and hydrogen releases is important to prevent fires. In some processes, care is needed to ensure that explosive concentrations of catalytic dust do not form during recharging.

(b) *Safety*: Inspection and testing of safety relief devices are important due to the very high pressures in this unit. Proper process control is needed to protect against plugging reactor beds. Unloading coked catalyst requires special precautions to prevent iron sulfide-induced fires. The coked catalyst should either be cooled to below 120° F before dumping, or be placed in nitrogen-inerted containers until cooled.

Because of the operating temperatures and presence of hydrogen, the hydrogen-sulfide content of the feedstock must be strictly controlled to a minimum to reduce the possibility of severe corrosion. Corrosion by wet carbon dioxide in areas of condensation also must be considered. When processing high-nitrogen feedstock, the ammonia and hydrogen sulfide form ammonium hydrosulfide, which causes serious corrosion at temperatures below the water dew point. Ammonium hydrosulfide is also present in sour water stripping.

(c) *Health*: Because this is a closed process, exposures are expected to be minimal under normal operating conditions. There is a potential for exposure to hydrocarbon gas and vapor emissions, hydrogen and hydrogen sulfide gas due to high-pressure leaks. Large quantities of carbon monoxide may be released during catalyst regeneration and changeover. Catalyst steam stripping and regeneration create waste streams containing sour water and ammonia. Safe work practices and/or the use of appropriate personal protective equipment may be needed for exposure to chemicals and other hazards such as noise and heat, during process sampling, inspection, maintenance, and turnaround activities, and when handling spent catalyst.

Catalytic Reforming

(a) *Fire prevention and protection*: This is a closed system; however, the potential for fire exists should a leak or release of reformate gas or hydrogen occur.

(b) *Safty*: Operating procedures should be developed to ensure control of hot spots during start-up. Safe catalyst handling is very important. Care must be taken not to break or crush the catalyst when loading the beds, as the small fines will plug up the reformer screens. Precautions against dust when regenerating or replacing catalyst should also be considered. Also, water wash should be considered where stabilizer fouling has occurred due to the formation of ammonium chloride and iron salts. Ammonium chloride may form in pretreater exchangers and cause corrosion and fouling. Hydrogen chloride from the hydrogenation of chlorine compounds may form acid or ammonium chloride salt.

(c) *Health*: Because this is a closed process, exposures are expected to be minimal under normal operating conditions. There is potential for exposure to hydrogen sulfide and benzene should a leak or release occur.

Small emissions of carbon monoxide and hydrogen sulfide may occur during regeneration of catalyst. Safe work practices and/ or appropriate personal protective equipment may be needed for exposures to chemicals and other hazards such as noise and heat during testing, inspecting, maintenance and turnaround activities, and when handling regenerated or spent catalyst.

Catalytic Hydrotreating

(a) *Fire prevention and protection*: The potential exists for fire in the event of a leak or release of product or hydrogen gas.

(b) *Safty*: Many processes require hydrogen generation to provide for a continuous supply. Because of the operating temperatures and presence of hydrogen, the hydrogen sulfide content of the feedstock must be strictly controlled to a minimum to reduce corrosion. Hydrogen chloride may form and condense as hydrochloric acid in the lower-temperature parts of the unit. Ammonium hydrosulfide may form in high-temperature, high-pressure units. Excessive contact time and/or temperature will create coking. Precautions need to be taken when unloading coked catalyst from the unit to prevent iron sulfide fires. The coked catalyst should be cooled to below 120° F before removal, or dumped into nitrogen-inerted bins where it can be cooled before further handling. Special antifoam additives may be used to prevent catalyst poisoning from silicone carryover in the coker feedstock.

(c) *Healthy*: Because this is a closed process, exposures are expected to be minimal under normal operating conditions. There is a potential for exposure to hydrogen sulfide or hydrogen gas in the event of a release, or to ammonia should a sour-water leak or spill occur. Phenol also may be present if high boiling-point feedstocks are processed. Safe work practices and/or appropriate personal protective equipment may be needed for exposures to chemicals and other hazards such as noise and heat; during process sampling, inspection, maintenance, and turnaround activities; and when handling amine or exposed to catalyst.

Isomerization

(a) *Fire protection and prevention*: Although this is a closed process, the potential for a fire exists should a release or leak contact a source of ignition such as the heater.

(b) *Safety*: If the feedstock is not completely dried and desulfurized, the potential exists for acid formation leading to catalyst poisoning and metal corrosion. Water or steam must not be allowed to enter areas where hydrogen chloride is present. Precautions are needed to prevent HCl from entering sewers and drains.

(c) *Health*: Because this is a closed process, exposures are expected to be minimal during normal operating conditions. There is a potential for exposure to hydrogen gas, hydrochloric acid, and hydrogen chloride and to dust when solid catalyst is used. Safe work practices and/or appropriate personal protective equipment may be needed for exposures to chemicals and other hazards such as heat and noise, and during process sampling, inspection, maintenance, and turnaround activities.

Polymerization

(a) *Fire prevention and protection*: Polymerization is a closed process where the potential for a fire exists due to leaks or releases reaching a source of ignition.

(b) *Safety*: The potential for an uncontrolled exothermic reaction exists should loss of cooling water occur. Severe corrosion leading to equipment failure will occur should water make contact with the phosphoric acid, such as during water washing at shutdowns. Corrosion may also occur in piping manifolds, reboilers, exchangers, and other locations where acid may settle out.

(c) *Health*: Because this is a closed system, exposures are expected to be minimal under normal operating conditions. There is a potential for exposure to caustic wash (sodium hydroxide), to phosphoric acid used in the process or washed out during turnarounds, and to catalyst dust. Safe work practices and/or appropriate personal protective equipment may be needed for exposures to chemicals and other hazards such as noise and heat, and during process sampling, inspection, maintenance, and turnaround activities.

Alkylation

(a) *Fire protection and prevention*: Alkylation units are closed processes; however, the potential exists for fire should a leak or release occur that allows product or vapor to reach a source of ignition.

(b) *Safety*: Sulfuric acid and hydrofluoric acid are potentially hazardous chemicals. Loss of coolant water, which is needed to maintain process temperatures, could result in an upset. Precautions are necessary to ensure that equipment and materials that have been in contact with acid are handled carefully and are thoroughly cleaned before they leave the process area or refinery. Immersion wash vats are often provided for neutralization of equipment that has come into contact with hydrofluoric acid. Hydrofluoric acid units should be thoroughly drained and chemically cleaned prior to turnarounds and entry to remove all traces of iron fluoride and hydrofluoric acid. Following shutdown, where water has been used the unit should be thoroughly dried before hydrofluoric acid is introduced.

Leaks, spills, or releases involving hydrofluoric acid or hydrocarbons containing hydrofluoric acid can be extremely hazardous. Care during delivery and unloading of acid is essential. Process unit containment by curbs, drainage, and isolation so that effluent can be neutralized before release to the sewer system is considered. Vents can be routed to soda-ash scrubbers to neutralize hydrogen fluoride gas or hydrofluoric acid vapors before release. Pressure on the cooling water and steam side of exchangers should be kept below the minimum pressure on the acid service side to prevent water contamination.

Some corrosion and fouling in sulfuric acid units may occur from the breakdown of sulfuric acid esters or where caustic is added for neutralization. These esters can be removed by fresh acid treating and hot-water washing. To prevent corrosion from hydrofluoric acid, the acid concentration inside the process unit should be maintained above 65 per cent and moisture below 4 per cent.

(c) *Health*: Because this is a closed process, exposures are expected to be minimal during normal operations. There is a potential for exposure should leaks, spills, or releases occur. Sulfuric acid and (particularly) hydrofluoric acid are potentially hazardous chemicals. Special precautionary emergency preparedness measures and protection appropriate to the potential hazard and areas possibly affected need to be provided. Safe work practices and appropriate skin and respiratory personal protective equipment are needed for potential exposures to hydrofluoric and sulfuric acids during normal operations such as reading gauges, inspecting, and process sampling, as well as during emergency response, maintenance, and turnaround activities. Procedures should be in place to ensure that protective equipment and clothing worn in hydrofluoric acid activities are decontaminated and inspected before reissue. Appropriate personal protection for exposure to heat and noise also may be required.

Sweetening and Treating Processes

(a) *Fire protection and prevention*: The potential exists for fire from a leak or release of feedstock or product. Sweetening processes use air or oxygen. If excess oxygen enters these processes, it is possible for a fire to occur in the settler due to the generation of static electricity, which acts as the ignition source.

(b) *Health*: Because these are closed processes, exposures are expected to be minimal under normal operating conditions. There is a potential for exposure to hydrogen sulfide, caustic (sodium hydroxide), spent caustic, spent catalyst (Merox), catalyst dust and sweetening agents (sodium carbonate and sodium bicarbonate). Safe work practices and/or appropriate personal protective equipment may be needed for exposures to chemicals and other hazards such as noise and heat, and during process sampling, inspection, maintenance, and turnaround activities.

Unsaturated Gas Plants

(a) *Fire prevention and protection*: The potential of a fire exists should spills, releases, or vapors reach a source of ignition.

(b) *Safety*: In unsat gas plants handling FCC feedstock, the potential exists for corrosion from moist hydrogen sulfide and cyanides. When feedstocks are from the delayed coker or the TCC, corrosion from hydrogen sulfide and deposits in the high pressure sections of gas compressors from ammonium compounds is possible.

(c) *Health*: Because these are closed processes, exposures are expected to be minimal under normal operating conditions. There is a potential for exposures to amine compounds such as monoethanolamine (MEA), diethanolamine (DEA) and methyldiethanolamine (MDEA) and hydrocarbons. Safe work practices and/or appropriate personal protective equipment may be needed for exposures to chemicals and other hazards such as noise and heat, and during process sampling, inspection, maintenance, and turnaround activities.

Amine Plants

(a) *Fire protection and prevention*: The potential for fire exists where a spill or leak could reach a source of ignition.

(b) *Safety*: To minimize corrosion, proper operating practices should be established and regenerator bottom and reboiler temperatures controlled. Oxygen should be kept out of the system to prevent amine oxidation.

(c) *Health*: Because this is a closed process, exposures are expected to be minimal during normal operations. There is potential for exposure to amine compounds (i.e. monoethanolamine, diethanolamine, methyldiethanolamine), hydrogen sulfide and carbon dioxide. Safe work practices and/or appropriate personal protective equipment may be needed for exposures to chemicals and other hazards such as noise and heat, and during process sampling, inspection, maintenance and turnaround activities.

Saturate Gas Plants

(a) *Fire protection and prevention*: There is potential for fire if a leak or release reaches a source of ignition such as the unit reboiler.

(b) *Safety*: Corrosion could occur from the presence of hydrogen sulfide, carbon dioxide, and other compounds as a result of prior treating. Streams containing ammonia should be dried before processing. Antifouling additives may be used in absorption oil to protect heat exchangers. Corrosion inhibitors may be used to control corrosion in overhead systems.

(c) *Health*: Because this is a closed process, exposures are expected to be minimal during normal operations. There is potential for exposure to hydrogen sulfide, carbon dioxide, and other products such as diethanolamine or sodium hydroxide carried over from prior treating. Safe work practices and/or appropriate personal protective equipment may be needed for exposures to chemicals and other hazards such as noise and heat, and during process sampling, inspection, maintenance, and turnaround activities.

Asphalt Production

The potential for a fire exists if a product leak or release contacts a source of ignition such as the process heater. Condensed steam from the various asphalt and deasphalting processes will contain trace amounts of hydrocarbons. Any disruption of the vacuum can result in the entry of atmospheric air and subsequent fire. In addition, raising the temperature of the vacuum tower bottom to improve efficiency can generate methane by thermal cracking. This can create vapors in asphalt storage tanks that are not detectable by flash testing but are high enough to be flammable.

(a) *Safety*: Deasphalting requires exact temperature and pressure control. In addition, moisture, excess solvent, or a drop in operating temperature may cause foaming, which affects the product temperature control and may create an upset.

(b) *Health*: Because these are closed processes, exposures are expected to be minimal during normal operations. Should a spill or release occur, there is a potential for exposure to residuals and asphalt. Air blowing can create some polynuclear aromatics. Condensed steam from the air-blowing asphalt process may also contain contaminants. The potential for exposure to hydrogen sulfide and sulfur dioxide exists in the production of asphalt. Safe work practices and/or appropriate personal protective equipment may be needed for exposures to chemicals and other hazards such as noise and heat, and during process sampling, inspection, maintenance, and turnaround activities.

Hydrogen Production

(a) *Fire protection and prevention*: The possibility of fire exists should a leak or release occur and reach an ignition source.

(b) *Safety*: The potential exists for burns from hot gases and superheated steam should a release occur. Inspections and testing should be considered where the possibility exists for valve failure due to contaminants in the hydrogen. Carryover from caustic scrubbers should be controlled to prevent corrosion in preheaters. Chlorides from the feedstock or steam system should be prevented from entering reformer tubes and contaminating the catalyst.

(c) *Health*: Because these are closed processes, exposures are expected to be minimal during normal operating conditions. There is a potential for exposure to excess hydrogen, carbon monoxide, and/or carbon dioxide. Condensate can be contaminated by process materials such as caustics and amine compounds, with resultant exposures. Depending on the specific process used, safe work practices and/or appropriate personal protective equipment may be needed for exposures to chemicals and other hazards such as noise and heat, and during process sampling, inspection, maintenance, and turnaround activities.

Blending

(a) *Fire prevention and protection*: Ignition sources in the area need to be controlled in the event of a leak or release.

(b) *Health*: Safe work practices and/or appropriate personal protective equipment may be needed for exposures to chemicals and other hazards such as noise and heat; when handling additives; and during inspection, maintenance, and turnaround activities.

Lubricant, Wax, and Grease Manufacturing Processes

(a) *Fire protection and prevention*: The potential for fire exists if a product or vapor leak or release in the lube blending and wax processing areas reaches a source of ignition. Storage of finished products, both bulk and packaged, should be in accordance with recognized practices.
While the potential for fire is reduced in lube oil blending, care must be taken when making metal-working oils and compounding greases due to the use of higher blending and compounding temperatures and lower flash point products.

(b) *Safety*: Control of treater temperature is important as phenol can cause corrosion above 400° F. Batch and in-line blending operations require strict controls to maintain desired product quality. Spills should be cleaned and leaks repaired to avoid slips and falls. Additives in drums and bags need to be handled properly to avoid strain. Wax can clog sewer or oil drainage systems and interfere with wastewater treatment.

(c) *Health*: When blending, sampling, and compounding, personal protection from steam, dusts, mists, vapors, metallic salts, and other additives is appropriate. Skin contact with any formulated grease or lubricant should be avoided. Safe work practices and/or appropriate personal protection may be needed for exposures to chemicals and other hazards such as noise and heat; during inspection, maintenance, and turnaround activities; and while sampling and handling hydrocarbons and chemicals during the production of lubricating oil and wax.

Syngas manufacture in SMDS is relatively expensive; between 50 and 60 per cent of total process capital costs are related to syngas production.

Chapter 4

UOP Processes

NEXOCTANE? TECHNOLOGY FOR ISOOCTANE PRODUCTION

Introduction

Environmental issues are threatening the future use of MTBE (methyl-tert-butyl ether) in gasoline in the United States. Since the late 1990s, concerns have arisen over ground and drinking water contamination with MTBE due to leaking of gasoline from underground storage tanks and the exhaust from two-cycle engines. In California a number of cases of drinking water pollution with MTBE have occurred. As a result, the elimination of MTBE in gasoline in California was mandated, and legislation is now set to go in effect by the end of 2003. The U.S. Senate has similar law under preparation, which would eliminate MTBE in the 2006 to 2010 time frame.

With an MTBE phase-out imminent, U.S. refiners are faced with the challenge of replacing the lost volume and octane value of MTBE in the gasoline pool. In addition, utilization of idled MTBE facilities and the isobutylene feedstock result in pressing problems of unrecovered and/or underutilized capital for the MTBE producers. Isooctane has been identified as a cost-effective alternative to MTBE. It utilizes the same isobutylene feeds used in MTBE production and offers excellent blending value. Furthermore, isooctane production can be achieved in a low-cost revamp of an existing MTBE plant. However, since isooctane is not an oxygenate, it does not replace MTBE to meet the oxygen requirement currently in effect for reformulated gasoline.

The NExOCTANE technology was developed for the production of isooctane. In the process, isobutylene is dimerized to produce isooctene, which can subsequently be hydrogenated to produce isooctane. Both products are excellent gasoline blend stocks with significantly higher product value than alkylate or polymerization gasoline.

History of MTBE

During the 1990s, MTBE was the oxygenate of choice for refiners to meet increasingly stringent gasoline specifications. In the United States and in a limited number of Asian countries, the use of oxygenates in gasoline was mandated to promote cleaner-burning fuels. In addition, lead phase-down programs in other parts of the world have resulted in an increased demand for high-octane blend stock. All this resulted in a strong demand for high-octane fuel ethers, and significant MTBE production capacity has been installed since 1990.

Today, the United States is the largest consumer of MTBE. The consumption increased dramatically with the amendment of the Clean Air Act in 1990 which incorporated the 2 per cent oxygen mandate. The MTBE production capacity more than doubled in the 5-year period from 1991 to 1995. By 1998, the MTBE demand growth had leveled off, and it has since tracked the demand growth for reformulated gasoline (RFG). The United States consumes about 300,000 BPD of MTBE, of which over 100,000 BPD is consumed in California. The U.S. MTBE consumption is about 60 per cent of the total world demand.

MTBE is produced from isobutylene and methanol. Three sources of isobutylene are used for MTBE production:

- On-purpose butane isomerization and dehydrogenation,
- Fluid catalytic cracker (FCC) derived mixed C_4 fraction,
- Steam cracker derived C_4 fraction.

The majority of the MTBE production is based on FCC and butane dehydrogenation derived feeds.

Nexoctane Background

Fortum Oil and Gas Oy, through its subsidiary Neste Engineering, has developed the NExOCTANE technology for the production of isooctane. NExOCTANE is an extension of Fortum's experience in the development and licensing of etherification technologies. Kellogg Brown & Root, Inc. (KBR) is the exclusive licenser of NExOCTANE. The technology licensing and process design services are offered through a partnership between Fortum and KBR.

The technology development program was initialized in 1997 in Fortum's Research and Development Center in Porvoo, Finland, for the purpose of producing high-purity isooctene, for use as a chemical intermediate. With the emergence of the MTBE pollution issue and the pending MTBE phase-out, the focus in the development was shifted in 1998 to the conversion of existing MTBE units to produce isooctene and isooctane for gasoline blending.

The technology development has been based on an extensive experimental research program in order to build a fundamental understanding of the reaction kinetics and key product separation steps in the process. This research has resulted in an advanced kinetic modeling capability, which is used in the design of the process for licensees. The process has undergone extensive pilot testing, utilizing a full range of commercial feeds. The first commercial NExOCTANE unit started operation in the third quarter of 2002.

Process Chemistry

The primary reaction in the NExOCTANE process is the dimerization of isobutylene over acidic ion-exchange resin catalyst. This dimerization reaction forms two isomers of trimethylpentene (TMP), or isooctene, namely, 2, 4, 4-TMP-1 and 2, 4, 4-TMP-2, according to the following reactions:

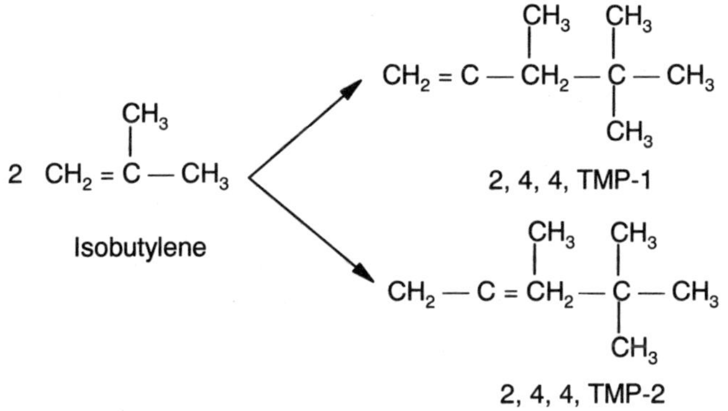

TMP further reacts with isobutylene to form trimers, tetramers, etc. Formation of these oligomers is inhibited by oxygen-containing polar components in the reaction mixture. In the NExOCTANE process, water and alcohol are used as inhibitors. These polar components block acidic sites on the ion-exchange resin, thereby controlling the catalyst activity and increasing the selectivity to the formation of dimers. The process conditions in the dimerization reactions are optimized to maximize the yield of high-quality isooctene product.

A small quantity of C_7 and C_9 components plus other C_8 isomers will be formed when other olefin components such as propylene, n-butenes, and isoamylene are present in the reaction mixture. In the NExOCTANE process, these reactions are much slower than the isobutylene dimerization reaction, and therefore only a small fraction of these components is converted.

Isooctene can be hydrogenated to produce isooctane, according to the following reaction:

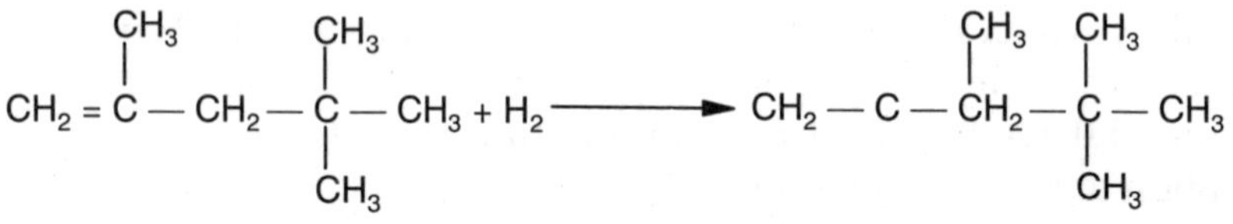

NExOCTANE PROCESS DESCRIPTION

The NExOCTANE process consists of two independent sections. Isooctene is produced by dimerization of isobutylene in the dimerization section, and subsequently, the isooctene can be hydrogenated to produce isooctane in the hydrogenation section. Dimerization and hydrogenation are independently operating sections.

STRATCO EFFLUENT REFRIGERATED H_2SO_4 ALKYLATION PROCESS

Introduction

Alkylation, first commercialized in 1938, experienced tremendous growth during the 1940s as a result of the demand for high-octane aviation fuel during World War II. During the mid-1950s, refiners' interest in alkylation shifted from the production of aviation fuel to the use of alkylate as a blending component in automotive motor fuel. Capacity remained relatively flat during the 1950s and 1960s due to the comparative cost of other blending components. The U.S. Environmental Protection Agency's lead phase-down program in the 1970s and 1980s further increased the demand for alkylate as a blending component for motor fuel. As additional environmental regulations are imposed on the worldwide refining community, the importance of alkylate as a blending component for motor fuel is once again being emphasized. Alkylation unit designs (grassroots and revamps) are no longer driven only by volume, but rather by a combination of volume, octane, and clean air specifications. Lower olefin, aromatic, sulfur, Reid vapor pressure (RVP), and drivability index (DI) specifications for finished gasoline blends have also become driving forces for increased alkylate demand in the United States and abroad. Additionally, the probable phase-out of MTBE in the United States will further increase the demand for alkylation capacity.

The alkylation reaction combines isobutane with light olefins in the presence of a strong acid catalyst. The resulting highly branched, paraffinic product is a low-vapor-pressure, high-octane blending component. Although alkylation can take place at high temperatures without catalyst, the only processes of commercial importance today operate at low to moderate temperatures using either sulfuric or hydrofluoric acid catalysts. Several different companies are currently pursuing research to commercialize a solid alkylation catalyst. The reactions occurring in the alkylation process are complex and produce an alkylate product that has a wide boiling range. By optimizing operating conditions,

The majority of the product is within the desired gasoline boiling range with motor octane numbers (MONs) up to 95 and research octane numbers (RONs) up to 98.

Process Description

A block flow diagram of the STRATCO effluent refrigerated H_2SO_4 alkylation project is shown in Figure. Each section of the block flow diagram is described below:

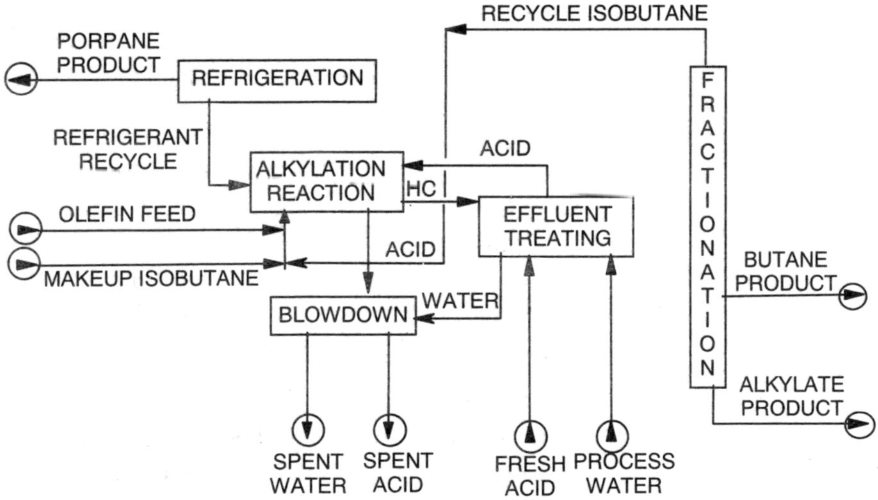

Figure 4.1: Block Flow Diagram of STRATCO Inc. Effluent Refrigerated Alkylation Process

Reaction section

Here the reacting hydrocarbons are brought into contact with sulfuric acid catalyst under controlled conditions.

Refrigeration Section

Here the heat of reaction is removed, and light hydrocarbons are removed from the unit.

Effluent Treating Section

Here the free acid, alkyl sulfates, and dialkyl sulfates are removed from the net effluent stream to avoid downstream corrosion and fouling.

Fractionation Section

Here isobutane is recovered for recycle to the reaction section, and remaining hydrocarbons are separated into the desired products.

Blowdown Section

Here spent acid is degassed, wastewater pH is adjusted, and acid vent streams are neutralized before being sent off-site.

The blocks are described in greater detail below:

Reaction Section

In the reaction section, olefins and isobutane are alkylated in the presence of sulfuric acid catalyst. As shown in Figure the olefin feed is initially combined with the recycle isobutane. The olefin and recycle isobutane mixed stream is then cooled to approximately 60°F (15.6°C) by exchanging heat with the net effluent stream in the feed/effluent exchangers.

Since the solubility of water is reduced at lower temperatures, water is freed from the hydrocarbon to form a second liquid phase. The feed coalescer removes this free water to minimize dilution of the sulfuric acid catalyst.

The feed stream is then combined with the refrigerant recycle stream from the refrigeration section. The refrigerant recycle stream provides additional isobutane to the reaction zone. This combined stream is fed to the STRATCO Contactor reactors.

The use of separate Contactor reactors in the STRATCO process allows for the segregation of different olefin feeds to optimize alkylate properties and acid consumption. In these cases, the unit will have parallel trains of feed/effluent exchangers and feed coalescers.

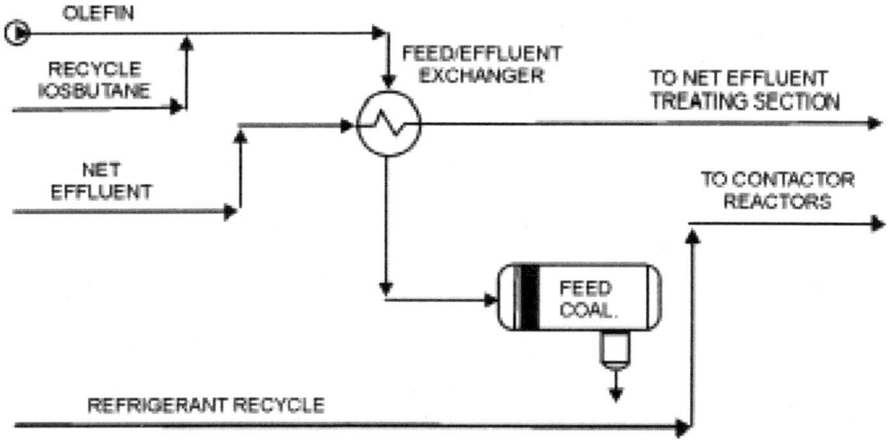

Figure 4.2: Feed Mixing and Cooling

At the "heart" of STRATCO's effluent refrigerated alkylation technology is the Contactor reactor. The Contactor reactor is a horizontal pressure vessel containing an inner circulation tube, a tube bundle to remove the heat of reaction, and a mixing impeller. The hydrocarbon feed and sulfuric acid enter on the suction side of the impeller inside the circulation tube. As the feeds pass across the impeller, an emulsion of hydrocarbon and acid is formed. The emulsion in the Contactor reactor is continuously circulated at very high rates.

The superior mixing and high internal circulation of the Contactor reactor minimize the temperature difference between any two points in the reaction zone to within 1°F (0.6°C). This reduces the possibility of localized hot spots that lead to degraded alkylate product and increased chances for corrosion. The intense mixing in the Contactor reactor also provides uniform distribution of the hydrocarbons in the acid emulsion. This prevents localized areas of nonoptimum isobutane/olefin ratios and acid/olefin ratios, both of which promote olefin polymerization reactions.

The typical Contactor reactor and acid settler arrangement. A portion of the emulsion in the Contactor reactor, which is approximately 50 LV per cent acid and 50 LV per cent hydrocarbon, is withdrawn from the discharge side of the impeller and flows to the acid settler. The hydrocarbon phase (reactor effluent) is separated from the acid emulsion in the acid settlers. The acid, being the heavier of the two phases, settles to the lower portion of the vessel. It is returned to the suction side of the impeller in the form of an emulsion, which is richer in acid than the emulsion entering the settlers.

The STRATCO alkylation process utilizes an effluent refrigeration system to remove the heat of reaction and to control the reaction temperature. With effluent refrigeration, the hydrocarbons in contact with the sulfuric acid catalyst are maintained in the liquid phase.

OP ALKYLENE™ PROCESS FOR MOTOR FUEL ALKYLATION

Introduction

The UOP Alkylene process is a competitive and commercially available alternative to liquid acid technologies for alkylation of light olefins and isobutane. Alkylate is a key blending component for gasoline having high octane, low Reid vapor pressure (RVP), low sulfur, and low volatility. It is composed of primarily highly branched paraffinic hydrocarbons. Changing gasoline specifications in response to legislation will increase the importance of alkylate, making it an ideal "clean fuels" blend stock. Existing liquid acid technologies, while well proven and reliable, are increasingly under political and regulatory pressure to reduce environmental and safety risks through increased monitoring and risk mitigation. A competitive solid catalyst alkylation technology, such as the Alkylene process, would be an attractive alternative to liquid acid technologies.

UOP developed the Alkylene process during the late 1990s, in response to the industry's need for an alternative to liquid acid technologies. Early attempts with solid acid catalysts found some to have good alkylation properties, but the catalysts also had short life, on the order of hours. In addition, these materials could not be regenerated easily, requiring a carbon burn step. Catalysts with acid incorporated on a porous support had been investigated but not commercialized. UOP invented the novel HAL-100 catalyst that has high alkylation activity and long catalyst stability and easily regenerates without a high-temperature carbon burn. Selectivity of the HAL-100 is excellent, and product quality is comparable to that of the product obtained from liquid acid technologies.

Alkylene Process

Olefins react with isobutane on the surface of the HAL-100 catalyst to form a complex mixture of isoalkanes called alkylate. The major constituents of alkylate are highly branched trimethylpentanes (TMP) that have high-octane blend values of approximately 100. Dimethyl hexanes (DMH) have lower-octane blend values and are present in alkylate at varying levels.

Alkylation proceeds via a carbenium ion mechanism, as shown in Figure.

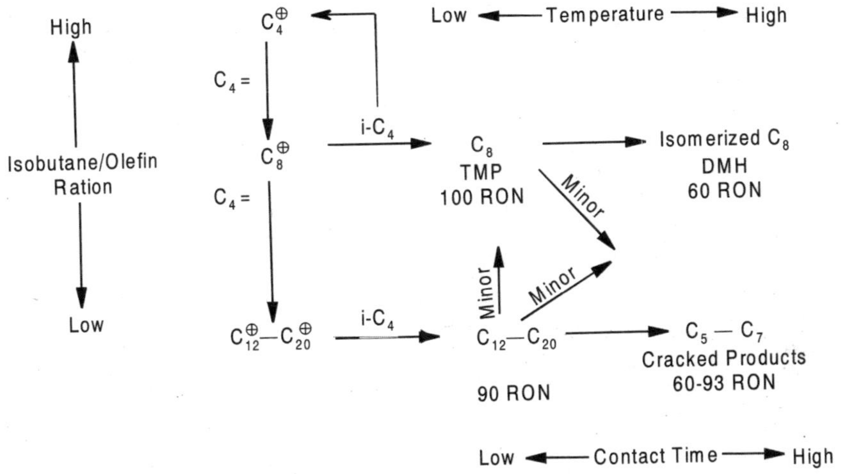

Figure 4.3: Reaction Mechanism

The complex reaction paths include an initiation step, a propagation step, and hydrogen transfer. Secondary reactions include polymerization, isomerization, and cracking to produce other isoalkanes including those with carbon numbers which are not multiples of 4. The primary reaction products are formed via simple addition of isobutane to an olefin such as propylene, butenes, and amylenes. The key reaction step is the protonation of a light olefin on the solid catalyst surface followed by alkylation of an olefin on the C_4 carbocation, forming the C_8 carbocation. Hydride transfer from another isobutane molecule forms the C_8 paraffin product. Secondary reactions result in less desirable products, both lighter and heavier than the high-octane C8 products. Polymerization to acid-soluble oil (ASO) is found in liquid acid technologies and results in additional catalyst consumption and yield loss. The Alkylene process does not produce acid-soluble oil. The Alkylene process also has minimal polymerization, and the alkylate has lighter distillation properties than alkylate from HF or H_2SO_4 liquid acid technologies.

Alkylation conditions that favor the desired high-octane trimethylpentane include low process temperature, high localized isobutane/olefin ratios, and short contact time between the reactant and catalyst. The Alkylene process is designed to promote quick, intimate contact of short duration between hydrocarbon and catalyst for octane product, high yield, and efficient separation of alkylate from the catalyst to minimize undesirable secondary reactions. Alkylate produced from the Alkylene process is comparable to alkylate produced from traditional liquid acid technologies without the production of heavy acid-soluble oil. The catalyst is similar to other hydroprocessing and conversion catalysts used in a typical refinery. Process conditions are mild and do not require expensive or exotic metallurgy.

Reactor temperature, isobutene/olefin ratio, contact time, and catalyst/ olefin ratios are the key operating parameters.

Feeds to the Alkylene unit are dried and treated to move impurities and contaminants such as diolefins, oxygenates, nitrogen, and sulfur. These contaminants also cause higher acid consumption, higher acid-soluble oil formation, and lower acid strength in liquid acid technologies. Diolefin saturation technology, such as the Huels Selective Hydrogenation Process technology licensed by UOP LLC, saturates diolefins to the corresponding monoolefin and isomerizes the 1-butene to 2-butene. The alkylate formed by alkylating isobutane with 2-butene is the preferred 2,2,3-TMP compared to the 2,2-DMH formed by alkylating isobutane with 1-butene.

The olefin and isobutane (Figure) are combined and injected into a carbon-steel riser reactor with continuous catalyst reactivation to maintain a constant catalyst activity and minimize catalyst inventory. This provides constant product quality, high yield, and high on-stream efficiency.

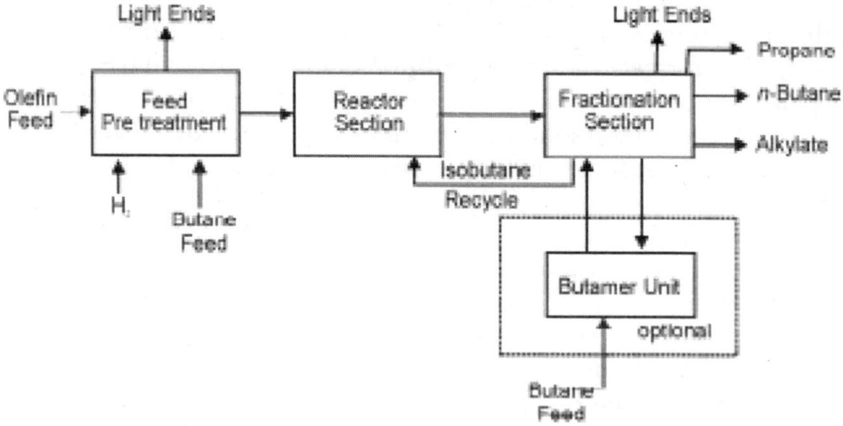

Figure 4.4: Alkylene Process Flow Scheme

Liquid-phase hydrocarbon reactants transport the catalyst around the reactor circuit where velocities are low relative to those of other moving catalyst processes. The reaction time is on the order of minutes for the completion of the primary reactions and to minimize secondary reactions. The catalyst and hydrocarbon are intimately mixed during the reaction, and the catalyst is easily disengaged from the hydrocarbon product at the top of the reactor. The catalyst is reactivated by a simple hydrogenation of the heavier alkylate on the catalyst in the reactivation wash zone. Hydrogen consumption is minimal as the quantity of heavy alkylate on the HAL-100 catalyst is very small. The reactivation process is highly effective, restoring the activity of the catalyst to nearly 100 per cent of fresh. The liquid-phase operation of the Alkylene process results in less abrasion than in other catalyst circulation processes due to the lubricating effect of the liquid. Furthermore, the catalyst and hydrocarbon velocities are low relative to those in other moving catalyst processes. This minimizes the catalyst replacement requirements. Catalyst circulation is maintained to target catalyst/olefin ratios. A small catalyst slipstream flows into a separate vessel for reactivation in vapor phase with relatively mild conditions to remove any last traces of heavy material and return the catalyst activity to essentially the activity of fresh catalyst.

Alkylate from the reactor is sent to a downstream fractionation section, which is similar to fractionation sections in liquid acid process flow schemes. The fractionation section recycles the unconverted isobutane back to the reactor and separates out the final alkylate product.

UOP HF ALKYLATION TECHNOLOGY

Introduction

The UOP HF Alkylation process for motor fuel production catalytically combines light olefins, which are usually mixtures of propylene and butylenes, with isobutane to produce a branched-chain paraffinic fuel. The alkylation reaction takes place in the presence of hydrofluoric (HF) acid under conditions selected to maximize alkylate yield and quality. The alkylate product possesses excellent antiknock properties and high-octane because of its high content of highly branched paraffins. Alkylate is a clean-burning, low-sulfur, low-RVP gasoline blending component that does not contain olefinic or aromatic compounds.

The HF Alkylation process was developed in the UOP laboratories during the late 1930s and early 1940s. The process was initially used for the production of high-octane aviation fuels from butylenes and isobutane. In the mid-1950s, the development and consumer acceptance of more-sophisticated high-performance automotive engines placed a burden on the petroleum refiner both to increase gasoline production and to improve motor fuel quality. The advent of catalytic reforming techniques, such as the UOP Platforming* process, provided an important tool for the production of high-quality gasolines available to refiners. However, the motor fuel produced in such operations is primarily aromaticbased and is characterized by high sensitivity (that is, the spread between research and motor octane numbers). Because automobile performance is more closely related to road octane rating (approximately the average of research and motor octanes), the production of gasoline components with low sensitivity was required. A natural consequence of these requirements was the expansion of alkylation operations. Refiners began to broaden the range of olefin feeds to both existing and new alkylation units to include propylene and occasionally amylenes as well as butylenes. By the early 1960s, the HF Alkylation process had virtually displaced motor fuel polymerization units for new installations, and refiners had begun to gradually phase out the operation of existing polymerization plants.

The importance of the HF Alkylation process in the refining situation of the 2000s has been increased even further by the scheduled phase-out of MTBE and the increased emphasis on low-sulfur gasoline. The contribution of the alkylation process is critical in the production of quality motor fuels including many of the "environmental" gasoline blends. The process provides refiners with a tool of unmatched economy and efficiency, one that will assist refiners in maintaining or strengthening their position in the production and marketing of gasolines.

Process Chemistry

General

In the HF Alkylation process, HF acid is the catalyst that promotes the isoparaffin-olefin reaction. In this process, only isoparaffins with tertiary carbon atoms, such as isobutane or isopentane, react with the olefins. In practice, only isobutane is used because isopentane has a high octane number and a vapor pressure that has historically allowed it to be blended directly into finished gasolines. However, where environmental regulations have reduced the allowable vapor pressure of gasoline, isopentane is being removed from gasoline, and refiner interest in alkylating this material with light olefins, particularly propylene, is growing.

The actual reactions taking place in the alkylation reactor are many and are relatively complex. The equations in Figure illustrate the primary reaction products that may be expected for several pure olefins.

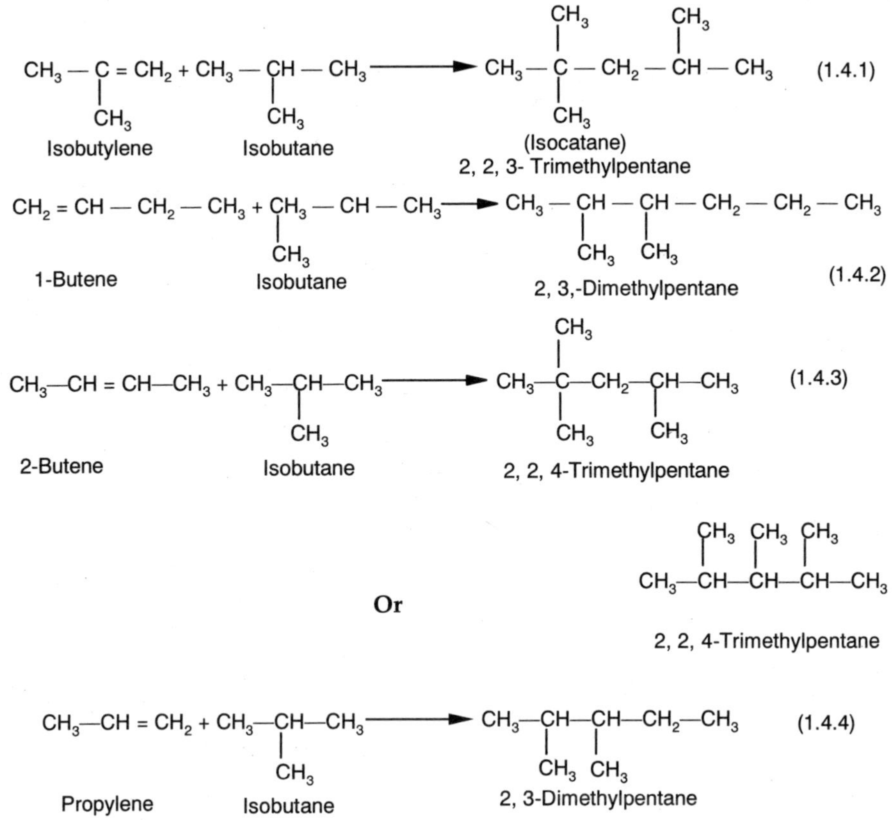

Figure 4.5: HF Alkylation Primary Reactions for Monoolefins

In practice, the primary product from a single olefin constitutes only a per centage of the alkylate because of the variety of concurrent reactions that are possible in the alkylation environment.

- *Reaction Mechanism*: Alkylation is one of the classic examples of a reaction or reactions proceeding via the carbenium ion mechanism. These reactions include an initiation step and a propagation step and may include an isomerization step. In addition, polymerization and cracking steps may be involved. However, these side reactions are generally undesirable.
- *Initiation*: The initiation step generates the tertiary butyl cations that will subsequently carry on the alkylation reaction.
- *Propagation*: Propagation reactions involve the tertiary butyl cation reacting with an olefin to form a larger carbenium ion, which then abstracts a hydride from an isobutane molecule. The hydride

abstraction generates the isoparaffin plus a new tertiary butyl cation to carry on the reaction chain.

- *Isomerization*: Isomerization is very important in producing good octane quality from a feed that is high in 1-butene. The isomerization of 1-butene is favored by thermodynamic equilibrium. Allowing 1-butene to isomerize to 2-butene reduces the production of dimethylhexanes (research octane number of 55 to 76) and increases the production of trimethylpentanes. Many recent HF Alkylation units, especially those processing only butylenes, have upstream olefin isomerization units that isomerize the 1-butene to 2-butene.
- *Other Reactions*: The polymerization reaction [Eq. (1.4.14), shown in Figure 1.4.2d] results in the production of heavier paraffins, which are undesirable because they reduce alkylate octane and increase alkylate endpoint. Minimization of this reaction is achieved by proper choice of reaction conditions.

The larger polymer cations are susceptible to cracking or disproportionation reactions which form fragments of various molecular weights.

LINEAR ALKYLBENZENE (LAB) MANUFACTURE

Introduction

The detergent industry originated in the late 1940s with the advent of sodium alkylbenzene sulfonates, which had detergency characteristics far superior to those of natural soaps. Natural soaps are sodium salts of fatty acids obtained by the alkaline saponification of naturally occurring triglycerides from either vegetal or animal sources. The early alkylbenzene sulfonates (ABSs) were essentially sodium dodecylbenzene sulfonates (DDBSs), also known as branched alkylbenzene sulfonates (BABSs) obtained by the Friedel-Crafts alkylation of benzene with propylene tetramer, a mixture of branched $C1_2$ olefins. Dodecylbenzenes (DDBs) are then sulfonated with oleum or sulfur trioxide (SO_3) and neutralized with sodium hydroxide or soda ash.

Because of their lower cost and high effectiveness in a wide range of detergent formulations, DDBSs rapidly displaced natural soaps in household laundry and dishwashing applications. However, although excellent from a performance viewpoint, BABS exhibited slow rates of biodegradation in the environment and, in the early 1960s, started to be replaced by linear alkylbenzene sulfonate (LAS or LABS). The linear alkyl chains found in LAS biodegrade at rates that are comparable to those observed in the biodegradation of natural soaps and other natural and semisynthetic detergent products.

The use of DDBS has never been formally banned in the United States, but by the late 1960s, its use had been largely phased out in the United States, Japan, and several European countries. By the late 1970s, the use of LAS had become more generalized, and new facilities were added in developing countries around the world. Currently, LAS accounts for virtually the entire worldwide production of alkylbenzene sulfonates. The demand for linear alkylbenzene increased from about 1.0 million metric tons per year (MTA) in 1980 to about 1.7 million in 1990. The demand for LAB is approximately 2.5 million MTA and is growing at an annual rate of 3.5 per cent as of 2002. Worldwide LAB production capacity is approximately 2.8 million MTA as of 2002.

Technology Background

Various routes were developed and used in the production of LAB. The first hurdle to be overcome was the recovery, typically from kerosene or gas oil fractions, of linear paraffins (n-paraffins) in the C_{10} to C_{14} range. Initial recovery attempts were based on the use of urea adducts, which were soon replaced by adsorptive separation and recovery techniques, in either the vapor or the liquid phase. These techniques used a variety of adsorbents and desorbents. Adsorptive separation techniques based on the molecular sieve action of 5-Å zeolites have dominated this industry since the mid-1960s. Typical commercial process technologies for this separation include the UOP Molex process in the liquid phase with a hydrocarbon desorbent that makes use of UOP's Sorbex simulated moving-bed technology; the UOP IsoSiv* process (formerly Union Carbide's), which operates in the vapor phase also with a hydrocarbon desorbent; Exxon's Ensorb process, which is also in the vapor phase but has an ammonia desorbent; or a similar technology developed in the former German Democratic Republic (East Germany) and known as the GDR Parex process, which also operates in the vapor phase with ammonia desorbent. The GDR Parex process is not to be confused with UOP's Parex process for the selective recovery of high-purity p-xylene from aromatic streams using the Sorbex simulated moving-bed technology.

Once the linear paraffins have been recovered at sufficient purity, typically in excess of about 98 per cent, they have to be alkylated with benzene to produce LAB. To date, attempts to alkylate n-paraffins with benzene directly have failed, thus necessitating the activation of the n-paraffins to a more reactive intermediate before the alkylation with benzene can take place.

The following routes for the production of LAB emerged during the 1960s:

- Chlorination of n-paraffins to form primarily monochloroparaffins. Benzene is then alkylated with monochloroparaffins using an aluminum chloride ($AlCl_3$) catalyst. An example of this route was developed and commercialized by ARCO Technology Inc.1
- Chlorination of n-paraffins followed by dehydrochlorination and alkylation of the resulting olefins with benzene typically using hydrofluoric (HF) acid as catalyst. Shell's CDC process (for chlorination/dehydrochlorination) is an example of such a process. This type of technology was still used commercially until the mid-1980s by, among others, Hüls AG in Germany.
- Alkylation of linear olefins with benzene also using an HF catalyst. The olefins are usually linear alpha-olefins (LAOs) from wax cracking (now discontinued), alpha-olefins from ethylene oligomerization, or linear internal olefins (LIOs) from olefin disproportionation. Various companies, such as BP, Chevron (formerly Gulf), and Shell, offer technologies for the oligomerization of ethylene to LAO; Shell also produces linear internal olefins by disproportionation in its Shell Higher Olefins process (SHOP).
- Dehydrogenation of linear paraffins to a fairly dilute mixture of LIO in unconverted n-paraffins, followed by the alkylation of the olefins with benzene also using HF acid catalyst but without the separation and concentration of the LIO. UOP's Pacol process for the catalytic dehydrogenation of n-paraffins and UOP's HF Detergent Alkylate* process for the alkylation of the LIO with benzene are prime examples of this approach. A similar approach is also practiced by Huntsman Corp. (formerly Monsanto's).

During the early days of LAB production, paraffin chlorination followed by alkylation over AlCl3 gained some prominence. However, since the late 1960s, the dehydrogenation and HF alkylation route has been the most prominent because ofand higher-quality product. Although LAO and LIO obtained from sources other than dehydrogenation can equally be used, n-paraffin dehydrogenation routes have usually prevailed because of the lower cost of the starting kerosene fractions. Table shows an approximate 2001 distribution of world LAB production employing these technologies. The dehydrogenation followed by alkylation route accounts for 81 per cent of world LAB production.

Table 4.1: 2001 World LAB Production by Technology Route

Technology route	Production, %
Chlorination and alkylation	10
Dehydrogenation and alkylation	81
High-purity olefins to alkylation	9
Total	100

The Detal process, which replaces HF with a solid heterogenous acid catalyst, was introduced in 1995. The various routes for the production of LAB are illustrated schematically on Figure.

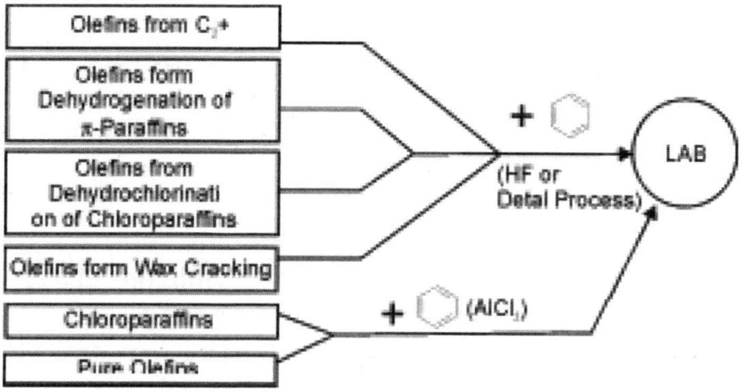

Figure 4.5: Routes to LAB its Economic Advantages

Commercial Experience

The first commercial operations of UOP's dehydrogenation and alkylation technologies were in Japan and Spain at the end of 1968. Almost all the units built since then throughout the world employ UOP technology. Over the years, UOP has continued research and development and has introduced numerous improvements that resulted in improved economics of LAB manufacture as well as consistently improved product quality. More than 30 LAB units now operate around the world with this process technology.

Q-MAX™ PROCESS FOR CUMENE PRODUCTION

Introduction

The Q-Max™ process converts benzene and propylene to high-quality cumene by using a regenerable zeolitic catalyst. The Q-Max process represents a substantial improvement over older cumene technologies and is characterized by its exceptionally high yield, superior product quality, low investment and operating costs, reduction in solid waste, and corrosion-free environment.

Cumene is produced commercially through the alkylation of benzene with propylene over an acid catalyst. Over the years, many different catalysts have been proposed for this alkylation reaction, including boron trifluoride, hydrogen fluoride, aluminum chloride, and phosphoric acid. In the 1930s, UOP introduced the UOP catalytic condensation process, which used a solid phosphoric acid (SPA) catalyst to oligomerize light olefin by-products from petroleum thermal cracking into heavier paraffins that could be blended into gasoline. During World War II, this process was adapted to produce cumene from benzene and propylene to make a high-octane blending component for military aviation gasoline. Today, cumene is no longer used as a fuel, but it has grown in importance as a feedstock for the production of phenol.

Although SPA is a highly efficient and economical catalyst for cumene synthesis, it has two important limitations:

1. Cumene yield is limited to about 95 per cent, because of the oligomerization of propylene and the formation of heavy alkylate by-products
2. The catalyst is not regenerable and must be disposed of at the end of each catalyst cycle.

In recent years, producers have been under increasing pressure to improve cumene product quality so that the quality of the phenol produced downstream (as well as acetone and alpha-methylstyrene, which are coproduced with phenol) could be improved. Twenty-five years ago, most phenol was used to produce phenolic resins, and acetone was used primarily as a solvent. Today, both phenol and acetone are used increasingly in the production of polymers such as polycarbonates and nylon. Over the years, improvements to the SPA process managed to keep pace with the demand for higher cumene product quality, but producers still sought an improved cumene process that would produce a better-quality product at higher yield.

Because zeolites are known to selectively perform many acid-catalyzed reactions, UOP began searching for a new cumene catalyst that would overcome the limitations of SPA. UOP's objective was to develop a regenerable catalyst that would increase the yield of cumene and lower the cost of production. More than 100 different catalyst materials were screened, including mordenites, MFIs, Y-zeolites, amorphous silicaaluminas, and beta-zeolite. The most promising materials were modified to improve their selectivity and then subjected to more-rigorous testing. By 1992, UOP had selected the most promising catalyst based on beta-zeolite for cumene production and then began to optimize the process design around this new catalyst. The result of this work is the Q-Max process and the QZ-2000 catalyst system.

Process Chemistry

The synthesis of cumene from benzene and propylene is a modified Friedel-Crafts alkylation, which can be accomplished by many different acid catalysts. The basic alkylation chemistry and reaction mechanism are shown in Figure. The olefin forms a carbonium ion intermediate, which attacks the benzene ring in an electrophilic substitution. The addition to the olefin double bond is at the middle carbon of propylene, in accordance with Markovnikov's rule. The addition of the isopropyl group to the benzene ring weakly activates the ring toward further alkylation, producing di-isopropyl-benzene (DIPB) and heavier alkylate by-products.

The QZ-2000 catalyst functions as strong acid. In the QZ-2000 catalyst, the active surface sites of the silica-alumina structure act to donate the proton to the adsorbed olefin. Because the QZ-2000 catalyst is a strong acid, it can be used at a very low temperature.

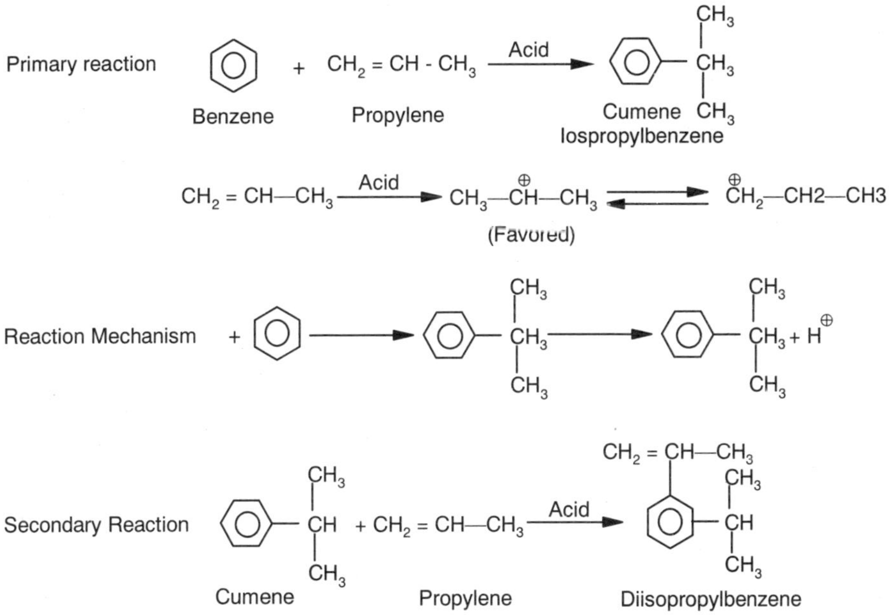

Figure 4.6: Alkylation chemistry

Low reaction temperature reduces the rate of competing olefin oligomerization reactions, resulting in higher selectivity to cumene and lower production of heavy by-products.

Transalkylation of DIPB

Transalkylation is the acid-catalyzed transfer of one isopropyl group from DIPB to a benzene molecule to form two molecules of cumene (Figure).

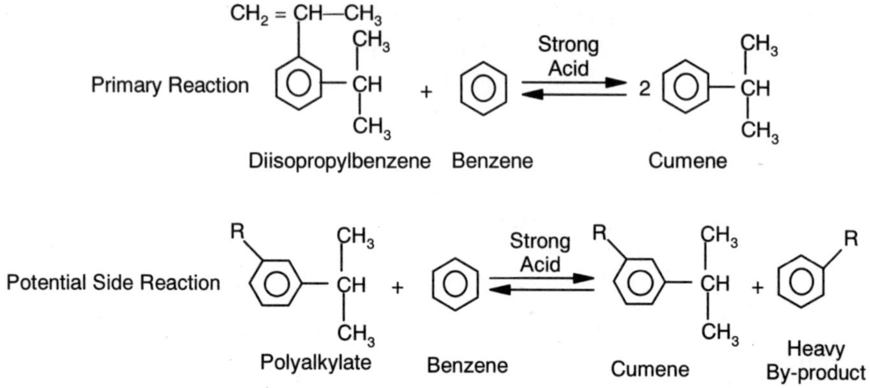

Figure 4.7: Transalkylation chemistry

The Q-Max process is designed with an alkylation reactor section, which produces about 85 to 95 wt per cent cumene and 5 to 15 wt per cent DIPB. After recovery of the cumene product by fractionation, the DIPB is reacted with recycle benzene at optimal conditions for transalkylation to produce additional cumene. With the alkylation and transalkylation reactors working together to take full advantage of the QZ-2000 catalyst, the overall yield of cumene is increased to 99.7 wt per cent.

Side Reactions

In addition to the principal alkylation reaction of benzene with propylene, all acid catalysts promote the following undesirable side reactions to some degree:

- *Oligomerization of olefins*: The model for acid-catalyzed alkylation is diffusion of the olefin to an active site saturated with benzene followed by adsorption and reaction. One possible side reaction is the combination of the propyl carbonium ion with propylene to form a C_6 olefin or even further reaction to form C_9, C_{12}, or heavier olefins.

- *Alkylation of benzene with heavy olefins*: Once heavy olefins have been formed through oligomerization, they may react with benzene to form hexylbenzene and heavier alkylated benzene by-products.
- *Polyalkylation*: The addition of an isopropyl group to the benzene ring to produce cumene weakly activates the ring toward further substitution, primarily at the meta and para positions, to make DIPB and heavier alkylates.
- *Hydride-transfer reactions*: Transfer of a hydrogen to an olefin by the tertiary carbon on cumene can form a cumyl carbonium ion that may react with a second benzene molecule to form diphenylpropane.

CONOCOPHILLIPS REDUCED VOLATILITY ALKYLATION PROCESS (REVAP)

Introduction

During the late 1930s, Phillips Petroleum Company researchers discovered the benefits of using hydrofluoric acid to catalyze the synthesis of high-octane fuels from a broad range of low-value C_3, C_4, and C_5 feedstocks. This research, as well as pilot-plant data, led to the commercialization of the HF Alkylation process at Phillips' Borger, Texas, refinery in 1942 to provide aviation gasoline during World War II. Since that time, alkylate has been, and continues to be, a valuable high-octane blending component for gasoline, as evidenced by its importance in refineries around the world. ConocoPhillips has built 11 HF Alkylation units in its own refineries and has licensed over 100 grassroots units.

Today, worldwide alkylation capacity exceeds 1.81 million bbl/day, with HF-based processes accounting for approximately 57 per cent of the total. Isobutane alkylate is an important component of modern fuels, due to the high-octane, clean-burning characteristics as well as the low vapor pressure and absence of sulfur, olefins, or aromatics. Alkylation is receiving renewed attention by refiners contemplating the replacement of MTBE in gasoline.

Chemistry

Alkylation occurs when isobutane reacts with olefins in the presence of hydrofluoric acid as the catalyst to produce branched paraffins. In simplest terms, those reaction are:

Propylene + isobutane ® 2, 3-dimethylpentane

Isobutylene + isobutane ® 2, 2, 4-trimethylepentane

1-Butene + isobutane ® 2, 2-dimethylhexane

2-Butene + isobutene ® 2, 2, 4- and 2, 3, 4-trimethylpentane

Amylene + isobutane ® C_9H_{20} (various isomers)

The trimethylpentanes are the preferred reaction products because they generally have the highest octane value. In practice, however, the reactions are not so simple. Reactions involving isomerization, hydrogen transfer, dimerization, polymerization, b-scission (or cracking), and disproportionation lead to a range of products. Furthermore, these side reactions produce substantial quantities of trimethylpentanes even when propylene or amylenes are the olefin feed. Polymerization produces conjunct polymers, which are complex, cyclic molecules, and this material is known as acid-soluble oil (ASO).

The reactions are also affected by the dispersion of hydrocarbons in the acid, the reaction temperature, the ratio of isobutane to olefin in the reaction zone, and the presence of water and ASO in the circulating acid. Since the hydrocarbon feeds are only slightly soluble in the HF acid, the reaction is enhanced by dispersing the hydrocarbons in the acid. Improved dispersion, i.e., smaller droplets of hydrocarbon, results in an alkylate product with more of the desired trimethylpentanes and lower amounts of the undesirable lighter and heavier compounds. Lower reaction temperature also favors the desired reaction products. A large excess of isobutene—above the stoichiometric amount—also favors the production of higher amounts of trimethylpentanes. Thus, the purity of isobutane in the recycle stream has an effect on alkylate quality, and buildup of C5+ components in this stream should be avoided. Small amounts of water enter the alkylation unit in the olefin and isobutane feeds. The water is allowed to accumulate in the acid phase and is found to be beneficial in that it produces alkylate with higher concentration of C_8 components and thus higher octane. Water in the HF is beneficial at levels up to about 3 to 4 wt per cent. However, water contents above about 2.0 per cent generally have a detrimental effect on corrosion rates in the unit and are avoided.

Description of the Conocophillips HF Alkylation Process

Isobutane reacts with propylene, butenes, and/or amylenes in the presence of hydrofluoric acid to produce a high-octane alkylate for motor gasoline. The reactions produce a variety of products, primarily C_8 branched paraffins, with lesser amounts of C_7 and C_9 branched paraffins and small amounts of lighter and heavier paraffins. For best operation, the following feedstock contaminant levels are recommended:

Sulfur — 20 wt ppm maximum

Water — 20 wt ppm maximum

Butadiene — 3000 wt ppm maximum

C6 + —0.1 LV % maximum

Oxygenates (MTBE, dimethyl enther, etc.)—30 wt ppm maximum

An alkylation unit will operate with feed contaminant at higher than the levels indicated above, but the adverse consequences are higher acid consumption, higher production of unwanted by-products, and possible lower octane number of the alkylate product. One feed treatment to remove butadiene is hydroisomerization, such as the ConocoPhillips Hydrisom Process. Hydroisomerization reduces butadiene (and pentadienes) to very low levels and also isomerizes 1-butene to cis- and trans-2-butene. The 2-butene isomers give higher-octane alkylate in the HF Alkylation Process.

We see that the olefin and makeup isobutane are typically mixed and then dried. The combined olefin and makeup isobutane are mixed with the recycle isobutane and sent to the differential gravity reactor of Phillips' proprietary design. This low-pressure reactor has no moving parts, such as impellers or stirrers, nor are there any pumps to circulate the acid. The feed mixture is highly dispersed into a moving bed of liquid acid, which circulates because of the difference in density between the acid and the hydrocarbon. Total conversion of olefins to alkylate occurs very quickly.

Operating conditions in the reactor are relatively mild. The temperature will typically be about 80 to 110°F (27 to 43°C), or only 5 to 15°F (2.5 to 8°C) above the cooling-water temperature. The pressure will be only slightly above that required to maintain the hydrocarbons in the liquid phase—usually in the range of 85 to 120 lb/in^2 gage (590 to 820 kPa). Each alkylation design case is carefully studied in order to maximize heat recovery and minimize the isobutene/olefin ratio, while producing alkylate of sufficient octane quality to meet the refiner's needs. Isobutane/olefin ratios in the range of 8:1 to 13:1 are typically used.

From the reaction zone, the hydrocarbons and catalyst flow upward to the settling zone. Here, the catalyst separates as a bottom phase and flows, by gravity, on a return cycle through the acid cooler to the reaction zone, where the reaction cycle is continued. The hydrocarbon phase from the settling zone, containing propane, excess isobutane, normal butane, alkylate, and a small amount of HF, is charged to the fractionation section. Recycle isobutane, essential for favorable control of reaction mechanisms, is returned to the reactor from the fractionator either as a liquid or as a vapor. In the latter case, the latent heat of vaporization is recovered in nearby exchangers.

Propane and HF are produced overhead in the fractionator. The HF phase separates in the overhead accumulator, which is shared with the HF stripper, and is returned to the acid settler. What HF remains in the propane from the fractionator is removed in the HF stripper, separates in the overhead accumulator, and is returned to the acid settler. The propane product from the HF stripper contains traces of propyl fluoride, which are removed in the propane defluorinators. The propane stream is heated and passed over alumina to remove the fluoride, yielding primarily aluminum fluoride and water with a trace of HF. The propane is sent through the KOH treater to remove the trace of HF and then to storage. A similar set of equipment may be used to treat n-butane, if it is produced as a separate product stream. The n-butane product may be blended with gasoline for vapor pressure control.

Alkylate is produced as a bottoms product from the fractionation section. The alkylate product is suitable for blending in motor gasoline, but may require additional fractionation for use in aviation gasoline.

To regenerate the system acid, a small slipstream of acid is fed to the acid rerun column to remove the ASO. The HF is stripped from the ASO with hot isobutane. The ASO is washed in the ASO caustic washer to remove free HF, and the ASO is disposed of, typically by burning in the reboiler furnace or blending with fuel oil. Excess water is also removed from the system acid in the acid rerun column.

Auxiliary systems within the alkylation unit include:

1. Relief-gas neutralizer to remove HF from gases before being sent to the refinery flare,
2. Storage for anhydrous HF during periods when the unit is down for maintenance,
3. A neutralizing system for surface drainage and sewer drainage in the acid area.

AROMATICS COMPLEXES

Introduction

An aromatics complex is a combination of process units that can be used to convert petroleum naphtha and pyrolysis gasoline (pygas) into the basic petrochemical intermediates: benzene, toluene, and xylenes (BTX). Benzene is a versatile petrochemical building block used in the production of more than 250 different products. The most important benzene derivatives are ethylbenzene, cumene, and cyclohexane (Figure).

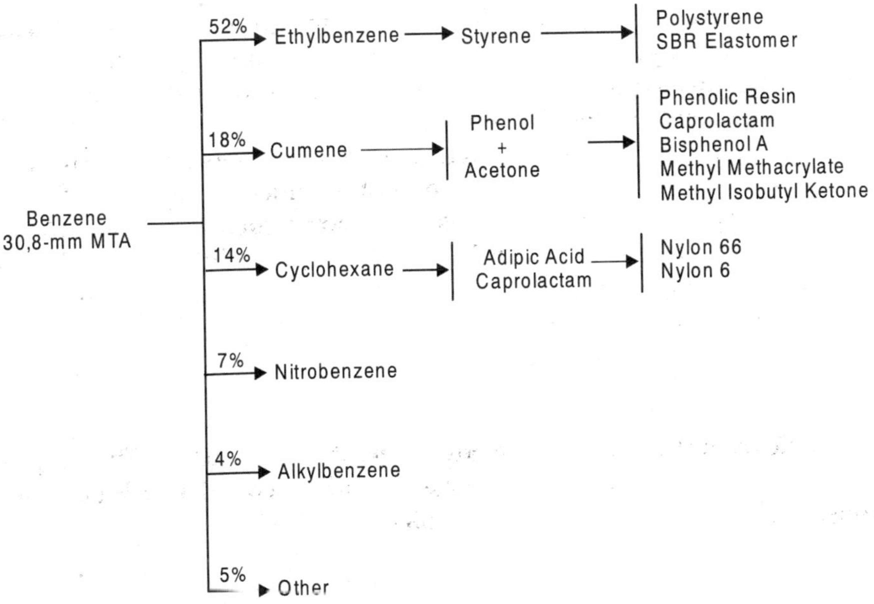

Figure 4.8: World Benzene Consumption, 2001

The xylenes product, also known as mixed xylenes, contains four different C_8 aromatic isomers: para-xylene, ortho-xylene, meta-xylene, and ethylbenzene. Small amounts of mixed xylenes are used for solvent applications, but most xylenes are processed further within the complex to produce one or more of the individual isomers. The most important C_8 aromatic isomer is para-xylene, which is used almost exclusively for the production of polyester fibers, resins, and films (Figure 4.9).

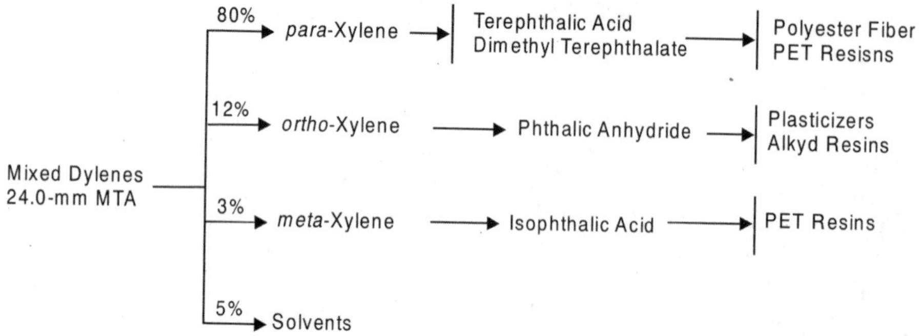

Figure 4.9: World Xylenes Consumption, 1999

In recent years, polyester fibers have shown growth rates of 5 to 6 per cent per year as synthetics are substituted for cotton. Resins have shown growth rates of 10 to 15 per cent per year, corresponding to the emergence of PET (polyethylene terephthalate) containers. Note that benzene can be a significant by-product of para-xylene production, depending on the type of technology being used. A small amount of toluene is recovered for use in solvent applications and derivatives, but most toluene is used to produce benzene and xylenes. Toluene is becoming increasingly important for the production of xylenes through toluene disproportionation and transalkylation with C_9 aromatics.

Configurations

Aromatics complexes can have many different configurations. The simplest complex produces only benzene, toluene, and mixed xylenes (Figure) and consists of the following major process units:

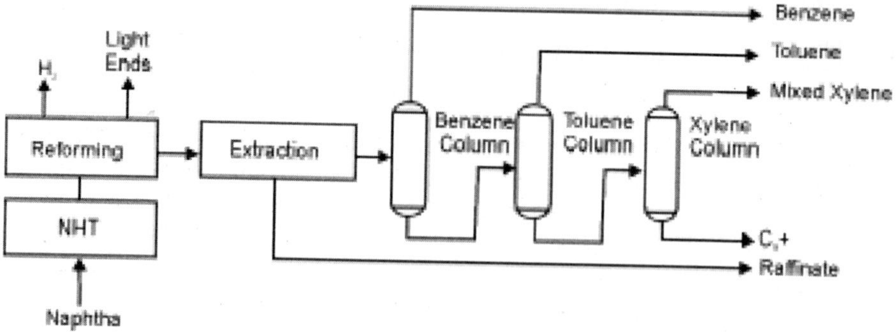

Figure 4.10: Simple Aromatics Complex

- Naphtha hydrotreating for the removal of sulfur and nitrogen contaminants

- Catalytic reforming for the production of aromatics from naphtha
- Aromatics extraction for the extraction of BTX

Most new aromatics complexes are designed to maximize the yield of benzene and para-xylene and sometimes ortho-xylene. The configuration of a modern, integrated UOP aromatics complex is shown in Figure 4.11.

This complex has been configured for maximum yield of benzene and para-xylene and includes the following UOP process technologies:

- CCR Platforming for the production of aromatics from naphtha at high severity,
- Sulfolane, Carom, on extractive distillation for the recovery of benzene and toluene,
- Parex for the recovery of para-xylene by continuous adsorptive separation,
- Isomar for the isomerization of xylenes and the conversion of ethylbenzene,
- Tatoray for the conversion of toluene and heavy aromatics to xylenes and benzene.

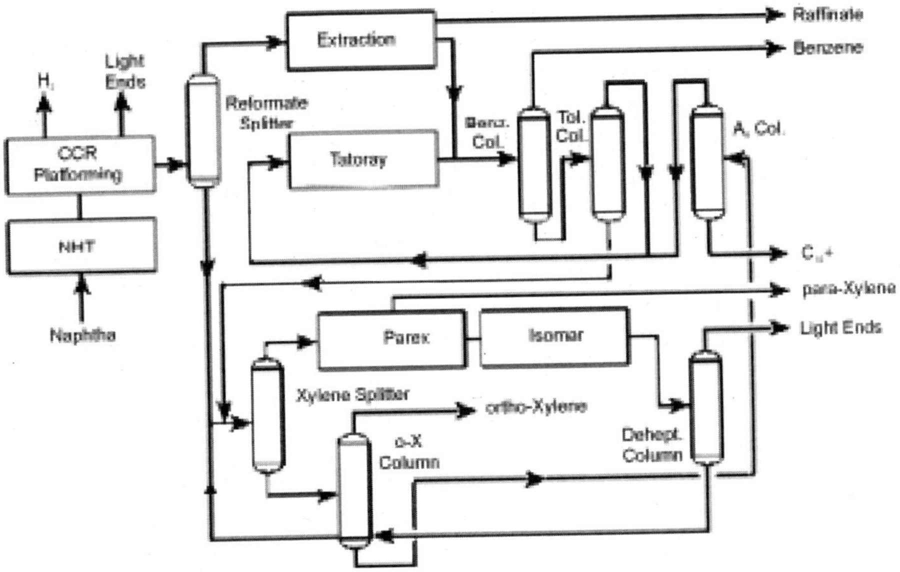

Figure 4.11: Integrated UOP Aromatics Complex

The Tatoray process is used to produce additional xylenes and benzene by toluene disproportionation and transalkylation of toluene plus C9 aromatics. The incorporation of a Tatoray unit into an aromatics complex can more than double the yield of para-xylene from a given amount of naphtha feedstock. Thus, the Tatoray process is used when paraxylene is the principal product. If there is significant need for benzene production, it can be achieved by adjusting the boiling range of the naphtha feed to include more benzene and toluene precursors. In such cases, technologies such as PX-Plus or even thermal hydrodealkylation (THDA) can be used to maximize benzene production. The cost of production is highest for THDA, so it is being used only in situations where benzene supply is scarce.

About one-half of the existing UOP aromatics complexes are configured for the production of both para-xylene and ortho-xylene. Figure shows an ortho-Xylene (o-X) column for recovery of ortho-xylene by fractionation. If ortho-xylene production is not required, the o-X column is deleted from the configuration, and all the C8 aromatic isomers are recycled through the Isomar unit until they are recovered as para-xylene.

UOP SULFOLANE PROCESS

Introduction

The UOP Sulfolane process is used to recover high-purity aromatics from hydrocarbon mixtures, such as reformed petroleum naphtha (reformate), pyrolysis gasoline (pygas), or coke-oven light oil.

The Sulfolane process takes its name from the solvent used: tetrahydrothiophene 1,1-dioxide, or Sulfolane. Sulfolane was developed as a solvent by Shell in the early 1960s and is still the most efficient solvent available for the recovery of aromatics. Since 1965, UOP has been the exclusive licensing agent for the Sulfolane process. Many of the process improvements incorporated in a modern Sulfolane unit are based on design features and operating techniques developed by UOP.

The Sulfolane process can be applied as a combination of liquid-liquid extraction (LLE) and extractive distillation (ED) or, with an appropriate feed, ED alone. The choice is a function of the feedstock and the processing objectives, as explained below.

The Sulfolane process is usually incorporated in an aromatics complex to recover high-purity benzene and toluene products from reformate. In a modern, fully integrated UOP aromatics complex (Figure), the Sulfolane unit is located downstream of the reformate splitter column. The C_6-C_7 fraction from the overhead of the reformate splitter is fed to the Sulfolane unit. The aromatic extract from the Sulfolane unit is clay-treated to remove trace olefins, and individual benzene and toluene products are recovered by simple fractionation. The paraffinic raffinate from the Sulfolane unit is usually blended into the gasoline pool or used in aliphatic solvents.

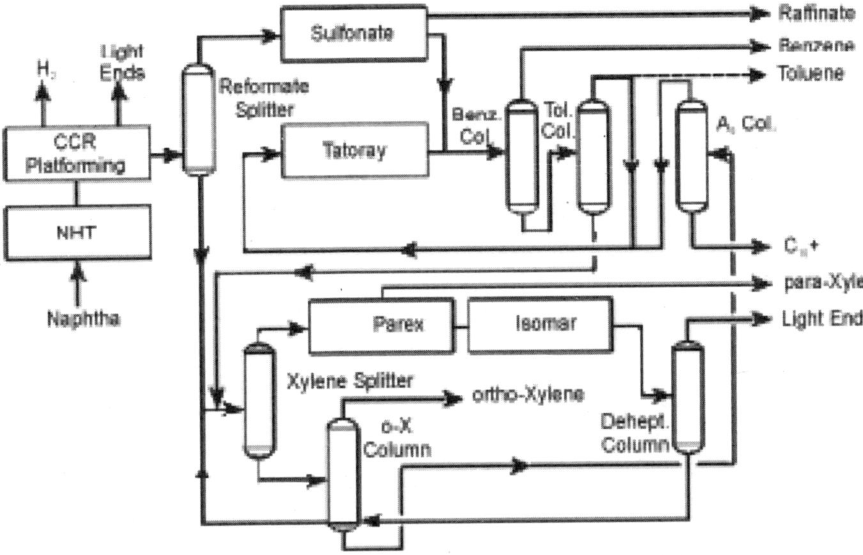

Figure 4.12: Integrated UOP Aromatics Complex

The Sulfolane process can also be an attractive way to reduce the benzene concentration in a refinery's gasoline pool so that it meets new reformulated gasoline requirements. In a typical benzene-reduction application (Figure), a portion of the debutanized reformate is sent to a reformate splitter column. The amount of reformate sent to the splitter is determined by the degree of benzene reduction required. Bypassing some reformate around the splitter and recombining it with splitter bottoms provide control of the final benzene concentration. The benzene-rich splitter overhead is sent to the Sulfolane unit, which produces a high-purity benzene product that can be sold to the petrochemical market. The raffinate from the Sulfolane unit can be blended back into the gasoline pool or upgraded in an isomerization unit.

Improvements in the Sulfolane process have allowed the application of extractive distillation alone to feeds that have traditionally been sent to a combination LLE/ED unit. For the same feed rate, an extractive distillation unit is about 80 per cent of the installed cost of a combined LLE/ED unit. The economics of one versus the other is largely a question of the utilities required to achieve the high-purity product specifications at satisfactory recoveries of BTX. The application of extractive distillation is favored when the Sulfolane feed is rich in aromatics. In such a case, there is less raffinate to boil overhead in the extractive distillation column, which consumes energy.

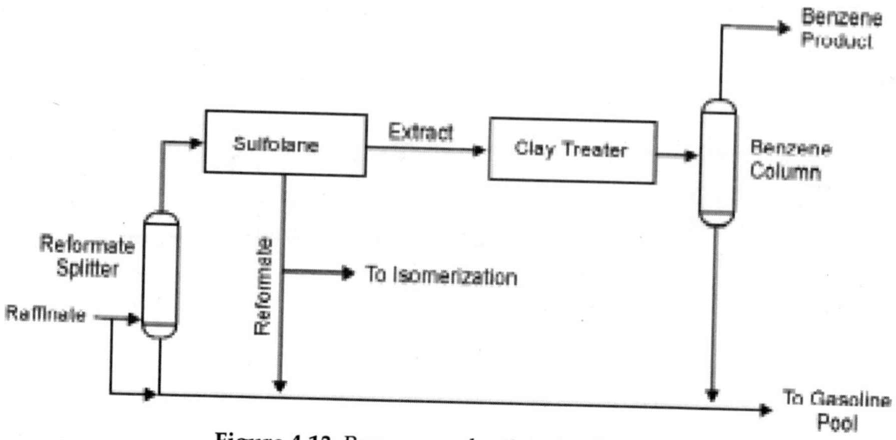

Figure 4.13: Benzene-reduction Application

The more economical choice is an economic and engineering decision. Factors to consider include:

- New versus revamp equipment
- Cost of utilities
- Feed composition (boiling range, nonaromatics, impurities)
- Product specifications

Solvent Selection

The suitability of a solvent for aromatics extraction involves the relationship between the capacity of the solvent to absorb aromatics (solubility) and the ability of the solvent to differentiate between aromatics and nonaromatics (selectivity). A study of the common polar solvents used for aromatic extraction reveals the following qualitative similarities:

- When hydrocarbons containing the same number of carbon atoms are compared, solubilities decrease in this order: aromatics>naphthenes>olefins>paraffins.
- When hydrocarbons in the same homologous series are compared, solubility decreases as molecular weight increases.
- The selectivity of a solvent decreases as the hydrocarbon content, or loading, of the solvent phase increases.

In spite of these general similarities, various commercial solvents used for aromatics recovery have significant quantitative differences. Sulfolane demonstrates better aromatic solubilities at a given selectivity than any other commercial solvent. The practical consequence of these differences is that an extraction unit designed to use Sulfolane solvent requires a lower solvent circulation rate and thus consumes less energy.

In addition to superior solubility and selectivity, Sulfolane solvent has three particularly advantageous physical properties that have a significant impact on plant investment and operating cost:

- High specific gravity (1.26). High specific gravity allows the aromatic capacity of Sulfolane to be fully exploited while maintaining a large density difference between the hydrocarbon and solvent phases in the extractor. This large difference in densities minimizes the required extractor diameter. The high density of the liquid phase in the extractive distillation section also minimizes the size of the equipment required there.
- Low specific heat—0.4 cal/(g•°C) [0.4 Btu/(lb•°F)]. The low specific heat of Sulfolane solvent reduces heat loads in the fractionators and minimizes the duty on solvent heat exchangers.
- High boiling point [287°C (549°F)]. The boiling point of Sulfolane is significantly higher than that of the heaviest aromatic hydrocarbon to be recovered, facilitating the separation of solvent from the aromatic extract.

Process Concept

The Sulfolane process combines both liquid-liquid extraction and extractive distillation in the same process unit. This mode of operation has particular advantages for aromatic recovery

UOP THERMAL HYDRODEALKYLATION (THDA) PROCESS

Introduction

The importance of benzene as an intermediate in the production of organic-based materials is exceeded only by that of ethylene. Benzene represents the basic building block for direct or indirect manufacture of well over 250 separate products or product classifications.

Historically, the major consumption of benzene has been in the production of ethylbenzene (for polystyrene), cumene (for phenol and acetone), and cyclohexane (for nylon). Significant quantities of benzene are also consumed in the manufacture of aniline, detergent alkylate, and maleic anhydride.

At present, approximately 92 per cent of the benzene produced worldwide comes directly from petroleum sources. Catalytic reforming supplies most of the petroleumderived petrochemical benzene. However, toluene is produced in greater quantities than benzene in the reforming operation, and in many areas, low market demand for toluene can make its conversion to benzene via dealkylation economically attractive. Approximately 13 per cent of the petrochemical benzene produced in the world is derived from toluene dealkylation.

The thermal hydrodealkylation (THDA) process provides an efficient method for the conversion of alkylbenzenes to high-purity benzene. In addition to producing benzene, the THDA process can be economically applied to the production of quality naphthalene from suitable feedstocks.

Process Description

The UOP THDA process converts alkylbenzenes and alkylnaphthalenes to their corresponding aromatic rings, benzene and naphthalene. The relation between product distribution and operating severity is such that, for both benzene and naphthalene operations, the conversion per pass of fresh feed is maintained at somewhat less than 100 per cent. A simplified process flow diagram for benzene manufacture is presented in Figure.

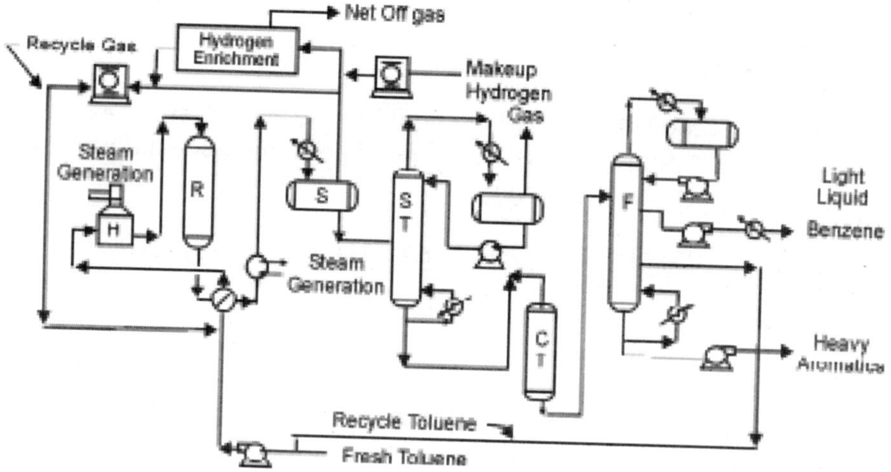

H = Heater, R = Reactor, S = Separator, ST = Stripper, CT = Clay Tower, F = Fractionator

Figure 4.14: UOP THDA Process for Benzene Production

The alkyl-group side chains of the alkyl-aromatic feed as well as nonaromatics that may be present in the unit feed are converted to a light paraffinic coproduct gas consisting mainly of methane. The basic hydrodealkylation reaction enables the process to produce a high-purity benzene or naphthalene product without applying extraction or superfractionation techniques, even when charging a mixture of alkyl aromatics and nonaromatic hydrocarbons. Excessive nonaromatics in the charge significantly add to hydrogen consumption.

Product yields approach stoichiometric with benzene yield from toluene approximating 99 per cent on a molal basis. A small amount of heavy-aromatic material consisting of biphenyl-type compounds is coproduced.

In a benzene unit, fresh toluene feedstock is mixed with recycle toluene and recycle and fresh hydrogen gases, heated by exchange in a fired heater, and then charged to the reactor. Alkyl aromatics are hydrodealkylated to benzene and nonaromatics, and paraffins and naphthalenes are hydrocracked. The effluent from the reactor is cooled and directed to the product separator, where it separates into a liquid phase and gas phase. The hydrogen-rich gas phase is recycled to the reactor, and the separator liquid is charged to a stripper for the removal of light ends. Stripper bottoms are percolated through a clay treater to the fractionation section, where high-purity benzene is obtained as an upper sidecut from a benzene fractionation column. Unconverted toluene is recycled to the reactor from the lower sidecut of the benzene column. Heavy-aromatic by-product is withdrawn from the bottom of the column to storage.

The reactor-section process flow in a naphthalene THDA unit is similar to that described for the benzene unit. Fresh feed is mixed with unconverted recycle alkyl aromatics and makeup and recycle hydrogen. The mixture is then heated and charged to the reactor. Materials in the feedstock materials that boil close to naphthalene would make the recovery of high-purity product either impossible or uneconomic if they remain unconverted. Process conditions are set to ensure that these materials are hydrocracked or dealkylated, or both, to products easily separated by fractionation.

In the case of naphthalene, the aromatic-splitter bottoms are charged to a naphthalene splitter, where the small amount of heavy-aromatic coproduct is rejected as a bottoms product. Naphthalene splitter overhead is directed to the naphthalene fractionator, where high-purity naphthalene is recovered as an overhead product. Naphthalene fractionator bottoms are recycled to the reactor section. In both benzene and naphthalene THDA units, clay treating of the product is generally required to meet the usual acid-wash color specifications.

Several design options are available for optimizing hydrogen usage in both types of units. Coproduct light ends, primarily methane, must be removed from the reaction section to maintain hydrogen purity. When the supply of makeup hydrogen is limited, consideration also must be given to the elimination of C3 and heavier nonaromatic hydrocarbons from the makeup gas. If present, these materials hydrocrack and substantially increase hydrogen consumption. During the design stage of hydrodealkylation units, careful attention must be given to hydrogen consumption and availability as related to overall refinery operation.

Depending on the application, THDA units can process a wide variety of feedstocks. For the production of benzene, feedstocks could include extracted light alkylbenzene, suitably treated coke-oven light oil; and pyrolysis coproducts. Feedstocks to produce naphthalene could include heavy reformate, extracted cycle oils from the fluid catalytic cracking (FCC) process, and coal-tar-derived materials.

Benzene produced from commercial THDA units typically has a freeze point of 5.5°C, which exceeds the ASTM Benzene-545 benzene specifications.

Process Economics

Although THDA yields are about 99 per cent on a molar basis, they are considerably lower on a weight basis because of the change in molecular weight. Investment and utility requirements are shown in Table.

Table 4.1: THDA Yields

Benzene production	*Feeds, wt %*	*Product, wt %*
Hydrogen (chemical consumption)	2.3	
Methane		17.7
Ethane		0.6
Benzene		83.6
Toluene	100	
Heavy aromatics		0.4
Total	102.3	102.3

BP-UOP CYCLAR PROCESS

Introduction

In recent years, light hydrocarbons have become increasingly attractive as fuels and petrochemical feedstocks, and much effort has been devoted to improving the recovery, processing, and transportation of liquefied petroleum gas (LPG) and natural gas. Because production areas are often located in remote areas that are far removed from established processing plants or consumers, elaborate product transport infrastructures are required. Although natural gas can be moved economically through pipelines, condensation problems limit the amount of LPG that can be transported in this way. Thus, most LPG is transported by such relatively expensive means as special-purpose tankers or railcars. The high cost of transporting LPG can often depress its value at the production site. This statement is especially true for propane, which is used much less than butane for gasoline blending and petrochemical applications.

British Petroleum (BP) recognized the problem with transporting LPG and in 1975 began research on a process to convert LPG to higher-value liquid products that could be shipped more economically. This effort led to the development of a catalyst that was capable of converting LPG to petrochemical-grade benzene, toluene, and xylenes (BTX) in a single step. However, BP soon realized that the catalyst had to be regenerated often in this application and turned to UOP for its well-proven CCR technology, which continuously regenerates the catalyst. UOP developed a high-strength formulation of the BP catalyst that would work in CCR service and also applied the radial-flow, stacked-reactor design originally developed for the UOP Platforming* process. The result of this outstanding technical collaboration is the BP-UOP Cyclar process.

Process Chemistry

The Cyclar process converts LPG directly to a liquid aromatics product in a single operation. The reaction is best described as dehydrocyclodimerization and is thermodynamically favored at temperatures above 425°C (800°F). The dehydrogenation of light paraffins propane and butanes) to olefins is the rate-limiting step (Figure).

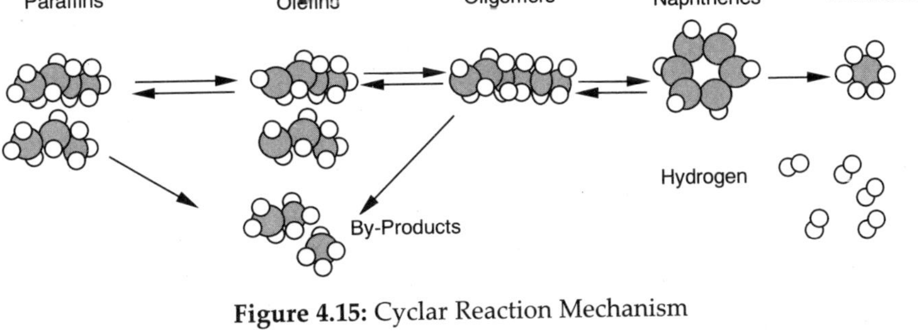

Figure 4.15: Cyclar Reaction Mechanism

Once formed, the highly reactive olefins oligomerize to form larger intermediates, which then rapidly cyclize to naphthenes. These reactions—dehydrogenation, oligomerization, and cyclization—are all acid-catalyzed.

The shape selectivity of the zeolite component of the catalyst also promotes the cyclization reaction and limits the size of the rings formed. The final reaction step is the dehydrogenation of the naphthenes to their corresponding aromatics. This reaction is highly favored at Cyclar operating conditions, and the result is virtually complete conversion of the naphthenes.

The reaction intermediates can also undergo a hydrocracking side reaction to form methane and ethane. This side reaction results in a loss of yield because methane and ethane are inert at Cyclar operating conditions.

Because olefins are a key reaction intermediate, they can of course be included in the feed to the Cyclar unit. Heavier paraffins, such as pentanes, can also be included in the feed. Olefins and pentanes are almost completely converted in the Cyclar unit, but the unit must be designed to handle them because they result in a higher catalyst-coking rate than pure butane and propane feedstocks.

Although the reaction sequence involves some exothermic steps, the preponderance of dehydrogenation reactions results in a highly endothermic overall reaction. Five moles of hydrogen are produced for every mole of aromatic components formed from propane or butane.

Because propane and butanes are relatively unreactive, the Cyclar process requires a catalyst with high activity. At the same time, the production of methane and ethane from unwanted hydrocracking side reactions must be minimized. An extensive joint effort by BP and UOP has resulted are a catalyst that combines several important features to ensure efficient commercial operation:

- At the conditions necessary for high selectivity to aromatics, the conversion performance of the catalyst declines slowly.
- The selectivity to aromatics is nearly constant over the normal range of conversion, resulting in stable product yield and quality. Thus, economic process performance can be maintained despite normal fluctuations in unit operation.
- At normal process conditions, the rate of carbon deposition on the catalyst is slow and steady, amounting to less than 0.02 wt per cent of the feed processed. Because the carbon levels on spent catalyst are low, regeneration requirements are relatively mild. Mild regeneration conditions extend the life of the catalyst and make it insensitive to process upsets and changes in feedstock composition.
- The catalyst exhibits high thermal stability and is relatively insensitive to common feedstock contaminants. Regeneration fully

restores the activity and selectivity of the catalyst to the performance seen with fresh catalyst.

- High mechanical strength and low attrition characteristics make the catalyst well suited for continuous catalyst regeneration.

Description Of The Process Flow

A Cyclar unit is divided into three major sections. The reactor section includes the radialflow reactor stack, combined feed exchanger, charge heater, and interheaters. The regenerator section includes the regenerator stack and catalyst transfer system. The product recovery section includes the product separators, compressors, stripper, and gas recovery equipment.

The flow scheme is similar to the UOP CCR Platforming process, which is used widely throughout the world for reforming petroleum naphtha.

Fresh feed and recycle are combined and heat exchanged against reactor effluent. The combined feed is then raised to reaction temperature in the charge heater and sent to the reactor section. Four adiabatic, radial-flow reactors are arranged in a vertical stack. Catalyst flows vertically by gravity down the stack, and the charge flows radially across the annular catalyst beds. Between each reactor, the vaporized charge is reheated to reaction temperature in an interheater.

The effluent from the last reactor is split into vapor and liquid products in a product separator. The liquid is sent to a stripper, where light saturates are removed from the C_6+ aromatic product. Vapor from the product separator is compressed and sent to a gas recovery section, typically a cryogenic unit, for separation into a 95 per cent pure hydrogen product stream, a fuel gas stream of light saturates, and a recycle stream of unconverted LPG. Hydrogen is not recycled.

Because coke builds up on the Cyclar catalyst over time at reaction conditions, partially deactivated catalyst is continually withdrawn from the bottom of the reactor stack for regeneration. Figure 4.15 shows additional details of the catalyst regeneration section. A discrete amount of spent catalyst flows into a lock hopper, where it is purged with nitrogen. The purged catalyst is then lifted with nitrogen to the disengaging hopper at the top of the regenerator. The catalyst flows down through the regenerator, where the accumulated carbon is burned off. Regenerated catalyst flows down into the second lock hopper, where it is purged with hydrogen and then lifted with hydrogen to the top of the reactor stack. Because the reactor and regenerator sections are separate, each operates at its own optimal conditions. In addition, the regeneration section can be temporarily shut down for maintenance without affecting the operation of the reactor and product recovery sections.

Feedstock Considerations

Propane and butanes should be the major components in the feedstock to a Cyclar unit. The C_1 and C_2 zsaturates should be minimized because these components act as inert diluents. Olefins should be limited to less than 10 per cent of the feed. Higher concentrations of olefins require hydrogenation of the feed. The C_5 and C_6 components increase the rate of coke formation in the process and should be limited to less than 20 and 2 wt per cent, respectively, in Cyclar units designed for LPG service. Cyclar units can be designed to process significantly higher amounts of C_5 and C_6 materials if necessary.

UOP ISOMAR PROCESS

Introduction

The UOP Isomar process is used to maximize the recovery of a particular xylene isomer from a mixture of C8 aromatic isomers. The Isomar process is most often applied to para-xylene recovery, but it can be used to maximize the recovery of ortho-xylene or meta-xylene. The term mixed xylenes is used to describe a mixture of C_8 aromatic isomers containing a near-equilibrium distribution of para-xylene, ortho-xylene, meta-xylene, and ethylbenzene (EB).

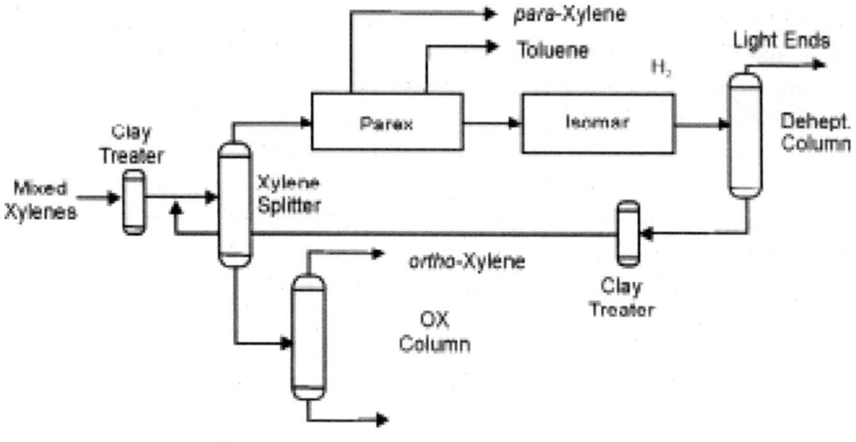

Figure 4.16: Typical Parex-Isomar Loop

In the case of para-xylene recovery, a mixed-xylenes feed is charged to a UOP Parex* unit where the para-xylene isomer is preferentially extracted at 99.9 wt per cent purity and 97wt per cent recovery per pass. The Parex raffinate is almost entirely depleted of para-xylene and is then sent to the Isomar unit (Figure). The Isomar unit reestablishes a near-equilibrium distribution of xylene isomers, essentially creating additional para-xylene from the remaining ortho and meta isomers. Effluent from the Isomar unit is then recycled to the Parex unit for recovery of additional para-xylene. In this way, the ortho and meta isomers and EB are recycled to extinction.

Process Chemistry

The two main categories of xylene isomerization catalysts are EB dealkylation catalysts and EB isomerization catalysts. The primary function of both catalyst types is to reestablish an equilibrium mixture of xylene isomers; however, they differ in how they handle the EB in the feed. An EB dealkylation catalyst converts EB to a valuable benzene coproduct. An EB isomerization catalyst converts EB to additional xylenes.

UOP offers both EB isomerization catalysts I-9, I-210 and I-400 and EB dealkylation catalysts I-300 and I-330. Both types are bifunctional catalysts that have a balance of catalytic sites between zeolitic (acid) and metal functions. The acid function on each catalyst serves the same function: isomerization of xylenes.

The EB isomerization catalyst systems I-9 and I-210 isomerize EB to xylenes through a naphthene intermediate (Figure).

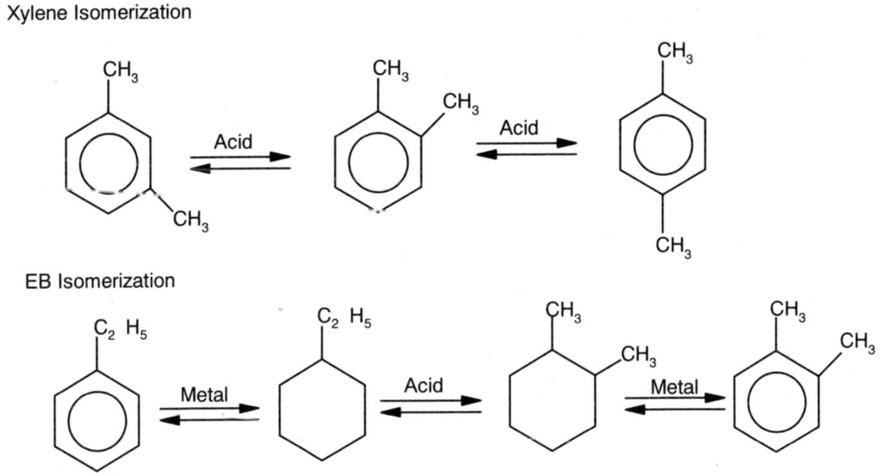

Figure 4.17: EB Isomerization Chemistry

The metal function first saturates the EB to ethylcyclohexane, then the acid function isomerizes it to dimethylcyclohexane, and finally the metal function dehydrogenates the naphthene to xylene. Because the isomerization of EB is an equilibrium-limited reaction, the conversion of EB is usually limited to about 30 to 35 wt per cent per pass. In a typical aromatics complex using the I-9 catalyst, naphthenes are recycled to the Isomar unit through the xylene column and Parex unit to suppress the formation of naphthenes in the Isomar unit and thereby increase the yield of para-xylene from the complex.

UOP will be introducing I-400, a new EB isomerization catalyst, in 2003. I-400 will provide enhanced EB conversion, increased xylene yield over I-9 and I-210, while allowing longer processing cycles between regenerations.

The EB dealkylation catalyst systems I-300 and I-330 use an EB dealkylation mechanism in which the ethyl group is cleaved from the aromatic ring by the acid function of the catalyst (Figure).

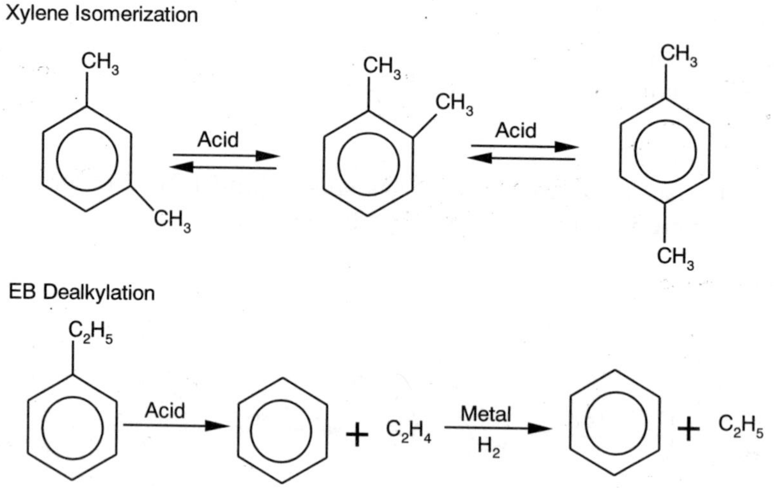

Figure 4.18: EB Dealkylation Chemistry

This reaction is not equilibrium-limited, thereby allowing EB conversion of up to 70 wt per cent or greater per pass. Because this reaction does not involve a naphthene intermediate, C8 naphthenes need not be recycled through the Parex-Isomar loop.

All xylene isomerization catalysts exhibit some by-product formation across the reactor. A large portion of the total feed to the Parex-Isomar loop is recycled from the Isomar unit. A typical Parex-Isomar loop is designed with a recycle/feed ratio of 2.5:3.5. Byproduct formation across the isomerization process is magnified accordingly. Therefore, a small reduction in the by-product formation across the

Isomar reactor translates to a large, overall yield advantage. In the Isomar process, the precise level of expected by-product formation varies with catalyst type and operating severity, but it is normally in the range of 1.0 to 4.0 wt per cent per pass of the feed. The lower end of the range is representative of operation with the later-generation catalysts I-100, I-210, and I-300. The upper end of the range is representative of operation with the I-9 catalyst. By-products are predominantly aromatic, such that overall ring retention is greater than 99 per cent. The proper selection of the isomerization catalyst type depends on the configuration of the aromatics complex, the composition of the feedstocks, and the desired product slate. The choice of isomerization catalyst must be based on an economic analysis of the entire aromatic complex. The C8 fraction of the reformate from a typical petroleum naphtha contains approximately 15 to 17 wt per cent EB, but up to 30 wt per cent EB may be in a similar pyrolysis gasoline (pygas) fraction. Using an EB isomerization catalyst maximizes the yield of para-xylene from an aromatics complex by converting EB to xylenes. An EB isomerization catalyst is usually chosen when the primary goal of the complex is to maximize the production of para-xylene from a limited supply of feedstock. The EB isomerization catalyst system will also minimize the quantity of benzene by-product produced.

Alternatively, an EB dealkylation catalyst can be used to debottleneck an existing Parex unit or crystallizer by converting more EB per pass through the isomerization unit and eliminating the requirement for naphthene intermediate circulating around the Parex-Isomar loop. For a new aromatics complex design, using an EB dealkylation catalyst minimizes the size of the xylene column and Parex and Isomar units required to produce a given amount of para-xylene. However, this reduction in size of the Parex-Isomar loop comes at the expense of lower para-xylene yields, because all the EB in the feed is being converted to benzene rather than to additional para-xylene. Lower para-xylene yield means that more feedstock will be required, which increases the size of the CCR* Platforming,* Sulfolane,* and Tatoray units in the front end of the complex, as well as most of the fractionators.

UOP PAREX PROCESS

Introduction

The UOP Parex process is an innovative adsorptive separation method for the recovery of para-xylene from mixed xylenes. The term mixed xylenes refers to a mixture of C_8 aromatic isomers that includes ethylbenzene, para-xylene, meta-xylene, and ortho-xylene. These isomers boil so closely together that separating them by conventional distillation is not practical. The Parex process provides an efficient means of recovering para-xylene by using a solid zeolitic adsorbent that is selective for para-xylene. Unlike conventional chromatography, the Parex process simulates the countercurrent flow of a liquid feed over a solid bed of adsorbent. Feed and products enter and leave the adsorbent bed continuously at nearly constant compositions. This technique is sometimes referred to as simulated moving-bed (SMB) separation.

In a modern aromatics complex (Figure), the Parex unit is located downstream of the xylene column and is integrated with a UOP Isomar unit. The feed to the xylene column consists of the C_8+ aromatics product from the CCR Platforming unit together with the xylenes produced in the Tatoray unit. The C_8 fraction from the overhead of the xylene column is fed to the Parex unit, where high-purity para-xylene is recovered in the extract. The Parex raffinate is then sent to the Isomar unit, where the other C_8 aromatic isomers are converted to additional para-xylene and recycled to the xylene column.

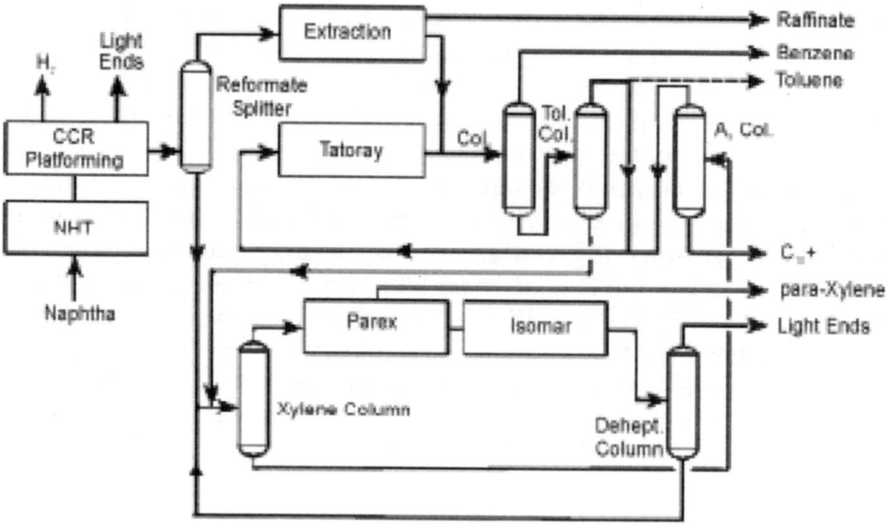

Figure 4.19: UOP Aromatics Complex—Maximum Para-xylene

UOP Parex units are designed to recover more than 97 wt per cent of the para-xylene from the feed in a single pass at a product purity of 99.9 wt per cent or better. The Parex design is energy-efficient, mechanically simple, and highly reliable. On-stream factors for Parex units typically exceed 95 per cent.

Parex Versus Crystallization

Before the introduction of the Parex process, para-xylene was produced exclusively by fractional crystallization. In crystallization, the mixed-xylenes feed is refrigerated to –75°C (–100°F), at which point the para-xylene isomer precipitates as a crystalline solid. The solid is then separated from the mother liquor by centrifugation or filtration. Final purification is achieved by washing the para- xylene crystals with either toluene or a portion of the para-xylene product.

Soon after it was introduced in 1971, the UOP Parex process quickly became the world's preferred technology for para-xylene recovery. Since that time, virtually all new para-xylene production capacity has been based on the UOP Parex process (Figure).

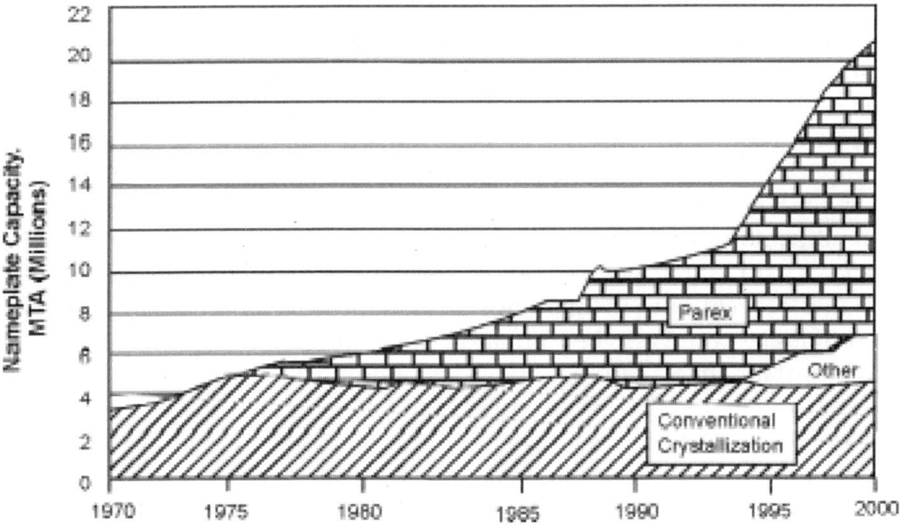

Figure 4.20: Total installed Parex capacity. In 2002, Parex represented 69 per cent of the total capacity.

The principal advantage of the Parex adsorptive separation process over crystallization technology is the ability of the Parex process to recover more than 97 per cent of the para-xylene in the feed per pass. Crystallizers must contend with a eutectic composition limit that restricts para-xylene recovery to about 65 per cent per pass. The implication of this difference is clearly illustrated in Figure.

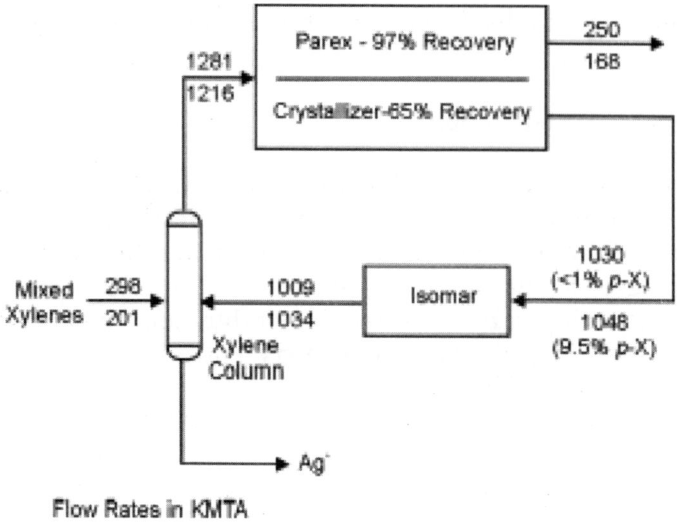

Figure 4.21: Comparison of Parex with Crystallization

A Parex complex producing 250,000 metric tons per annum (MTA) of para-xylene is compared with a crystallizer complex producing 168,000 MTA. The upper numbers in the figure indicate the flow rates through the Parex complex; the lower numbers indicate the flow rates through a comparable crystallizer complex. A Parex complex can produce about 50 per cent more para-xylene from a given-size xylene column and isomerization unit than a complex using crystallization. In addition, the yield of para-xylene per unit of fresh feed is improved because a relatively smaller recycle flow means lower losses in the isomerization unit. The technologies could also be compared by keeping the para-xylene, product rate constant. In this case, a larger xylene column and a larger isomerization unit would be required to produce the same amount of para-xylene, thus increasing both the investment cost and the utility consumption of the complex.

A higher para-xylene recycle rate in the crystallizer complex not only increases the size of the equipment in the recycle loop and the utility consumption within the loop, but also makes inefficient use of the xylene isomerization capacity. Raffinate from a Parex unit is almost completely depleted of para-xylene (less than 1 wt per cent), whereas mother liquor from a typical crystallizer contains about 9.5 wt per cent para-xylene. Because the isomerization unit cannot exceed an equilibrium concentration of para-xylene (23 to 24 wt per cent), any para-xylene in the feed to the isomerization unit reduces the amount of para-xylene produced in that unit per pass. Thus, the same isomerization unit produces about 60 per cent more para-xylene per pass when processing Parex raffinate than it does when processing crystallizer mother liquor.

In 1997, UOP, Washington Group International, and Niro Process Technology recognized the value that the three companies could bring to the marketplace by consolidating their combined 80+ years of process design and know-how to reevaluate para-xylene production from a multidiscipline perspective. In 1998, this alliance introduced the HySorb XP process, a simplified, single-chamber, light desorbent adsorption process coupled with single-stage crystallization and Niro wash column technology. This combination of technologies when integrated into existing multistage crystallization facilities can increase para-xylene production by as much as 500 per cent. The HySorb process produces a 95 wt per cent para-xylene concentrate, eliminating eutectic constraints and enabling single-stage crystallization recoveries above 90 per cent at much improved utilities' consumption. Para-xylene production utilizing single-stage crystallization is categorized as "Other" in Figure Economic studies indicate that the HySorb XP configuration does not provide any.

UOP TATORAY PROCESS

Introduction

An aromatics complex is a combination of process units which are used to convert petroleum naphtha and pyrolysis gasoline to the basic petrochemical intermediates: benzene, toluene, and xylenes (BTX). A fully integrated modern complex designed to produce benzene, para-xylene (PX), and sometimes ortho-xylene includes UOP's Tatoray process. A simplified flow diagram of a typical aromatics complex designed for maximum production of PX is shown in Figure. The Tatoray process is integrated between the aromatics extraction and xylene recovery sections of the plant (Figure).

Toluene (A_7) is fed to the Tatoray unit rather than being blended into the gasoline pool or sold for solvent applications. If the goal is to maximize the production of PX from the complex, the C_9 aromatic (A_9) by-product can also be fed to the Tatoray unit rather than blending it into the gasoline pool. Processing A_9-A_{10} in a Tatoray unit shifts the chemical equilibrium in the unit toward decreased benzene production and increased production of xylenes.

The Tatoray process provides an ideal way of producing additional mixed xylenes from low-value toluene and heavy aromatics. What is seldom recognized, however, is where the xylenes are produced within the complex. For an aromatics complex that includes transalkylation, approximately 50 per cent of the xylenes in the complex come from the transalkylation reaction, that is, the Tatoray process.

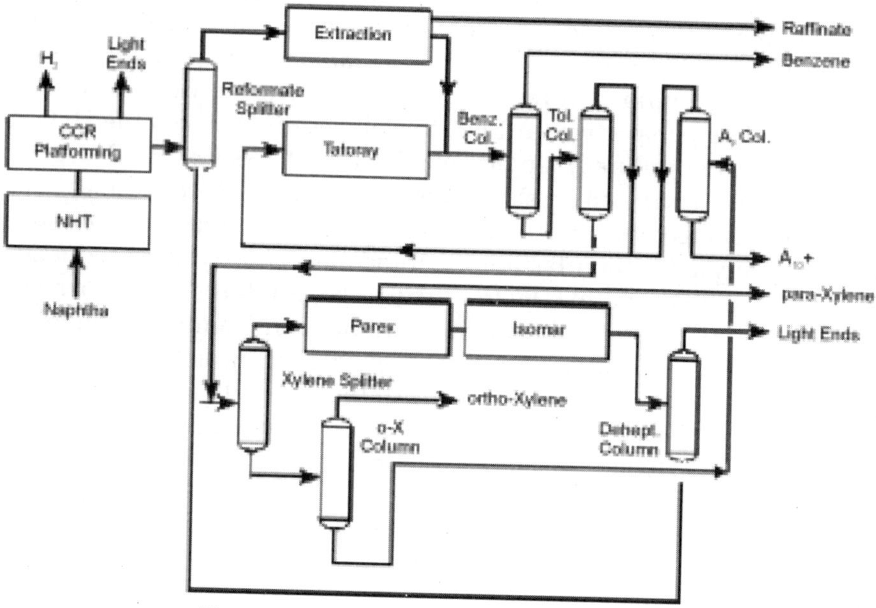

Figure 4.22: Typical UOP Aromatics Complex

The reformate provides approximately 45 per cent, and 5 per cent comes from C_8 aromatic isomerization such as the Isomar process, via conversion of ethylbenzenes to xylenes. The Tatoray process produces an equilibrium mixture of xylenes plus ethylbenzene. The xylenes are recovered and isomerized to PX while the ethylbenzene can also be converted to xylenes. The incorporation of a Tatoray unit to an aromatics complex can more than double the yield of PX from naphtha feedstock. It is extremely important, when looking at improving xylene production economics, that one focus not only on the reformer, but also on the transalkylation process and its performance. There are a number of different strategies that producers are pursuing to increase profitability. Two in particular are to reduce feedstock consumption (and the associated cost) and upgrade by-products to increase their sale value. Transalkylation can play a key role in both strategies.

Process Chemistry

The two major reactions in the Tatoray process, disproportionation and transalkylation, are illustrated in Figure.

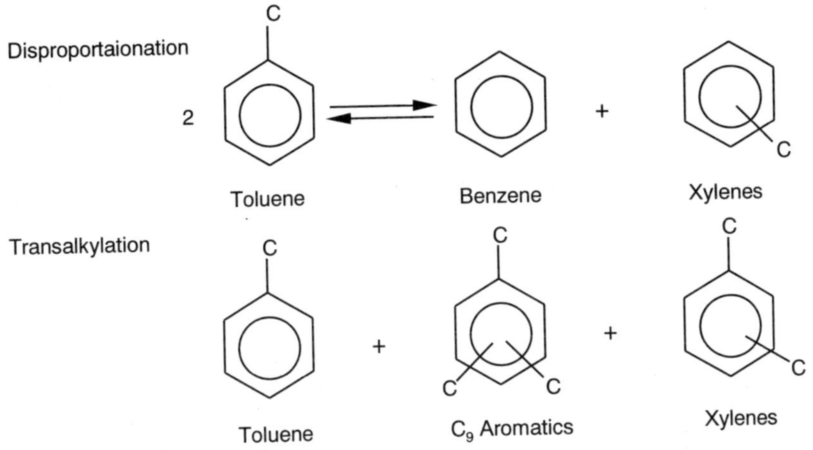

Figure 4.23: Major Tatoray Reactions

The conversion of toluene alone to an equilibrium mixture of benzene and xylenes is called disproportionation. The conversion of a blend of toluene and A_9 to xylenes through the migration of methyl groups between methyl-substituted aromatics is called transalkylation. In general, both reactions proceed toward an equilibrium distribution of benzene and alkyl-substituted aromatics. Methyl groups are stable at Tatoray reaction conditions, and thus the reaction equilibrium is easy to estimate when the feed consists of all methyl-substituted aromatics. The equilibrium distribution is illustrated in Figure. The reaction pathways involving A_9 for the Tatoray process have also been described elsewhere.

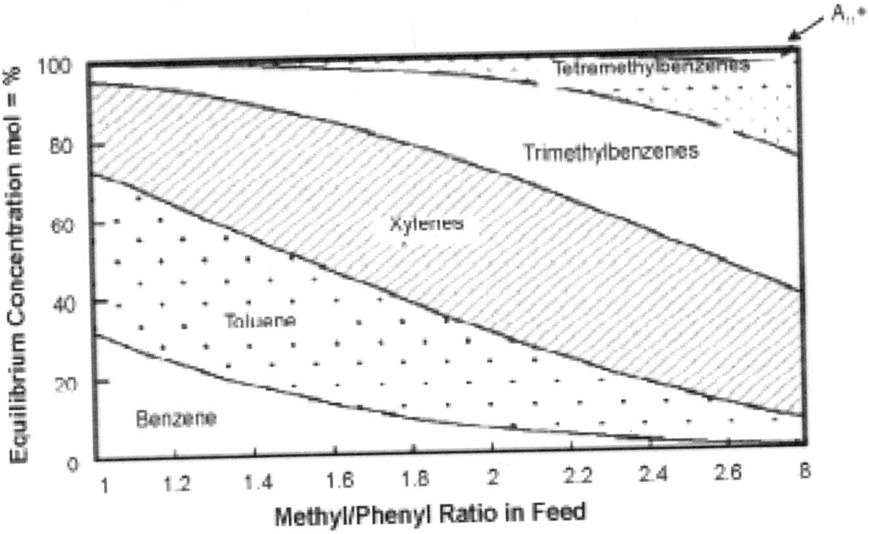

Figure 4.25: Equilibrium Distribution of Methyl Groups at 700 K

More complex reaction pathways occur when other alkyl groups are present in the feed. The Tatoray process effectively converts the ethyl, propyl, and higher alkyl group substituted A_9-A_{10} to lighter single-ring aromatics via dealkylation, while preserving the methyl groups. The lighter, mostly methyl-substituted, aromatics proceed with transalkylation to produce benzene and xylenes in a yield pattern governed by equilibrium. The dealkylation reactions involving propyl and higher substituted groups typically proceed to completion. It is also known that the diffusion coefficients of ethyl and higher alkyl group substituted rings in some aluminosilicates are much lower than those of the methyl-only substituted rings. The Tatoray catalyst enhances the transport properties of the reactants, thereby increasing the reaction efficiency.

The TA series of catalysts was first commercialized in 1969. A new generation has been introduced, on average, every 6 years. UOP introduced TA-4 in 1988. Tatoray licensees are very familiar with TA-4 catalyst and have experienced its ruggedness and ability to handle a wide variety of operating conditions while maintaining performance. The catalyst demonstrates superior selectivity and stability over a wide range of feed rates and feed compositions. High per pass conversion to benzene and mixed xylenes established TA-4 as economically attractive.

The most recent Tatoray catalyst to be successfully commercialized is called TA-5.3 This catalyst was designed as a "drop-in" reload catalyst for service in the Tatoray process and was commercialized in October 2000. The stability of TA-5 catalyst is more than double that of TA-4. This results in improved on-stream efficiency and a reduction in regeneration frequency. Commercial units have also shown that the activity of TA-5 is at least 50 per cent better than that of TA-4, allowing users to potentially increase throughput while maintaining current cycle lengths. Alternatively, users can use the TA-5 catalyst to process heavier aromatic feeds to produce higher-value benzene and mixed-xylenes products.

KBR FLUID CATALYTIC CRACKING PROCESS

Introduction

Fluid catalytic cracking (FCC) technology is a technology with more than 60 years of commercial operating experience. The process is used to convert higher-molecular-weight hydrocarbons to lighter, more valuable products through contact with a powdered catalyst at appropriate conditions. Historically, the primary purpose of the FCC process has been to produce gasoline, distillate, and C_3/C_4 olefins from low-value excess refinery gas oils and heavier refinery streams. FCC is often the heart of a modern refinery because of its adaptability to changing feedstocks and product demands and because of high margins that exist between the FCC feedstocks and converted FCC products. As oil refining has evolved over the last 60 years, the FCC process has evolved with it, meeting the challenges of cracking heavier, more contaminated feedstocks, increasing operating flexibility, accommodating environmental legislation, and maximizing reliability.

The FCC unit continuously circulates a fluidized zeolite catalyst that allows rapid cracking reactions to occur in the vapor phase. The KBR Orthoflow FCC unit consists of a stacked disengager-regenerator system that minimizes plot space requirements.

The cracking reactions are carried out in an up-flowing vertical reactor-riser in which a liquid oil stream contacts hot powdered catalyst. The oil vaporizes and cracks to lighter products as it moves up the riser and carries the catalyst along with it. The reactions are rapid, requiring only a few seconds of contact time. Simultaneously with the desired reactions, coke, a material having a low ratio of hydrogen to carbon, deposits on the catalyst and renders it less catalytically active. Catalyst and product vapors separate in a disengaging vessel with the catalyst continuing first through a stripping stage and second through a regeneration stage where coke is combusted to rejuvenate the catalyst and provide heat for operation of the process. The regenerated catalyst then passes to the bottom of the reactor-riser, where the cycle starts again. Hydrocarbon product vapors flow downstream for separation into individual products.

KBR, through its ancestry in The M.W. Kellogg Company, has been a leader in FCC technology developments since the inception of the process. In recent years, KBR has worked with its FCC partner, ExxonMobil, to create and refine FCC technology features that have led the industry. To date, KBR has licensed more than 120 grassroots FCC units throughout the world, including 13 grassroots units and more than 120 revamps since just 1990.

Feedstocks

The modern FCC unit can accept a broad range of feedstocks, a fact which contributes to FCC's reputation as one of the most flexible refining processes in use today. Examples of common feedstocks for conventional distillate feed FCC units are:

- Atmospheric gas oils
- Vacuum gas oils

- Coker gas oils
- Thermally cracked gas oils
- Solvent deasphalted oils
- Lube extracts
- Hydrocracker bottoms

Residual FCCU (RFCCU) processes Conradson carbon residue and metals-contaminated feedstocks such as atmospheric residues or mixtures of vacuum residue and gas oils. Depending on the level of carbon residue and metallic contaminants (nickel and vanadium), these feedstocks may be hydrotreated or deasphalted before being fed to an RFCCU. Feed hydrotreating or deasphalting reduces the carbon residue and metals levels of the feed, reducing both the coke-making tendency of the feed and catalyst deactivation.

Products

Products from the FCC and RFCC processes are typically as follows:

- Fuel gas (ethane and lighter hydrocarbons)
- C_3 and C_4 liquefied petroleum gas (LPG)
- Gasoline
- Light cycle oil (LCO)
- Fractionator bottoms (slurry oil)
- Coke (combusted in regenerator)
- Hydrogen Sulfide (from amine regeneration)

Although gasoline is typically the most desired product from an FCCU or RFCCU, design and operating variables can be adjusted to maximize other products. The three principal modes of FCC operation are (1) maximum gasoline production, (2) maximum light cycle oil production, and (3) maximum light olefin production, often referred to as maximum LPG operation. These modes of operation are discussed below:

Maximum Gasoline

The maximum gasoline mode is characterized by use of an intermediate cracking temperature (510 to 540°C), high catalyst activity, and a high catalyst/oil ratio. Recycle is nor

DEEP CATALYTIC CRACKING, THE NEW LIGHT OLEFIN GENERATOR

Basis

The fluid catalytic cracking (FCC) unit is the most important and widely used heavy oil conversion process in the modern refinery. Historically, the FCC unit has operated in maximum gasoline and maximum distillate modes, depending on seasonal product demands and refinery locale. Recently, with the advent of reformulated gasoline requirements, the FCC unit has been increasingly required to operate in the maximum olefin mode. Light isoolefins, isobutylene and isoamylene, from the FCC unit are necessary feedstocks for methyl tertiary butyl ether (MTBE) and tertiary amyl methyl ether (TAME) oxygenated reformulated gasoline blending components. Increased alkylate demand to meet reformulated gasoline requirements also necessitates an increase in light olefins.

At the same time as these changes are occurring in the refining industry, the petro-chemical industry is experiencing increased demands for propylene for the manufacture of polypropylene products. Nearly one-half of the propylene used by the chemical industry is obtained from refineries, and the remainder comes from steam cracking (SC).1 As a result, the demand for propylene from both FCC units and SC units is rising. Since SC units produce ethylene as the primary product, a catalytic process is more suitable for making propylenes and butylenes.

The demand for propylene, both as an alkylation feed and for polypropylene production, is expected to continue growing well into the 21st century. More isoolefins are also needed for those locations where MTBE and TAME can be used in the gasoline pool. This places a considerable strain on the FCC unit and SC unit in order to meet the demand. Obviously, a need for an economical light olefin generating process is required to meet these demands for light olefins (C_3 through C_5).

To this end, Stone & Webster has entered into an agreement with the Research Institute of Petroleum Processing (RIPP) and Sinopec International, both located in the People's Republic of China, to exclusively license RIPP's Deep Catalytic Cracking (DCC) technology outside China. DCC is a fully commercialized process, similar to FCC, for producing light olefins (C3 to C5) from heavy feedstocks such as gas oils and paraffinic residuals. Stone & Webster's proven position in FCC technology and steam cracking is a natural complement to DCC technology.

Numerous DCC units have been put in commercial service. Table 4.2 is a list of all DCC units operating at present.

Process Description

DCC is a fluidized catalytic process for selectively cracking a variety of feedstocks to light olefins. A traditional reactor/regenerator unit design is employed with a catalyst having physical properties much like those of FCC catalyst. The DCC unit may be operated in one of two modes: maximum propylene (type I) and maximum isoolefins (type II). Each operational mode employs a unique catalyst and operating conditions. DCC reaction products are light olefins, high-octane gasoline, light cycle oil, dry gas, and coke. A small amount of slurry oil may also be produced.

DCC maximum propylene operation (type I) employs both riser and bed cracking at severe reactor conditions. Maximum isoolefin operation (type II) utilizes riser cracking, as does a modern FCC unit, at slightly milder conditions than a type I operation.

Fresh feed is finely atomized by steam and injected into the riser through Stone & Webster proprietary FCC feed injection nozzles over a dense phase of catalyst. The atomized oil intimately mixes with the catalyst and begins to crack into lighter, more valuable products. A good feed injection system is required for DCC, just as for FCC operations, to ensure rapid oil vaporization and selective catalytic cracking reactions.

Riser steam is injected just above the feed injection point to supplement feed dispersion and stripping steam in order to achieve optimal hydrocarbon partial pressure for the DCC operation. Simple steam injection nozzles are employed for riser steam injection. (Steam requirements for DCC type II operation are considerably less and may not need additional steam injection nozzles.)

Table 4.2:

Location	*Feed, MTA**	*Start-up*	*DCC type*
Jinan, China	60,000	1990	I
Jinan expansion	150,000	1994	I and II
Anqing, China	400,000	1995	I
Daqing, China	120,000	1995	I
Jinmen, China	800,000	1997	II
TPI, Thailand†	900,000	1997	I
Shenyang, China	400,000	1998	II
Jinzhou, China	300,000	1999	I
Urumchi, China	800,000	1999	II

*MTA = metric tons per year.

Slurry recycle is injected, if required, just above the riser steam nozzles. This recycle stream is not required to increase overall conversion but rather to optimize the unit heat balance, as a large slurry reaction product is coke.

At the top of the riser, catalyst, steam, and hydrocarbon pass through a riser terminator located below the reactor bed. Conversion of the DCC feedstock can be regulated by adjusting the catalyst bed height (hydrocarbon weight hourly space velocity) above the riser distributor, the catalyst circulation rate, and/or the reactor temperature. Two-stage high-efficiency reactor cyclones remove entrained catalyst from the reactor vapors. Products, inerts, steam, and a small amount of catalyst flow from the reactor into the bottom of the main fractionator to begin product separation.

UOP FLUID CATALYTIC CRACKING PROCESS

Introduction

The fluid catalytic cracking (FCC) process is a process for the conversion of straight-run atmospheric gas oils, vacuum gas oils, certain atmospheric residues, and heavy stocks recovered from other refinery operations into high-octane gasoline, light fuel oils, and olefin-rich light gases. The features of the FCC process are relatively low investment, reliable long-run operations, and an operating versatility that enables the refiner to produce a variety of yield patterns by simply adjusting operating parameters. The product gasoline has an excellent front-end octane number and good overall octane characteristics. Further, FCC gasoline is complemented by the alkylate produced from the gaseous olefinic by-products because alkylate has superior midrange octane and excellent sensitivity.

In a typical FCC unit, the cracking reactions are carried out in a vertical reactor riser in which a liquid oil stream contacts hot powdered catalyst. The oil vaporizes and cracks to lighter products as it moves up the riser and carries the catalyst powder along with it. The reactions are rapid, and only a few seconds of contact time are necessary for most applications. Simultaneously with the desired reactions, coke, a carbonaceous material having a low ratio of hydrogen to carbon (H/C), deposits on the catalyst and renders it less catalytically active. The spent catalyst and the converted products are then separated; and the catalyst passes to a separate chamber, the regenerator, where the coke is combusted to rejuvenate the catalyst. The rejuvenated catalyst then passes to the bottom of the reactor riser, where the cycle begins again.

Development History

With early development of the process taking place in the late 1930s, the first commercial FCC unit was brought on-stream in the United States in

May 1942. This design, Model I, was quickly followed by a Model II design. A total of 31 Model II units were designed and built. Although engineered by different organizations, these units were similar in concept because the technology came from the same pool, a result of wartime cooperative efforts.

Of those first units, several remain in operation today. The principal features of the Model II unit included a reactor vessel near ground level and the catalyst regenerator offset and above it. A rather short transfer line carried both catalyst and hydrocarbon vapor to a dense-bed reactor. Dual slide valves were used at various points in the unit, and this configuration resulted in a low-pressure regenerator with a higher-pressure reactor. Commercial evidence indicated that although conversions were rather low on these early units [40 to 55 liquid volume per cent (LV per cent)], a large portion of the cracking reactions actually took place in the short transfer line carrying both hydrocarbon and catalyst.

After the war, the stacked FCC design (Figure), which featured a low-pressure reactor stacked directly above a higher-pressure regenerator, was commercialized by UOP.

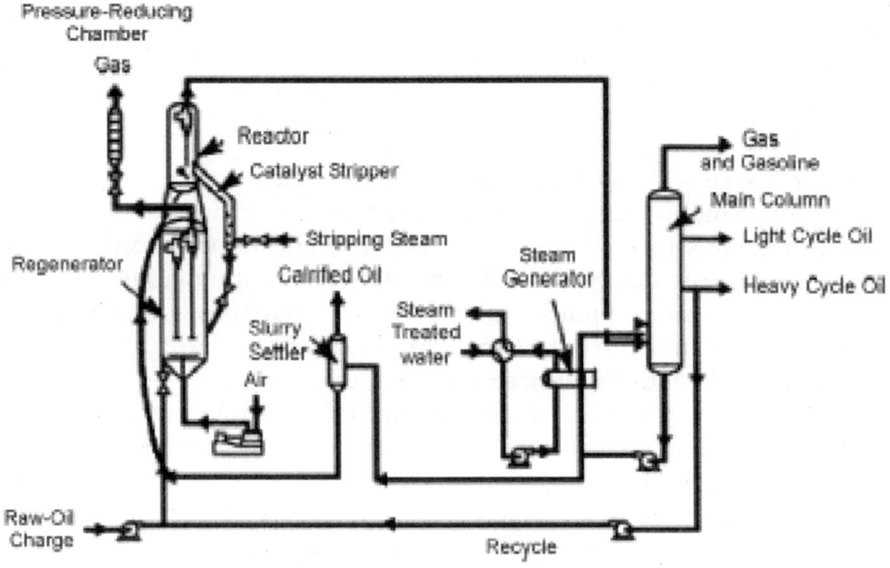

Figure 4.26: UOP Stacked FCC Unit

This design was a major step toward shifting the cracking reaction from the dense phase of the catalyst bed to the dilute phase of the riser. In the mid-1950s, the straight-riser design, also called the side-by-side design (Figure), was introduced. In this unit, the regenerator was located near ground level, and the reactor was placed to the side in an elevated position. Regenerated catalyst, fresh feed, and recycle were directed to the reactor by means of a long, straight riser located directly below the reactor. Compared with earlier designs, product yields and selectivity were substantially improved.

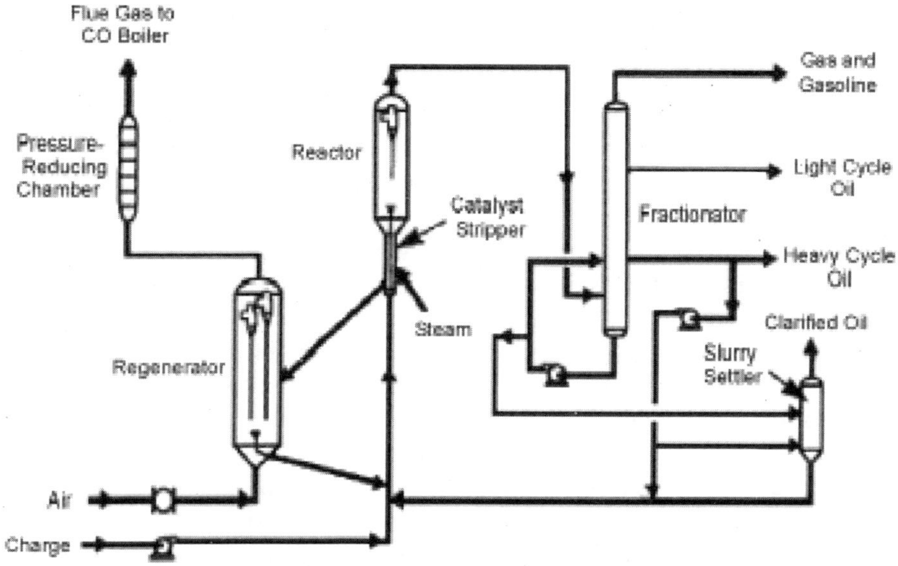

Figure 4.27: UOP Straight-riser FCC Unit

A major breakthrough in catalyst technology occurred in the mid-1960s with the development of zeolitic catalysts. These sieve catalysts demonstrated vastly superior activity, gasoline selectivity, and stability characteristics compared to the amorphous silica-alumina catalysts then in use. The availability of zeolitic catalysts served as the basis for most of the process innovations that have been developed in recent years.

The continuing sequence of advances first in catalyst activity and then in process design led to an emphasis on achieving more of the reactions within the dilute phase of the riser, or riser cracking, as it is commonly called. In 1971, UOP commercialized a new design based on this riser cracking concept, which was then quickly extended to revamps of many of the existing units. Commercial results confirmed the advantages of this system compared to the older designs. Riser cracking provided a higher selectivity to gasoline and reduced gas and coke production that indicated a reduction in secondary cracking to undesirable products.

This trend has continued throughout the years as process designs emphasize greater selectivity to desired primary products and a reduction of secondary by-products. When processing conditions were relatively mild, extended risers and rough-cut cyclones were adequate. As reaction severities were increased, vented risers and direct-connected cyclones were used to terminate the riser. To achieve even higher levels of hydrocarbon containment, further enhancements to prestrip, or displace, hydrocarbons that would otherwise be released from the cyclone diplegs into the reactor vessel now provide an even more selective operation. One example of such selective riser termination designs is the vortex separation system (VSS). Such designs have truly approached the concept of all-riser cracking, where almost all the reaction now takes place within the riser and its termination system.

The emphasis on improved selectivity with all-riser cracking has placed a premium on good initial contact of feed and catalyst within the riser. Thus, much attention over the years has been given to improving the performance of the feed distributor as well as to properly locating it. The quantity of dispersant and the pressure drop required as well as the mechanical characteristics of various feed nozzles have been carefully studied, leading to the development of the highly successful Optimix feed distributor. The feed nozzle, though important, is just one component of a complete feed distribution system. Again the push for higher reaction severities has placed an even greater emphasis on the characteristics of this complete feed distribution system in the design of a modern FCC unit.

Thus far, the discussion has centered on the reactor design; however, significant changes have taken place on the regeneration side. For the first 20 years or so of its history, the regenerator of the FCC unit was operated so that the flue gas contained substantial quantities of carbon monoxide (CO) and carbon dioxide (CO_2). In this partial combustion mode, the spent catalyst was regenerated to the point of leaving a few tenths of a per cent of carbon still remaining on the regenerated catalyst. A major improvement in FCC technology in the early 1970s was the development of catalysts and hardware to permit complete internal combustion of CO to CO_2. In 1973, an existing FCC unit was revamped to include a new high-efficiency concept in regeneration technology to achieve direct con.

Stone & Webster–Institut Français Du Pétrole Fluid RFCC Process

History

Stone & Webster (S&W), in association with Institut Français du Pétrole (IFP), is the licenser of the S&W-IFP residual fluid catalytic cracking (R2R) process. The original S&W-IFP R2R (reactor 2 regenerators) process was developed during the early 1980s by Total Petroleum Inc. at its Arkansas

City, Kansas, and Ardmore, Oklahoma, refineries. Because the development of this process saw heavy input from an operating company, unit operability and mechanical durability were incorporated into the design to ensure smooth operation and long run lengths. To process the heavy, viscous residual feedstocks, which can contain metals in high concentrations and produce relatively high amounts of coke, the design incorporates an advanced feed injection system, a unique regeneration strategy, and a catalyst transfer system which produces extremely stable catalyst circulation. Recent technology advances have been made in the areas of riser termination, reactant vapor quench, mix temperature control (MTC), and stripping.

Table 4.3: S&W-IFP Full RFCC Technology Units

Refinery	*Location*	*Capacity, BPSD*	*Start-up*
A	Kansas	20,000	1981
B	Oklahoma	25,000/40,000*	1982
C	Canada	19,000	1985
D	Japan	40,000	1987
E	Australia	25,000	1987
F	Canada	25,000	1987
G	China	23,000	1987
H	China	21,000	1989
I	China	28,000	1990
J	China	21,000	1990
K	China	21,000	1991
L	Japan	30,000	1992
M	Japan	31,600	1994
N	Uruguay	9,000	1994
O	Singapore	24,000	1995
P	Korea	50,000	1995
Q	Korea	30,000	1995
R	Thailand	37,000	1996
S	India	15,000	1997
T	Canada	65,000	2000
U	India	15,000	2001
V	India	26,000	2002
W	India	60,000	2003
X	India	65,000	2003
Y	Europe	30,000	2004
Z	Vietnam	65,000	2004

Note: BPSD = barrels per stream-day.

Today 26 full-technology S&W-IFP RFCC units have been licensed worldwide (revamp and grassroots), more than all other RFCC licensers combined. Within the Pacific Rim, S&W-IFP's 19 licensed units outnumber the competition by more than 2 to 1. From 1980 to 2001, there were 20 operating S&W-IFP FCC units totaling more than 190 years of commercial operation. A listing of all S&W-IFP (full-technology) licensed R2R units is shown in Table 4.3.

While the conception of this technology was based on processing residual feed, the technology has been proved and is widely accepted for processing lighter gas oil feedstocks. Stone & Webster and IFP have ample experience revamping gas oil FCC units to upgrade the feed injection system, combustion air distributor, riser termination device, etc. At present, more than 60 FCC units are processing over 2,400,000 barrels per day (BPD) of FCC feed employing the S&W-IFP feed injection technology. In fact, S&W-IFP systems have replaced feed injection systems of virtually every competing licenser and have always provided measurable benefits.

Process Description

R2R Converter

Two configurations of the grassroots R2R unit are offered. The first, and the most common, is the stacked regenerator version shown in Figure, which minimizes plot space

The second is side-by-side regenerator design configuration, which is discussed in "FCC Revamp to RFCC" below and is more typical of FCC units which have been revamped to RFCC or for units larger than 100,000 B/D.

The process flow will be presented by using the stacked version shown in Figure. The RFCC utilizes a riser-reactor, catalyst stripper, first-stage regeneration vessel, second-stage regeneration vessel, catalyst withdrawal well, and catalyst transfer lines. Process flow for the side-by-side configuration is identical except for the catalyst transfer between the first- and second-stage regenerators.

Fresh feed is finely atomized with dispersion steam and injected into the riser through the feed injection nozzles over a dense catalyst phase. The small droplets of feed contact the freshly regenerated catalyst and instantaneously vaporize. The oil molecules intimately mix with the catalyst particles and crack into lighter, more valuable products.

Mix temperature control nozzles inject a selected recycle stream which quenches the catalyst and feed vapor. This feature allows control of the critical feed-catalyst mix zone temperature independent of the riser outlet temperature and provides some cooling of the regenerator. Riser outlet temperature (ROT) is controlled by the regenerated catalyst slide valve.

As the reaction mixture travels up the riser, the catalyst, steam, and hydrocarbon product mixture pass through a riser termination device. S&W-IFP currently offers a number.

UOP Platforming Process

Process Evolution

The Platforming process is a UOP-developed and -engineered catalytic reforming process in widespread use today throughout the petroleum and petrochemical industries. The first UOP Platforming unit went on-stream in 1949. The Platforming process has since become a standard feature in refineries worldwide.

In the Platforming process, light petroleum distillate (naphtha) is contacted with a platinum-containing catalyst at elevated temperatures and hydrogen pressures ranging from 345 to 3450 kPa (50 to 500 lb/in^2 gage). Platforming produces a high-octane liquid product that is rich in aromatic compounds. Chemical-grade hydrogen, light gas, and liquefied petroleum gas (LPG) are also produced as reaction by-products.

Originally developed to upgrade low-octane-number straight-run naphtha to high-octane motor fuels, the process has since been applied to the production of LPG and high-purity aromatics. A wide range of specially prepared platinum-based catalysts permit tailored processing schemes for optimum operation. With proper feed preparation, Platforming efficiently handles almost any refinery naphtha.

Since the first Platforming unit was commercialized, UOP has been at the industry forefront in advancing reforming technology. UOP has made innovations and advances in process-variable optimization, catalyst formulation, equipment design, and maximization of liquid and hydrogen yields. Since higher yields and octane are obtained at low pressure and high severity, innovations at UOP were driven to meet these objectives while controlling the coke deposition and catalyst deactivation.

The first Platforming units were designed as semiregenerative (SR), or fixed-bed, units employing monometallic catalysts. Semiregenerative Platforming units are periodically shut down to regenerate the catalyst. This regeneration includes burning off catalyst coke and reconditioning the catalyst's active metals. To maximize the length of time (cycle) between regenerations, these early units were operated at high pressures in the range of 2760 to 3450 kPa (400 to 500 lb/in^2 gage).

A typical SR Platforming flow diagram is presented in Figure.

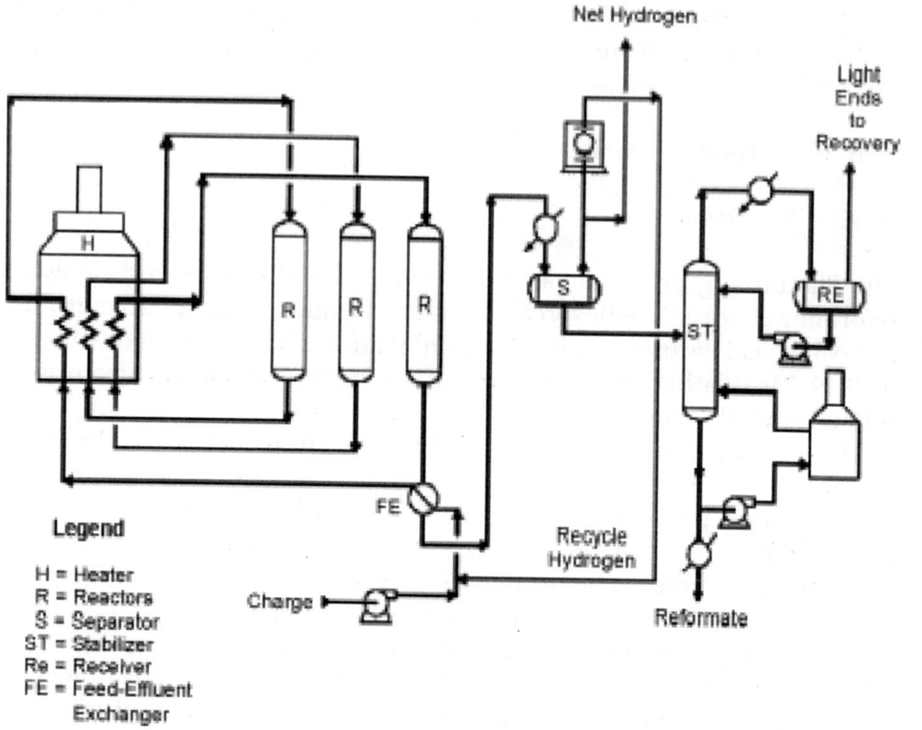

Figure 4.28: UOP Platforming Process

In the process flow, feed to the Platforming unit is mixed with recycled hydrogen gas, preheated by a feed-effluent exchanger, further heated to reaction temperature by a fired heater, and then charged to the reactor section. Because most of the reactions that occur in the Platforming process are endothermic, the reactor section is separated into several stages, or reactors. Fired heaters are installed between these reactors to reheat the process stream up to the correct temperature for the next stage. Effluent from the last reactor is cooled by exchanging heat with the feed for maximum heat recovery. Additional cooling to near-ambient temperature is provided by air or water cooling. The effluent is then charged to the separation section, where the liquid and gas products are separated. A portion of the gas from the separator is compressed and recycled to the reactor section. The net hydrogen produced is sent to hydrogen users in the refinery complex or to the fuel header. The separator liquid is pumped to a product stabilizer, where the more-volatile light hydrocarbons are fractionated from the high-octane liquid product.

UOP initially improved the Platforming process by introducing bimetallic catalysts to SR Platforming units. These catalysts enabled a lower-pressure, higher-severity operation: about 1380 to 2070 kPa (200 to 300 lb/in^2 gage), at 95 to 98 octane with typical cycle lengths of 1 year. The increased coking of the catalyst at the higher severity limited the operating run length and the ability to further reduce pressure. Catalyst development alone could not solve these problems; process innovations were needed. In the 1960s, cyclic reforming was developed to sidestep this barrier. Cyclic reforming employs fixed-bed reforming, but the reactors can be individually taken off-line, regenerated, and then put back into service without shutting down the unit and losing production.

UOP recognized the limitations of fixed-bed catalyst stability and so commercialized Platforming with continuous regeneration, the CCR* Platforming process, in 1971. The process employs continuous catalyst regeneration in which catalyst is continuously removed from the last reactor, regenerated in a controlled environment, and then transferred back to the first reactor (Figure).

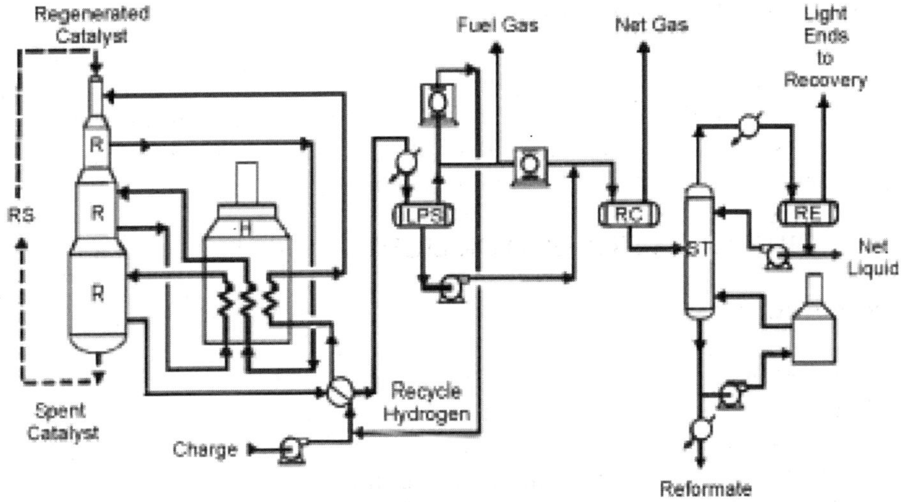

Legend

H = Heater	LPS = Rlow Pressure Separator
R = Reactor	RC = Recontractor
RE = Receiver	RS = Regeneration Section
ST = Stabilizer	

Figure 4.29: UOP CCR Platforming Process

The CCR Platforming process represents a step change in reforming technology. With continuous regeneration, coke laydown is no longer an issue because the coke is continuously burned off and the catalyst is reconditioned to its original performance. The CCR Platforming process has enabled ultralow-pressure operations at 345 kPa (50 lb/in^2 gage) and produced product octane levels as high as 108. The continuous regeneration approach has been very successful with more than 95 per cent of the new catalytic reformers being designed as CCR Platforming units. In addition, many units that were originally built as SR Platforming units have been revamped to CCR Platforming units.

In summary, the UOP Platforming process has evolved continuously throughout its history. The operating pressure has been lowered by more than 2760 kPa (400 lb/in^2 gage), and hydrogen yield has doubled. Product octane was increased by more than 12 numbers along with a C_5+ yield increase of 2 liquid volume per cent (LV %). The evolution of UOP Platforming performance is depicted in Figure, which shows the increase in both C_5+ yield and octane through time and innovation compared to the theoretical limit.

Process Chemistry

Feed and Product Compositions

The Platforming naphtha charge typically contains C_6 through C_{11} paraffins, naphthenes, and aromatics. The primary purpose of the Platforming process is to produce aromatics from the paraffins and naphthenes. The product stream is a premium-quality gasoline

UOP OLEFLEX PROCESS FOR LIGHT OLEFIN PRODUCTION

Introduction

The UOP Oleflex process is catalytic dehydrogenation technology for the production of light olefins from their corresponding paraffin. An Oleflex unit can dehydrogenate propane, isobutane, normal butane, or isopentane feedstocks separately or as mixtures spanning two consecutive carbon numbers. This process was commercialized in 1990, and by 2002 more than 1,250,000 metric tons per year (MTA) of propylene and more than 2,800,000 MTA of isobutylene were produced from Oleflex units located throughout the world.

Process Description

The UOP Oleflex process is best described by separating the technology into three different sections:

- Reactor section
- Product recovery section
- Catalyst regeneration section

Reactor Section

Hydrocarbon feed is mixed with hydrogen-rich recycle gas (Figure). This combined feed is heated to the desired reactor inlet temperature and converted at high monoolefin selectivity in the reactors.

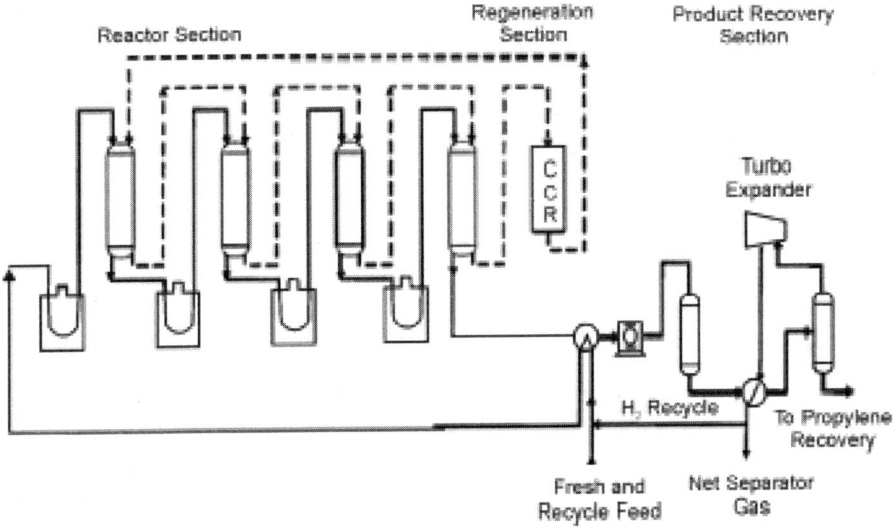

Figure 4.30: Oleflex Process Fow

The reactor section consists of several radial-flow reactors, charge and interstage heaters, and a reactor feed-effluent heat exchanger. The diagram shows a unit with four reactors, which would be typical for a unit processing propane feed. Three reactors are used for butane or isopentane dehydrogenation. Three reactors are also used for blends of C_3-C_4 or C_4-C_5 feeds.

Because the reaction is endothermic, conversion is maintained by supplying heat through interstage heaters. The effluent leaves the last reactor, exchanges heat with the combined feed, and is sent to the product recovery section.

Product Recovery Section

The reactor effluent is cooled, compressed, dried, and sent to a cryogenic separation system. The dryers serve two functions: (1) to remove trace amounts of water formed from the catalyst regeneration and (2) to remove hydrogen sulfide. The treated effluent is partially condensed in the cold separation system and directed to a separator.

Two products come from the Oleflex product recovery section: separator gas and separator liquid. The gas from the cold high-pressure separator is expanded and divided into two streams: recycle gas and net gas. The net gas is recovered at 90 to 93 mol per cent hydrogen purity. The impurities in the hydrogen product consist primarily of methane and ethane. The separator liquid, which consists primarily of the olefin product and unconverted paraffin, is sent downstream for processing.

Catalyst Regeneration Section

The regeneration section, shown in Figure, is similar to the CCR* unit used in the UOP Platforming* process. The CCR unit performs four functions:

- Burns the coke off the catalyst
- Redistributes the platinum
- Removes the excess moisture
- Reduces the catalyst prior to returning to the reactors

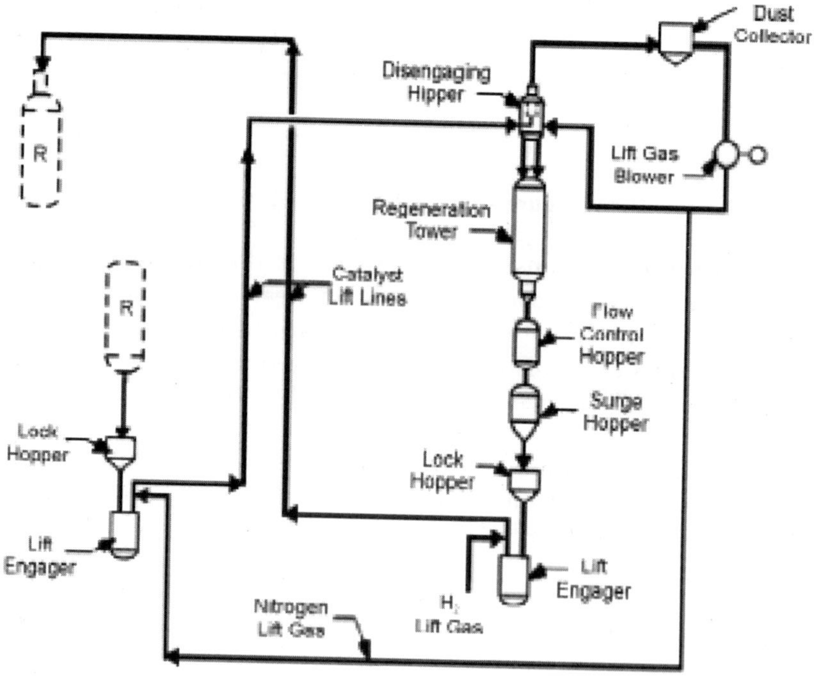

Figure 4.31: Oleflex Regeneration Section

The slowly moving bed of catalyst circulates in a loop through the reactors and the regenerator. The cycle time around the loop can be adjusted within broad limits but is typically anywhere from 5 to 10 days, depending on the severity of the Oleflex operation and the need for regeneration. The regeneration section can be stored for a time without interrupting the catalytic dehydrogenation process in the reactor and recovery sections.

Dehydrogenation Plants

Propylene Plant

Oleflex process units typically operate in conjunction with fractionators and other process units within a production plant. In a propylene plant (Figure), a propane-rich liquefied petroleum gas (LPG) feedstock is sent to a depropanizer to reject butanes and heavier hydrocarbons. The depropanizer overhead is then directed to the Oleflex unit. The once-through conversion of propane is approximately 40 per cent, which closely approaches the equilibrium value defined by the Oleflex process conditions. Approximately 90 per cent of the propane conversion reactions are selective to propylene and hydrogen; the result is a propylene mass selectivity in excess of 85 wt per cent. Two product streams are created within the C_3 Oleflex unit: a hydrogen-rich vapor product and a liquid product rich in propane and propylene.

Trace levels of methyl acetylene and propadiene are removed from the Oleflex liquid product by selective hydrogenation. The selective diolefin and acetylene hydrogenation step is accomplished with the Hüls SHP process, which is available for license through UOP. The SHP process selectively saturates diolefins and acetylenes to monoolefins

UOP PACOL DEHYDROGENATION PROCESS

Introduction

Paraffins can be selectively dehydrogenated to the corresponding monoolefins by using suitable dehydrogenation catalysts. Iron catalysts have long been used for the dehydrogenation of ethylbenzene to styrene, and catalysts made of chromia (chrome oxide) supported on alumina have long been used for the dehydrogenation of light paraffins (for example, n-butane to n-butene) and the deeper dehydrogenation of olefins to diolefins (for example, n-butene to 1,3-butadiene). However, newer commercial processes for the dehydrogenation of light and heavy paraffins are based on the use of noble-metal catalysts because of the superior stability and selectivity of these catalyst systems.

In the late 1940s and through the 1950s, the pioneering work done at UOP by Vladimir Haensel on platinum catalysis for the catalytic reforming of naphthas for the production of high-octane gasolines and high-purity aromatics showed that platinum catalysts have interesting dehydrogenation functions. This research area was later pursued by Herman Bloch and others also within UOP. In 1963-64, UOP started development work on heterogeneous platinum catalysts supported on an alumina base for the dehydrogenation of heavy n-paraffins. The resulting successful process, known as the Pacol* process (for paraffin conversion to olefins), was first commercialized in 1968. The advent of the UOP Pacol process marked a substantial transformation in the detergent industry and contributed to the widespread use of linear alkylbenzene sulfonate (LAS or LABS) on an economical, cost-effective basis. As of mid-2003, more than 40 Pacol units have been built, or are under design or construction; practically all new linear alkylbenzene (LAB) capacity built on a worldwide basis over the last two decades makes use of UOP's Pacol catalytic dehydrogenation process.

Maintaining technological superiority over some 30-odd years requires continued innovation and improvement, principally of the dehydrogenation catalyst, the reactor design, and operating conditions because these have the greatest impact on the overall process economics. The first commercial Pacol dehydrogenation catalysts, denoted DeH-3 and DeH-4, came on-stream in the mid-1960s. They were soon superseded by a newer catalyst, DeH-5, that was commercialized in 1971 and dominated the market for several years. In 1983, DeH-7 catalyst was introduced. This new catalyst exhibited about 1.75 times the stability of its predecessor, DeH-5, and soon replaced it as the dominant Pacol catalyst. Development efforts continued, and in 1998, DeH-11 was commercialized. This catalyst is the first "layered sphere" catalyst to be offered by UOP in which a thin reactive layer is coated onto an inert one. The result is an advantage in selectivity to mono-olefins. In 2001 DeH-201 was introduced. This catalyst, also a layered sphere, allows for higher conversion operation than previous Pacol catalysts. All these various generations of paraffin dehydrogenation catalysts have resulted in improved yields at higher conversion and higher operating severities, thus allowing for smaller and more economical units for a given production capacity.

Since 1980, UOP has adapted similar catalysts to the selective catalytic dehydrogenation of light olefins (propane to propylene and isobutane to isobutylene) in the Oleflex process; a number of large-capacity units have been built for this application. Because of the higher severity, light paraffin dehydrogenation units make use of UOP's proprietary CCR continuous catalyst regeneration technology, which was originally developed and commercialized for the catalytic reforming of naphthas at high severity. Because the Pacol process operates at a lower severity, catalyst runs are significantly longer, and CCR technology is not needed.

Process Description

The catalytic reaction pathways found in the dehydrogenation of n-paraffins to n-monoolefins [linear internal olefins (LIO)] in addition to other thermal cracking reactions are illustrated in Figure. A selective catalyst is required if only LIO is to be the main product.

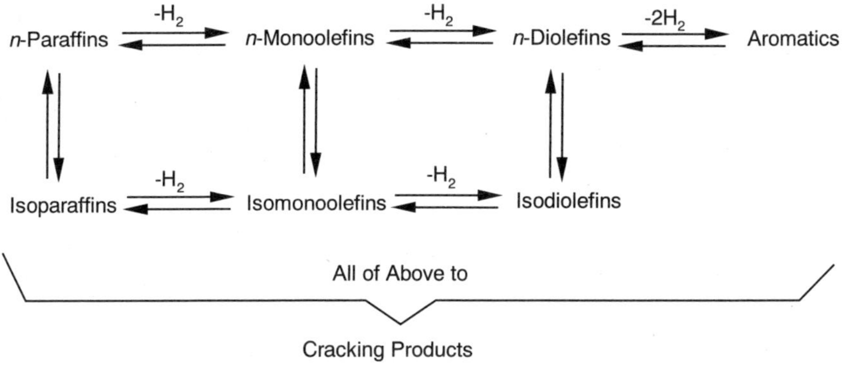

Figure 4.32: Dehydrogenation Reaction Pathways

In the Pacol reaction mechanism, the conversion of n-paraffins to monoolefins is near equilibrium, and therefore a small but significant amount of diolefins and aromatics is produced. In the alkylation process, the diolefins consume 2 moles of benzene to yield heavier diphenylalkane compounds or form heavier polymers that become part of the heavy alkylate and the bottoms by-products of the hydrofluoric (HF) acid regenerator. Thus, diolefin formation represents a net loss of alkylate yield. In 1984, UOP developed the DeFine process, a highly selective catalytic hydrogenation process to convert diolefins back to monoolefins. Detergent complexes licensed prior to 1986 included only Pacol and HF Detergent Alkylate units. The first DeFine unit came on-stream during the fourth quarter of 1986; all subsequent Pacol process units have also incorporated DeFine hydrogenation reactors, and DeFine reactors have also been retrofitted into a growing number of existing older Pacol units. Both Pacol and DeFine processes are also used in the latest process developed and commercialized by UOP, the Detal process, for the production of LAB using a heterogeneous solid catalyst instead of the older, traditional HF acid catalyst.

The dehydrogenation of n-paraffins is an endothermic reaction with a heat of reaction of about 125 kJ/g × mol (30 kcal/g × mol; 54,000 Btu/lb × mol). The equilibrium conversion for the dehydrogenation reaction is determined by temperature, pressure, and hydrogen partial pressure. As expected, the equilibrium conversion increases with temperature and decreases with pressure and with increasing hydrogen-to-hydrocarbon ratio. Kinetically, the overall conversion depends on space velocity (feed-to-catalyst ratio): excessively high space velocities do not allow for sufficient conversions, and space velocities that are too low lead to lower selectivities because of the onset of side and competitive reactions.

Figure illustrates the flow scheme of an integrated complex incorporating Pacol, DeFine, and HF Detergent Alkylate units or Pacol, DeFine, and Detal units. The main differences between the two flow schemes are in the alkylation section as a result of the elimination of the HF acid handling and neutralization facilities; for example, no alumina treater is used in conjunction with a Detal process unit.

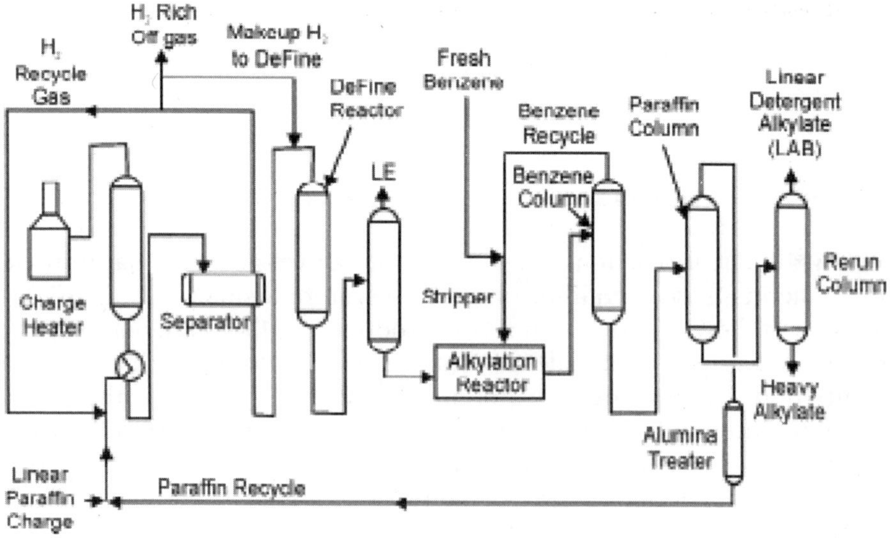

Figure 4.33: Integrated LAB Complex

In the Pacol process, linear paraffins are dehydrogenated to linear olefins in the presence of hydrogen over a selective platinum dehydrogenation catalyst. An adiabatic radial-flow reactor with feed preheat is normally used to compensate for the endothermic temperature drop and to minimize pressure drop within an efficient reactor volume.

Relatively high space velocities are used so that only a modest amount of catalyst is required. Hydrogen and some by-product light ends are separated from the dehydrogenation reactor effluent, and a part of this hydrogen gas is recycled back to the dehydrogenation reactor to minimize coking and enhance catalyst stability.

FW HYDROGEN PRODUCTION

Introduction

As hydrogen use has become more widespread in refineries, hydrogen production has moved from the status of a high-technology specialty operation to an integral feature of most refineries. This has been made necessary by the increase in hydrotreating and hydrocracking, including the treatment of progressively heavier feedstocks.

Steam reforming is the dominant method for hydrogen production. This is usually combined with pressure-swing adsorption (PSA) to purify the hydrogen to greater than 99 vol per cent.

As hydrogen production grows, a better understanding of the capabilities and requirements of the modern hydrogen plant becomes ever more useful to the refiner. This will help the refiner get the most from existing or planned units and make the best use of hydrogen supplies in the refinery.

Uses of Hydrogen

Overview

There has been a continual increase in refinery hydrogen demand over the last several decades. This is a result of two outside forces acting on the refining industry: environmental regulations and feedstock shortages. These are driving the refining industry to convert from distillation to conversion of petroleum. Changes in product slate, particularly outside the United States, are also important. Refiners are left with an oversupply of heavy, high-sulfur oil, and in order to make lighter, cleaner, and more salable products, they need to add hydrogen or reject carbon.

Within this trend there are many individual factors depending on location, complexity of the refinery, etc.

Hydrogen Demand

The early use of hydrogen was in naphtha hydrotreating, as feed pretreatment for catalytic reforming (which in turn was producing hydrogen as a by-product). As environmental regulations tightened, the technology matured and heavier streams were hydrotreated. These included light and heavy distillates and even vacuum residue.

Hydrotreating has also been used to saturate olefins and make more stable products. For example, the liquids from a coker generally require hydrotreating, to prevent the formation of polymers.

At the same time that demand for cleaner distillates has increased, the demand for heavy fuel oil has dropped. This has led to wider use of hydrocracking, which causes a further large increase in the demand for hydrogen.

Table shows approximate hydrogen consumption for hydrotreating or hydrocracking of various feedstocks.

Table 4.4: Hydrogen Consumption Data

Chemical consumption only			
Process	*Wt % on feed*	*SCF/bbl*	*Wt % on crude*
Hydrotreating:			
Straight-run naphtha	0.05	20	0.01
FCC naphtha/coker naphtha	1	500	0.05–0.01
Kerosene	0.1	50	0.1–0.02
Hydrodesulfurization:			
Low-sulfur gas oil to 0.2% S	0.1	60	0.03
High-sulfur gas oil to 0.2% S	0.3	170	0.04
Low-sulfur gas oil to 0.05% S	0.15	80	0.04
High-sulfur gas oil to 0.05% S	0.35	200	0.05
FCC gas oil/coker gas oil	1	600	0.1
Cycle oil hydrogenation	3	1700	0.3
Hydrocracking vacuum gas oil	2–3	1200–1800	0.5–0.8
Deep atmospheric residue conversion	2–3.5	1200–2200	1–2

Note: FCC = fluid catalytic cracker; SCF = standard cubic feet.

Hydrogen Production

Hydrogen has historically been produced in catalytic reforming, as a by-product of the production of the high-octane aromatic compounds used in gasoline. Changes in this process have had a large impact on the refinery hydrogen balance.

As reforming has changed from fixed-bed to cyclic to continuous regeneration, pressures have dropped and hydrogen production per barrel of reformate has increased. Recent changes in gasoline composition, due to environmental concerns, have tended to reduce hydrogen production, however. Besides limits on aromatics, requirements for oxygenates in gasoline have resulted in reduced reforming severity, as the high-octane oxygenates have displaced reformate from the gasoline pool. The only safe statement is that the situation will continue to change.

Where by-product hydrogen production has not been adequate, hydrogen has been manufactured by steam reforming. In some cases partial oxidation has been used, particularly where heavy oil is available at low cost. However, oxygen is then required, and the capital cost for the oxygen plant makes partial oxidation high in capital cost.

Figure shows a typical modern hydrogen plant. This unit produces 82 million SCFD (at 60°F and 14.7 lb/in^2 absolute) [92,000 (N) m^3/h (N represents normal temperature and pressure at 0°C and 1.0332 kg/cm^2 absolute)] of hydrogen from natural gas for a Far Eastern refinery, at a purity of 99.9 vol per cent. The Foster Wheeler Terrace Wall* steam reforming furnace is visible in the background, with the 12 absorbers and two surge drums of the pressure-swing adsorption unit in the foreground.

Chemistry

Steam Reforming. In steam reforming, light hydrocarbons such as methane are reacted with steam to form hydrogen

$$CH_4 + H_2O \rightarrow 3H_2 + CO$$
$$\Delta H = 97,400 \text{ Btu/(1b.mol) } 227 \text{ kj/(g.mol)}$$

where DH is the heat of reaction. The reaction equation can be generalized to

$$C_nH_m + (n)H_2O \rightarrow \left(n + \frac{m}{2}\right)H_2 + nCO.$$

Isocracking—Hydrocracking for Superior Fuels and Lubes Production

Hydrocracking technology plays the major role in meeting the need for cleaner-burning fuels, effective feedstocks for petrochemical operations, and more effective lubricating oils. Only through hydrocracking can heavy fuel oil components be converted to transportation fuels and lubricating oils whose quality will meet tightening environmental and market demands.

The Chevron Lummus Global Isocracking process, widely licensed for more than 40 years, has technological advantages for gasoline, middle-distillate, and lubricating oil production. Optimizing a refinery is always a matter of balance. Every benefit has a cost; every incremental gain in margin trades off against a loss somewhere else. Isocracking helps with this balance by delivering trade-off advantages with respect to product yield, quality, catalyst choice and run length, capital costs, operating costs, versatility, and flexibility. Through its families of amorphous and zeolitic catalysts, Isocracking provides refiners with essential flexibility in choices of crude to buy, products to sell, specifications to meet, configurations to use, and efficiency and profitability to achieve, all with the trade-off advantage. This chapter explains the process technology that provides these benefits.

Isocracking Chemistry

Chevron's Lummus Global's hydrocracking process was named Isocracking because of the unusually high ratio of isoparaffins to normal paraffins in its light products. A high per centage of isoparaffins increases light naphtha product octane numbers and produces outstanding middle-distillate cold flow properties—kerosene/jet fuel freeze point and diesel pour point. In 1992, Chevron enhanced its process capabilities in heavy paraffin isomerization by commercializing the Chevron Lummus Global Isodewaxing process. When combined with hydrocracking, Isodewaxing is the most efficient way to produce high-viscosity-index (VI), low-pour-point lube oil base stocks.

Isocracking provides a unique combination of aromatic saturation and paraffin isomerization which generates an attractive combination of product qualities.

Product Quality from Isocracking

Isocracking removes heavy aromatic compounds and creates isoparaffins to produce middle distillates with outstanding burning and cold flow properties.

- Kerosene with low freeze points and high smoke points
- Diesel fuels with low pour points and high cetane numbers
- Heavy naphthas with a high content of single-ring hydrocarbons
- Light naphthas with a high isoparaffin content
- Heavy products that are hydrogen-rich for feeding fcc units, ethylene plants, or lube oil dewaxing and finishing facilities

The process removes heavy aromatic compounds and produces middle distillates with outstanding burning qualities—high kerosene/jet fuel smoke points and high diesel cetane numbers. The heavy product is rich in hydrogen, making it a prime candidate for feedstock to lube oil facilities, ethylene crackers, or fluid catalytic cracking (FCC) plants.

When combined with other Chevron Lummus Global processes such as Isotreating (for the processing of light distillates), Isodewaxing or LC-Fining (for the processing of vacuum residuum), Isocracking can be used to process everything from residuum to cracked distillate stocks for the production of very high-quality middle distillates, LPG, finished lube base stocks, naphtha and low-sulfur fuel oils in addition to feeds for FCC or petrochemicals units.

The Importance of Hydrogen

Hydrocracking removes the undesirable aromatic compounds from petroleum stocks by the addition of hydrogen. The amount of hydrogen required depends on the character of the feedstock.

Isocracking produces cleaner fuels and more effective lubricants from a wide variety of petroleum stocks, different crude oil sources, and, in some cases, heavy oils generated by different processing routes. These technical challenges can be illustrated by focusing on feed and product hydrogen contents using a Stangeland diagram, as shown in Figure.

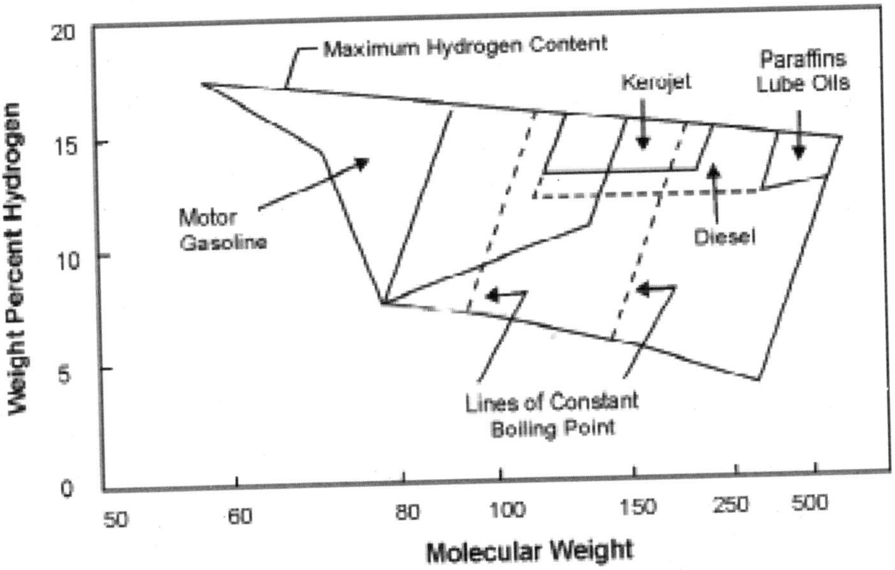

Figure 4.34: Stangeland diagram showing product hydrogen content. Regions that meet the specifications for jet/kerosene, diesel, and lube products all border the paraffin line. Aromatic compounds hurt the quality of these products.

This relates the hydrogen content of hydrocarbons to their molecular weight and provides a road map for all hydrocarbons present in petroleum stocks. By comparing the characteristics of feedstocks and products, the processing schemes required to go from one to the other can be represented.

The upper line of Figure represents the hydrogen content of pure paraffins, which have the highest hydrogen content of any hydrocarbon series. Aromatic compounds have much lower hydrogen content and fall considerably below the paraffin line. This diagram shows regions that meet the specifications for the most important refined products—motor gasoline, jet/kerosene, diesel, and lubricating oils. The regions for middle distillate and lubes all border the paraffin line. Aromatic compounds hurt the qualities of these products. The motor gasoline region is more complex because both hydrogen-rich isoparaffins and hydrogen-poor aromatics improve octane numbers.

The Isocracking process handles variations in feedstocks easily. Table shows some of the important properties of the straight-run distillates from four popular crude oils.

Table 4.6: Crude Oil Distillate Qualities

Each crude oil contains distillates of different sulfur levels and burning qualities.

	Boiling range						
°F	*400–500*	*500–650*		*650–800*		*800–1000*	
°C	*204–260*	*260–343*		*343–427*		*427–538*	
Inspection	*Smoke point, mm*	*Sulfur, wt %*	*Cetane index*	*Sulfur, wt %*	*Viscosit y index*	*Sulfur, wt %*	*Viscosit y index*
Arabian light	22	1.3	51	2.2	65	2.7	55
Sumatran light	27	0.1	60	0.1	75	0.1	60
Chinese (Shengli)	20	0.4	52	0.6	40	0.7	26
Russian (West Siberia)	20	1.1	49	1.8	46	2.3	35

Isocracking removes heavy aromatic compounds and creates isoparaffins to produce middle distillates with outstanding burning and cold flow properties.

- Kerosene with low freeze points and high smoke points
- Diesel fuels with low pour points and high cetane numbers
- Heavy naphthas with a high content of single-ring hydrocarbons
- Light naphthas with a high isoparaffin content
- Heavy products that are hydrogen-rich for feeding FCC units, ethylene plants, or lube oil dewaxing and finishing facilities

Arabian light, Sumatran light, Chinese (Shengli), and Russian (Western Siberia). Kerosene smoke point, diesel cetane, and vacuum gas oil V/s reflect the overall aromatic nature of the crude oil. Sumatran light is uniquely paraffinic and has the highest hydrogen content. The sulfur levels of the distillates from these different crude oils are also shown. Environmental pressures continue to push product sulfur levels down. Although the Stangeland diagram does not include this important contaminant, sulfur occurs primarily in the aromatic component of the feedstock. Isocracking effectively eliminates sulfur as it saturates and cracks the heavy aromatics.

Again, the paraffinic nature of the Indonesian crude oil is clearly shown. The Russian and Arabian distillates are much more deficient in hydrogen in the vacuum gas oil boiling range. Shengli distillates show less variation in hydrogen content from light to heavy distillates compared to the Russian and Arabian stocks.

UOP UNICRACKING PROCESS FOR HYDROCRACKING

Introduction

Hydrotreating and hydrocracking are among the oldest catalytic processes used in petroleum refining. They were originally employed in Germany in 1927 for converting lignite to gasoline and later used to convert petroleum residues to distillable fractions. The first commercial hydrorefining installation in the United States was at Standard Oil Company of Louisiana in Baton Rouge in the 1930s. Following World War II, growth in the use of hydrocracking was slow. The availability of Middle Eastern crude oils reduced the incentive to convert coal to liquid fuels, and new catalytic cracking processes proved more economical for converting heavy crude fractions to gasoline. In the 1950s, hydrodesulfurization and mild hydrogenation processes experienced a tremendous growth, mostly because large quantities of by-product hydrogen were made available from the catalytic reforming of low-octane naphthas to produce high-octane gasoline.

The first modern hydrocracking operation was placed on-stream in 1959 by Standard Oil Company of California. The unit was small, producing only 1000 barrels per streamday (BPSD). As hydrocracking units were installed to complement existing fluid catalytic cracking (FCC) units, refiners quickly recognized that the hydrocracking process had the flexibility to produce varying ratios of gasoline and middle distillate. Thus, the stage was set for rapid growth in U.S. hydrocracking capacity from about 3000 BPSD in 1961 to about 120,000 BPSD in just 5 years. Between 1966 and 1983, U.S. capacity grew eight-fold, to about 980,000 BPSD.

Outside the United States, early applications involved production of liquefied petroleum gas (LPG) by hydrocracking naphtha feedstocks. The excellent quality of distillate fuels produced when hydrocracking gas oils and other heavy feedstocks led to the choice of the hydrocracking process as a major conversion step in locations where diesel and jet fuels were in demand. Interest in high-quality distillate fuels produced by hydrocracking has increased dramatically worldwide. As of 2002, more than 4 million BPSD of hydrocracking capacity is either operating or is in design and construction worldwide

Process Applications

Hydrocracking is one of the most versatile of all petroleum refining processes. Any fraction from naphtha to nondistillables can be processed to produce almost any desired product with a molecular weight lower than that of the chargestock. At the same time that hydrocracking takes place, sulfur, nitrogen, and oxygen are almost completely removed, and olefins are saturated so that products are a mixture of essentially pure paraffins, naphthenes, and aromatics. Table illustrates the wide range of applications of hydrocracking by listing typical chargestocks and the usual desired products.

Table 4.7: Applications of the Unicracking Process

Chargestock	*Products*
Naphtha	Propane and butane (LPG)
Kerosene	Naphtha
Straight-run diesel	Naphtha and/or jet fuel
Atmospheric gas oil	Naphtha, jet fuel, and/or distillates
Natural gas condensates	Naphtha
Vacuum gas oil	Naphtha, jet fuel, distillates, lubricating oils
Deasphalted oils and demetallized oils	Naphtha, jet fuel, distillates, lubricating oils
Atmospheric crude column bottoms	Naphtha, distillates, vacuum gas oil, and low-sulfur residual fuel
Catalytically cracked light cycle oil	Naphtha
Catalytically cracked heavy cycle oil	Naphtha and/or distillates
Coker distillate	Naphtha
Coker heavy gas oil	Naphtha and/or distillates

The first eight chargestocks are virgin fractions of petroleum crude and gas condensates. The last four are fractions produced from catalytic cracking and thermal cracking. All these streams are being hydrocracked commercially to produce one or more of the products listed.

This flexibility gives the hydrocracking process a particularly important role as refineries attempt to meet the challenges of today's economic climate. The combined influences of low-quality feed sources, capital spending limitations, hydrogen limitations, environmental regulatory pressures, and intense competition have created a complex optimization problem for refiners. The hydrocracking process is uniquely suited, with proper optimization, to assist in solving these problems. UOP, with its broad background and research capabilities, has continued to develop both catalyst and process capabilities to meet the challenges.

Process Description

The UOP Unicracking process is carried out at moderate temperatures and pressures over a fixed catalyst bed in which the fresh feed is cracked in a hydrogen atmosphere. Exact process conditions vary widely, depending on the feedstock properties and the products desired. However, pressures usually range between 35 and 219 kg/cm^2 (500 and 3000 lb/in^2 gage) and temperatures between 280 and 475°C (536 and 887°F).

Chemistry

Hydrocracking chemistry is bifunctional catalytic chemistry involving acid-catalyzed isomerization and cracking reactions as well as metal-catalyzed hydrogenation reactions. The resulting products are lower in aromatics and contain naphthenes and highly branched paraffins due to the higher stability of the tertiary carbenium ion intermediate. For paraffins, the reaction network, shown in Figure, is postulated to begin with a dehydrogenation step at a metal site forming an olefin intermediate, which is quickly protonated at an acid site to yield a carbenium ion. This is quickly followed by a series of isomerization reactions to the most stable tertiary carbenium ions and subsequent cracking to smaller paraffin, which evolves off the catalyst surface and smaller carbenium ion intermediate. The carbenium ion can then eliminate a proton to form an olefinic intermediate, which gets hydrogenated at a metal site or directly abstract a hydride ion from a feed component to form a paraffin and desorb from the surface.

Formation of Olefin

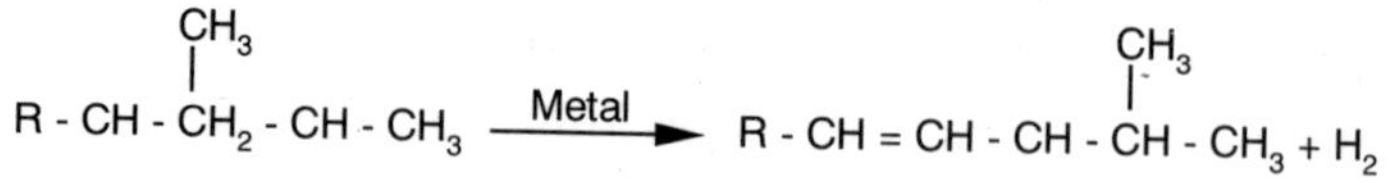

Formation of Teritiary Carbenlum lon

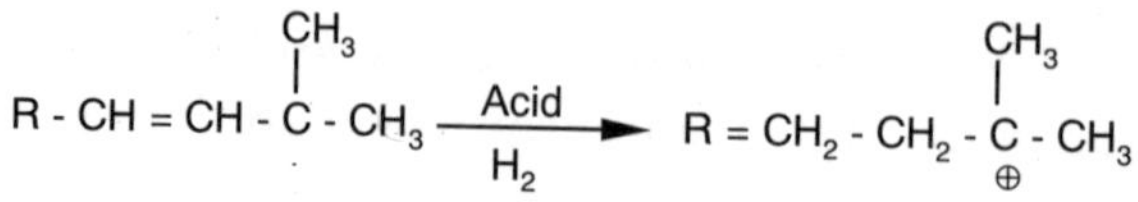

Cracking

Reaction of carbenium lon and Olefin

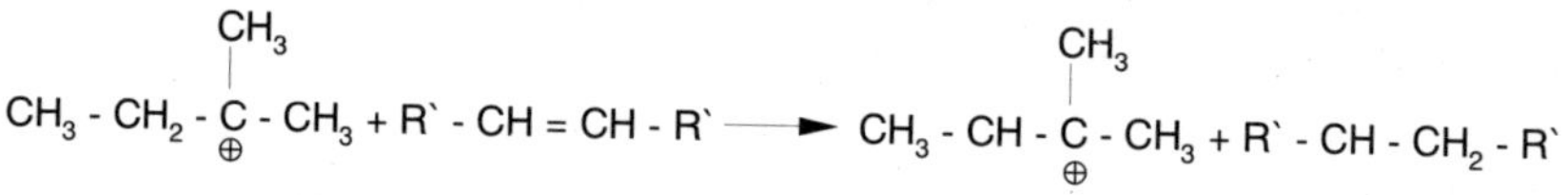

Olefin Hydrogenation

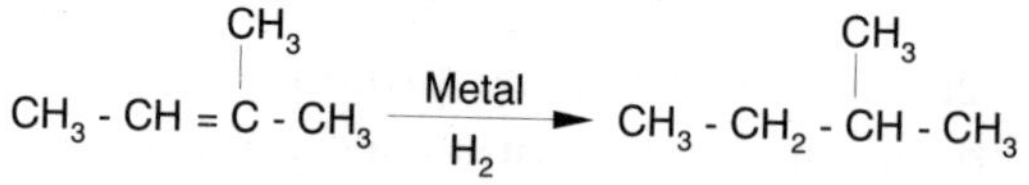

Figure 4.35: Postulated Paraffin-cracking Mechanism

A typical hydrocracking reaction for a cycloparaffin is known as a paring reaction, in which methyl groups are rearranged and then selectively removed from the cycloparaffin without severely affecting the ring itself. Normally the main acyclic product is isobutane. The hydrocracking of multiple-ring naphthene, such as decalin, is more rapid than that of a corresponding paraffin. Naphthenes found in the product contain a ratio of methylcyclopentane to methylcyclohexane that is far in excess of thermodynamic equilibrium.

Reactions during the hydrocracking of alkyl aromatics include isomerization, dealkylation, paring, and cyclization. In the case of alkylbenzenes, ring cleavage is almost absent, and methane formation is at a minimum.

CHEVRON LUMMUS GLOBAL RDS/VRDS HYDROTREATING—TRANSPORTATION FUELS FROM THE BOTTOM OF THE BARREL

Introduction

The Chevron Lummus Global (CLG) Residuum Desulfurization (RDS) and Vacuum Residuum Desulfurization (VRDS) Hydrotreating processes are used by refiners to produce low-sulfur fuel oils, and to prepare feeds for vacuum gas oil (VGO) fluid catalytic crackers (FCCs), residuum FCCs (RFCCs), visbreakers, and delayed cokers. Over half of the fixed-bed residuum hydrotreaters in operation use CLG's RDS/VRDS Hydrotreating technology.

RDS/VRDS Hydrotreaters upgrade residual oils by removing impurities and cracking heavy molecules in the feed to produce lighter product oils. Early applications of CLG's residuum hydroprocessing technology were used to remove sulfur from atmospheric residues (ARs) and vacuum residues (VRs), hence the term desulfurization. Today, RDS/VRDS Hydrotreaters perform equally well removing nitrogen, carbon residue (see "Process Chemistry" section), nickel, and vanadium from the oil and cracking heavy VR molecules to VGO, distillates, and naphtha products. The amount of impurities removed depends on the feed and on the product specifications desired by the refiner. Sulfur removal greater than 95 per cent, metal removal (primarily nickel and vanadium) greater than 98 per cent, nitrogen removal greater than 70 per cent, carbon residue reduction greater than 70 per cent, and cracking of vacuum residue (538°C+ material converted to 538°C–) as high as 60 liquid volume per cent (LV%) have been commercially demonstrated. RDS/VRDS Hydrotreating uses fixed beds of catalyst that typically operate at moderately high pressures [150 to 200 atm (2133 to 2850 lb/in2)] and temperatures [350 to 425°C (662–797°F)] in a hydrogen-rich atmosphere (80 to 95 mol per cent hydrogen at the reactor inlet) to process the oil feed. The feed to a VRDS Hydrotreater is generally the VR from a crude unit vacuum column with a typical starting true boiling point (TBP) cut point of 538°C (1000°F), although cut points of 575°C (1067°F) and higher are feasible. The feed to an RDS Hydrotreater is generally AR from a crude unit atmospheric column with a typical starting TBP cut point of 370°C (698°F). Other feeds (such as solvent deasphalted oil, solvent deasphalter pitch, vacuum gas oil, and cracked gas oils from visbreakers, FCCs, RFCCs, and cokers) can also be processed in either RDS or VRDS Hydrotreaters.

Residua from many crudes have been successfully processed in RDS and VRDS Hydrotreaters. Table shows a partial list of crudes that have been commercially processed in CLG RDS/VRDS Hydrotreaters.

Table 4.8: RDS/VRDS Hydrotreater Feedstocks that have been Commercially Processed

A.A. Bu Khoosh	Kuwait
Alaskan North Slope	Laguna
Algerian	Margham C.
Arabian Berri	Maya
Arabian Heavy	Mina Saud
Arabian Light	Minas
Arabian Medium	Murban
Basrah Light	Oguendjo
Cabinda	Oman
Colombian Limon	Qatar Land
Dubai	Qatar Marine
Duri	Russian
El Chaure	Shengli No. 2
Gipsland	Statfjord
Indonesian	Suez
Iranian Heavy	Tia Juana Pesado
Iranian Light	Umm Shaif
Isthmus	West Texas Intermediate
Khafji	West Texas Sour
Kirkuk	Zakum

The range of feeds which can be economically processed in RDS/VRDS Hydrotreaters expands significantly when On-Stream Catalyst Replacement (OCR) technology is added to the unit. OCR technology allows spent catalyst to be removed from a guard reactor and be replaced by fresh catalyst while the reactor remains in service. This enables the refiner to process heavy, high-metal feeds or to achieve deeper desulfurization from a fixed-bed residuum hydrotreater.

History

Hydrotreating of residual oils was a natural extension of hydrotreating distillate oils and VGOs to remove sulfur.1,2,3 CLG's first commercial RDS Hydrotreater was commissioned in 1969. Typical of many early residuum hydrotreaters, CLG's first RDS Hydrotreater was designed to remove sulfur to produce low-sulfur fuel oil (LSFO). CLG's first VRDS Hydrotreater, commissioned in 1977, was also designed to produce LSFO.

In 1984 Okinawa Sekiyu Seisei, a Japanese refiner, first reported4 the operation of a CLG RDS Hydrotreater in "conversion mode." In this operation, the reactor temperature was raised fairly high early in the run—

much higher than required to simply produce low-sulfur fuel oil—and held high until the end of run. This operation hydrocracked as much VR as possible to lighter boiling products (VR was "converted" to light products). It also shortened the run length because of higher catalyst deactivation from coke deposited on the catalyst through more of the run. Conversion mode operation has been favored by many RDS/VRDS Hydrotreater operators in recent years to minimize the production of fuel oil.

An alternative to destroying low-value fuel oil has been to convert it to higher value fuel oil. During the late 1970s and through the 1980s and 1990s, the demand for and value of high-sulfur (3 per cent) fuel oil and low-sulfur (1 per cent) fuel oil dropped. In some cases, power plant operators have been willing to pay higher prices for fuel oils with much lower sulfur content (0.1 to 0.5 wt %). RDS/VRDS Hydrotreating enabled refiners to produce these lower-sulfur fuel oils. The lowest-sulfur fuel oil commercially produced from sour crudes (about 3 wt % sulfur in the AR) was 0.1 wt per cent. This fuel oil was produced by the CLG RDS Hydrotreater at Idemitsu Kosan's refinery in Aichi, Japan (see Table).

Table 4.9: Production of Premium Low-Sulfur Fuel Oil

	RDS feed	*RDS 343°C+ product*
SULFUR, WT %	3.75	0.09
VISCOSITY, CST AT 50°C	248	84
SPECIFIC GRAVITY, $D^{15}/4$°C	0.9590	0.9105
Carbon residue, %	7.95	2.28
Ni/V, ppm	13/40	<1/<1
Nitrogen, ppm	2060	644
Distillation, °C		
IBP	257	272
5%	325	330
10%	353	358
20%	402	393
30%	435	425
40%	467	450
50%	502	480
60%	537	515
70%	—	555

The ability of residuum hydrotreaters to improve the economics of conversion units by pretreating their feeds has been understood for many years. The most noticeable economic impact of feed pretreatment is to lower the sulfur content of the feed to the conversion unit. For example, pretreatment of RFCC feed to reduce its sulfur to less than 0.5 wt per cent eliminates the need to install costly flue gas desulfurization facilities. Addition of hydrogen to the feed by the hydrotreater also improves the product yields and product qualities of the downstream conversion unit. In 1983, at Phillips' Borger, Texas, refinery, the first CLG RDS Hydrotreater was commissioned to pretreat residuum to feed an existing RFCC unit. Prior to the hydrotreater project, the RFCC had been feeding sweet domestic crude. The 50,000 barrel per day (BPD) RDS Hydrotreater was designed to achieve 92 per cent hydrodesulfurization (HDS) and 91 per cent hydrodemetallization (HDM) from a mixed domestic and Arabian Heavy AR for a 1-year cycle. In addition to its contribution toward meeting environmental requirements and reducing catalyst usage, the feed pretreatment significantly increased the gasoline yield from the RFCC.

An RDS Hydrotreater is used to pretreat residuum to feed a delayed coker unit at Chevron Texaco's refinery in Pascagoula, Mississippi.

SELECTIVE HYDROGENATION PROCESSES

The presence of dienes and acetylenes in light olefinic streams is often undesirable, and these reactive contaminants must be removed from the olefinic streams without affecting the nature or concentration of the olefins. The dienes and acetylenes are typically removed by selective hydrogenation to the corresponding monoolefins. Light olefin streams are produced by steam cracking, dehydrogenation of C_3/C_4 paraffins, or fluid catalytic cracking (FCC). There are a number of processes that selectively remove reactive components such as acetylenes or dienes from olefinic streams.

Introduction

Hydrogenation of dienes and acetylenic compounds can be accomplished selectively in the presence of monoolefins by using mild hydrogenation conditions. Chemische Werke Hüls developed the concept of selective hydrogenation in 1963. The first Hüls Selective Hydrogenation Process (SHP) was commercialized in 1980. The unit processed 160 kMTA of C4 feed derived from a steam cracking unit. The SHP unit is applicable to all C3–C5 feedstocks including sulfur-containing feeds from FCC units. The selectivity of the hydrogenation reaction is dependent on the nature of the catalyst and the operating conditions. Hydrogenation is carried out at mild conditions with a slight stoichiometric excess of hydrogen. The Hüls SHP unit is licensed by UOP for the selective hydrogenation of butadiene to butenes, for propadiene and methylacetylene to propene, and for pentadienes to pentenes. UOP also offers the KLP process to selectively remove acetylenes from crude butadiene feedstocks.

The KLP Process

1,3-Butadiene is an important petrochemical intermediate recovered from the C_4 fraction of a naphtha steam cracking unit. The C_4 cut from a steam cracking unit contains up to 60 per cent butadiene and also small amounts of C_4 acetylenes that need to be separated from the main butadiene product. Acetylenic compounds, such as vinyl acetylene and ethyl acetylene, can be selectively removed using the KLP process prior to butadiene extraction.6,7 The KLP process is used to convert essentially 100 per cent of the alpha-acetylenes to monoolefins and butadiene. The KLP process is highly selective, and there is no yield loss of butadiene; in fact, there may be a slight yield gain. With acetylenes no longer present, the extraction of butadiene can be accomplished in a single-stage unit. The KLP process can be readily integrated into existing extraction units and allows for a capacity increase or debottlenecking of the existing extraction unit. A process flow diagram of the KLP process integrated with a butadiene extraction unit is shown in Figure 4.35. The extracted butadiene is very high-purity, with acetylene levels typically less than 10 ppm. The raffinate I stream still contains low levels of butadiene that may be removed by the SHP unit.

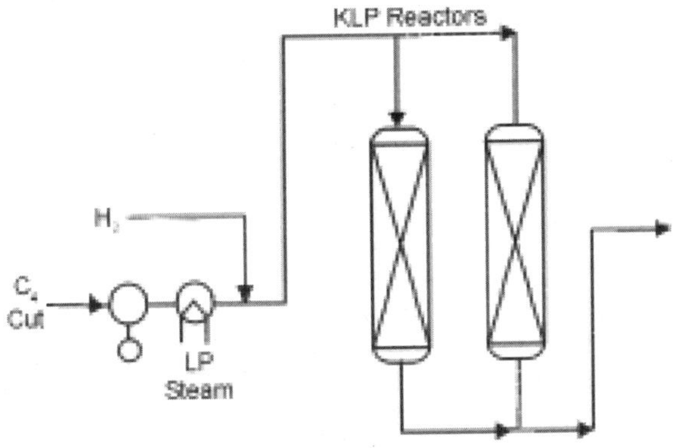

Figure 4.35: The KLP Process

The SHP Unit

The removal of butadiene from a C_4 olefin-rich stream can be readily accomplished using the SHP unit with a noble or non-noble-metal catalyst system. The process has been optimized to minimize losses of 1-butene through either hydrogenation or isomerization and to achieve levels of residual butadiene to as low as 10 ppm.

Hüls has successfully developed two modes of operation using a non-noble-metal or a noble metal catalyst depending on the processing objectives. In the first mode of operation, the dienes and acetylenes are selectively hydrogenated, and the n-butene isomer concentration moves toward the equilibrium concentration. At the low temperatures used in the SHP unit, the 2-butene is thermodynamically favored. This mode of operation is desirable for preparation of HF catalyzed alkylation feedstocks. The butene isomerization that occurs over the SHP catalyst will increase the ratio of 2-butene to 1-butene typically from about 2 to 8. The increased 2-butene content will increase alkylate octane by up to two research numbers in HF catalyzed alkylation units. In the second mode of operation the isomerization activity is suppressed while hydrogenation activity is maintained. This mode of operation is desirable for the production of high-purity comonomer-grade 1-butene. In addition, reducing the diene content in HF and H2SO4 catalyzed alkylation units results in lower acid consumption and increased alkylate yield. A flow diagram for the SHP unit is shown in Figure.

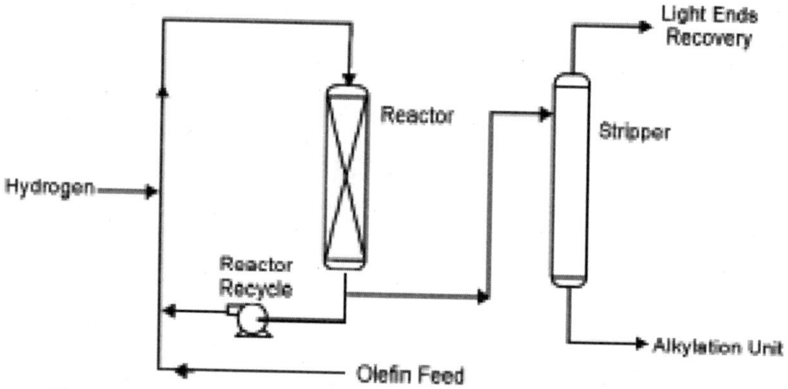

Figure 4.36: Hüls Selective Hydrogenation Process Flow Diagram

The feed is combined with hydrogen at near stoichiometric ratios to the diene and acetylene content of the feed. Hydrogenation then takes place in a fixed-bed reactor. In cases where the diene concentration of the feed is high, a portion of the reactor effluent is recycled. The hydrogen leaving the reactor is at a very low concentration and does not require removal unless a downstream process is sensitive to noncondensable gases. A Hüls SHP unit is easy to operate, and minimal utilities are required, especially if the feed and hydrogen are both available at suitable pressure.

Table illustrates the two different modes of operation. The same feedstock is used with the same hydrogen addition rate and space velocity. In the first mode, good hydrogenation is achieved with a high level of isomerization from 1-butene to 2-butene. In the second mode, the SHP unit is optimized for the production of 1-butene. Essentially complete removal of dienes and acetylenes is achieved in either case, and there is no loss of n-butenes.

Table 4.10: Typical Modes of SHP Unit Operation

Component	*Feed*	*Mode 1*	*Mode 2*
1-Butene, wt %	14.6	4.2	14.7
2-Butene, wt %	23.4	33.9	23.4
1,3-Butadiene, wt ppm	4460	20	<1
Butyne, wt ppm	68	6	<1

In refineries, light olefin streams are produced by FCC units. Removal of the acetylenes and dienes from C_3–C_5 streams is readily achieved with the SHP unit. For example, polymergrade propylene requires less than 5 wt ppm of dienes and acetylenes to meet specifications.

The C_4 and C_5 steams derived from an FCC unit typically contain low levels of sulfur. For sulfur-containing feedstocks in refinery applications, a non-noble-metal catalyst was developed that is robust and sulfur-tolerant. Catalyst life based on commercial experience is expected to be 4 years or more provided the feedstocks meet design specifications.

The C_5 olefin streams can be used as alkylation process feed or as feedstock for the production of tertiary amyl methyl ether (TAME). Typically, the feedstock to a TAME Ethermax unit requires the reduction of dienes in the C5 olefin stream. In addition, 3-methyl-1-butene can be isomerized in the SHP unit to 2-methyl-1-butene and 2-methyl-2-butene, which react to form TAME. This can increase the TAME yield by up to 8 wt per cent.

UOP UNIONFINING TECHNOLOGY

Introduction

Hydrotreating is one of the most mature technologies found in the refinery, rivaling the history and longevity of the thermal process. In 1952, UOP and Union Oil Co. of California began licensing hydrotreating under the name of the Unifining process. The partnerships and the development of this technology have gone through a series of changes over the years, and in 1995 the acquisition of the Unocal Process Technology and Licensing group by UOP resulted in the merger of two premier hydroprocessing companies and the combination of their expertise under the UOP Unionfining banner.

Generally speaking, the hydrotreating process removes objectionable materials from petroleum distillates by selectively reacting these materials with hydrogen in a catalyst bed at elevated temperature. These objectionable materials include sulfur, nitrogen, olefins, and aromatics. Lighter materials such as naphtha are generally treated for subsequent processing in catalytic reforming units, and the heavier distillates, ranging from jet fuel to heavy vacuum gas oils, are treated to meet strict product-quality specifications or for use as feedstocks elsewhere in the refinery. Many of the product-quality specifications are driven by environmental regulations that are becoming more stringent each year. This push toward more environmentally friendly products is resulting in the addition of hydroprocessing units in refineries throughout the world.

Process Chemistry

The chemistry behind the hydrotreating process can be divided into a number of reaction categories: (hydro) desulfurization, (hydro) denitrification, saturation of olefins, and saturation of aromatics. For each of these reactions, hydrogen is used to improve the quality of the petroleum fraction

Desulfurization

Desulfurization is by far the most common of the hydrotreating reactions. Sulfur-containing hydrocarbons come in a number of forms, and the ability to remove sulfur from the different types of hydrocarbons varies from one type to the next. The degree to which sulfur can be removed from the hydrocarbon varies from near-complete desulfurization for light straight-run naphthas to 50 to 70 per cent for heavier residual materials. Figure 4.37(a) lists several sulfur-containing compounds in order of the difficulty in removing the sulfur. The reaction of thiophenol, which is at the top of the list in Figure 4.37(a), proceeds quite rapidly; the reaction is shown schematically in Figure 4.37(b). Multiring thiophene-type sulfurs are more difficult to treat because the ring structure, which is attached to the sulfur on two sides, must be broken. Figure 4.37(c) is a schematic representation of the reaction for the desulfurization of dibenzothiophene.

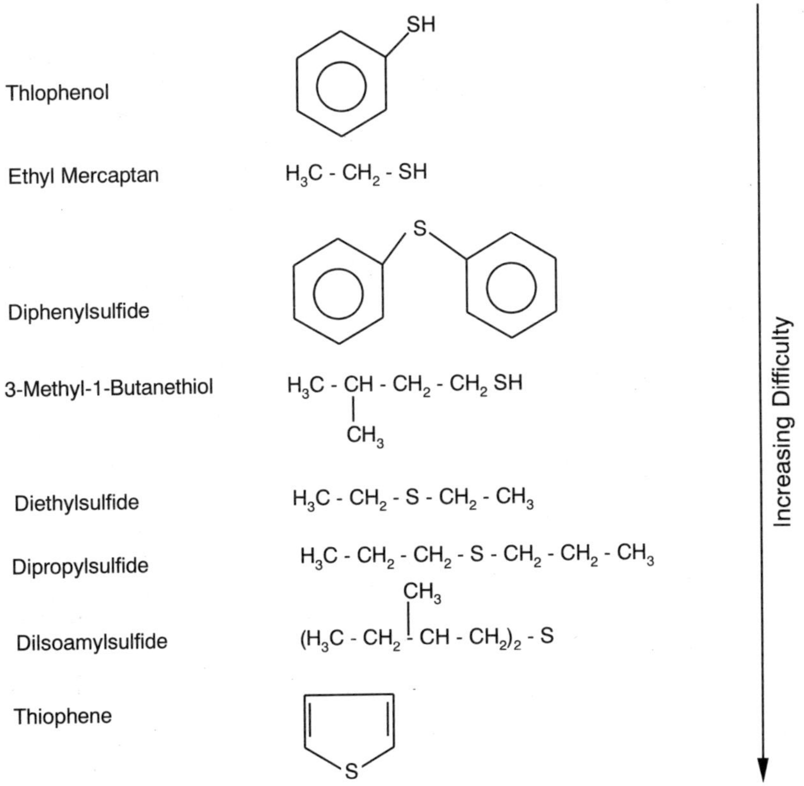

Figure 4.37(a): Relative Desulfurization Reactivities

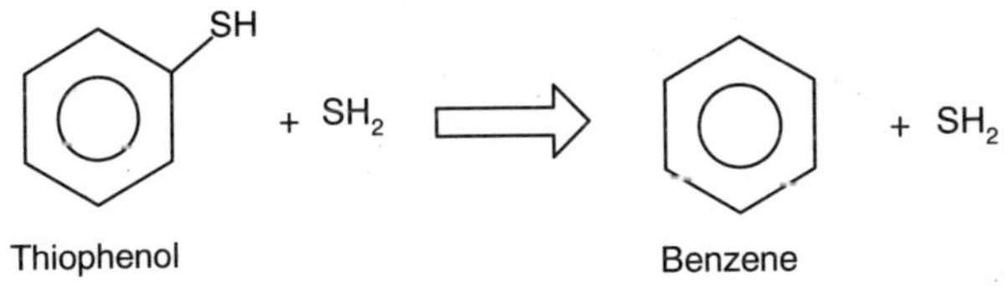

Figure 4.37(b): Desulfurization of Thiophenol

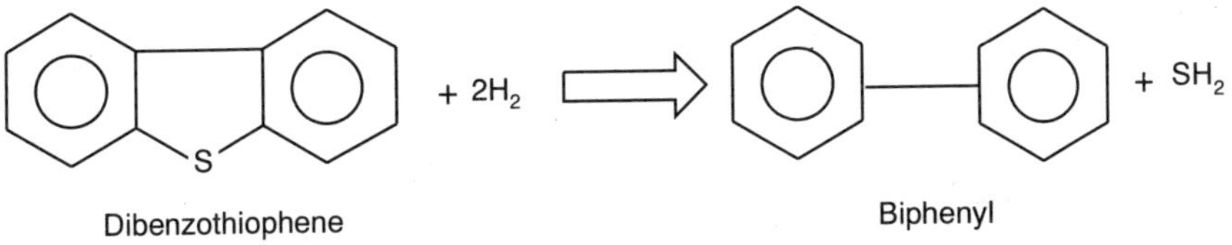

Figure 4.37(c): Desulfurization of Dibenzothiophene

In each case, the desulfurization reaction results in the production of hydrogen sulfide (H_2S) in the reactor section of the plant. To complete the desulfurization reaction, the H_2S must be removed in downstream fractionation.

Denitrification

The nitrogen compounds that occur naturally in crude oils and that would normally be found in the feed to a hydrotreater can be classified into two categories: basic nitrogen, which is generally associated with a six-member ring, and neutral nitrogen, which is generally associated with a five-member ring. Examples of these two types of nitrogen are shown in Figure 4.38. The complexity of the nitrogen compounds makes denitrification more difficult than desulfurization.

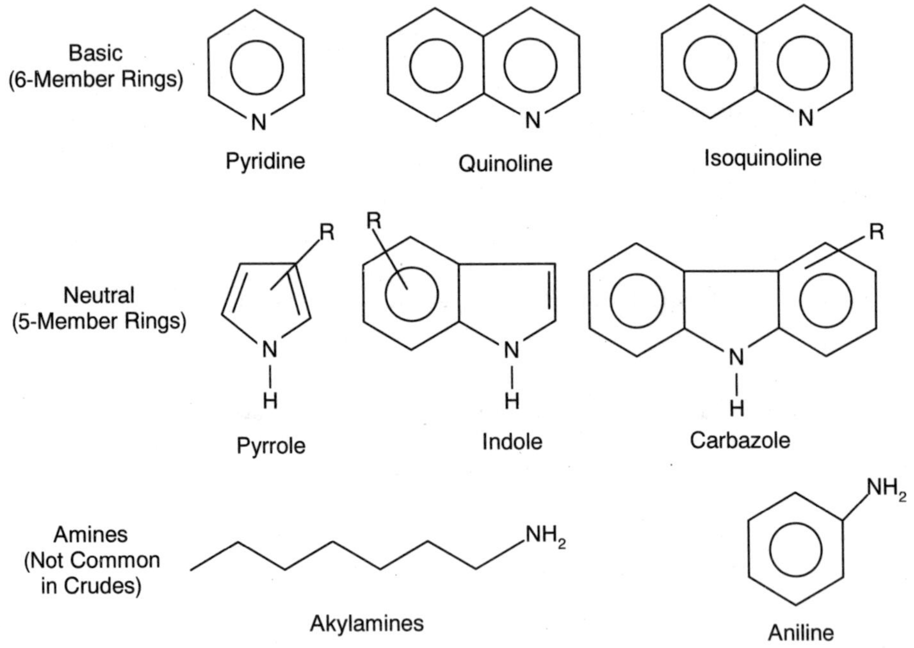

Figure 4.38: Types of Nitrogen Compounds

The denitrification reaction first proceeds through a step that saturates the aromatic ring. This saturation is an equilibrium reaction and normally sets the rate at which the denitrification reaction can occur.

UOP RCD UNIONFINING PROCESS

Introduction

The UOP RCD Unionfining reduced-crude desulfurization process represents the merger of three of the world's leaders in residual oil processing and catalyst technology. UOP's acquisition of the Unocal PTL Division in January 1995 resulted in the merging of UOP and Unocal's catalyst technology, commercial know-how, and design experience to create a new, improved residual hydrotreating process. Prior to this acquisition in 1993, UOP entered into an alliance with Catalyst & Chemicals Ind. Co. Ltd. (CCIC) in Japan that enabled UOP to offer CCIC's commercially proven portfolio of residual hydrotreating catalysts. In addition UOP has catalysts available from other leading catalyst manufacturers.

The RCD Unionfining process provides desulfurization, denitrification, and demetallization of reduced crude, vacuum-tower bottoms, or deasphalted oil (DAO). Contaminant removal is accompanied by partial conversion of nondistillables. The process employs a fixed bed of catalyst, operates at moderately high pressure, consumes hydrogen, and is capable of greater than 90 per cent removal of sulfur and metals. In addition to its role of providing low-sulfur fuel oil, the process is frequently used to improve feedstocks for downstream conversion units, such as cokers, fluid catalytic crackers (FCCs), and hydrocrackers.

Market Drivers for RCD Unionfining

The first commercial reduced-crude desulfurization unit, which came on-stream in 1967, was a licensed design from UOP. The residual hydrotreating units produced a low-sulfur fuel oil product that was in increasing demand as a result of the stringent laws relating to air pollution that were being enacted in the industrialized countries. Units were designed in the 1970s to produce fuel oil with a sulfur level as low as 0.3 wt per cent. The trend toward low-sulfur fuel oil has now extended around the world.

Other drivers have added to the need for the RCD Unionfining process. Fuel oil demand has been declining at a rate of about 0.2 per cent per year since the 1980s, and this decline has been coupled with a growth of about 1.4 per cent in the demand for refined products. As the demand for heavy fuel oil has fallen, the price differential between light and heavy crude oil has increased. This price differential has given the refiner an economic incentive to process heavy crude. However, heavy crude not only produces a disproportionate share of residual fuel, but also is usually high in sulfur content. Because heavy, high-sulfur crude is a growing portion of the worldwide crude oil reserves, refiners looking for future flexibility have an incentive to install substantial conversion and desulfurization capacity to produce the required product slate. In addition to providing low-sulfur fuel oil, the RCD Unionfining process provides excellent feedstock for downstream conversion processes producing more valuable transportation fuels.

Catalyst

Catalysts having special surface properties are required to provide the necessary activity and stability to cope with reduced-crude components. The cycle life of the catalyst used in the RCD Unionfining process is generally set by one of three mechanisms:

- Excessive buildup of impurities, such as scale or coke, that leads to unacceptable pressure drop in the reactor
- Coke formation from the decomposition and condensation of heavy asphaltic molecules.
- Metal deposition in catalyst pores from the hydrocracking of organometallic compounds in the feed.
- UOP provides a complete portfolio of catalysts to handle each of these three mechanisms (Table).

Table 4.11: CCIC Catalyst Portfolio

Catalyst name	*Application*	*Size and shape*	
CDS-NP1 CDS-NP5	ΔP relaxation and HDM	1/4 in shaped	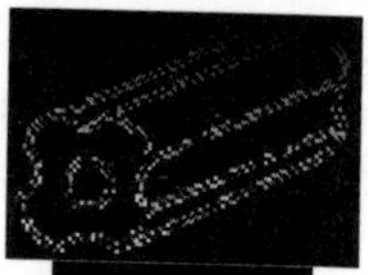
CDS-NPS1 CDS-NPS5	ΔP relaxation and HDM	1/6 in shaped	
CDS-DM1	HDM		
CDS-DM5	HDM		
CDS-R95	HDM/HDS	1/8, 1/12, 1/16, 1/22 in cylindrical or shaped	
CDS-R25H	HDS		
CDS-R55	HDS		

HDS = hydrodesulfurization; HDM = hydrodemetallization

Feed filtration for removing scale particulates is a standard part of the RCD Unionfining design. In most cases, this filtration satisfactorily prevents buildup of scale on the catalyst bed, and the resulting pressure drop does not limit cycle life. However, some feeds contain unfilterable components. Under these circumstances, the catalyst bed itself acts as a filter, and the impurities build up in the top section of catalyst to create unacceptable pressure drop. The CDS-NP series of catalyst helps prevent this problem by increasing the amount of void space in the top of the reactor for the impurities to collect. The CDS-NP catalysts are designed as a macaroni shape in 1/4-in and 1/6-in sizes. The macaroni shape maximizes the amount of void space and catalyst surface area available for deposition of the impurities. The catalysts are also loaded from the larger to smaller size. This kind of loading, which is called grading the catalyst bed, helps to maximize the void space available for deposition.

Coke formation reduces catalyst effectiveness by decreasing the activity of the reactive surface and decreasing the catalyst-pore volume needed for metals accumulation. For a given catalyst and chargestock operating under steady-state conditions, the amount of coke on the catalyst is a function of temperature and pressure. Successful hydrodesulfurization of reduced crudes requires that temperatures and pressures be selected to limit coke formation. When coke formation is limited, ultimate catalyst life is determined by the rate of metals removal from organometallic compounds in reduced crude and by the catalyst-pore volume available for the accumulation of metals.

The deposition rate of metals from organometallic compounds correlates with the level of desulfurization for a given catalyst. Thus, the rate of catalyst deactivation by metals deposition increases with the increasing level of desulfurization. Metals deposition results from the conversion of sulfur-bearing asphaltenes. Conversion exposes catalyst pores to the portion of the feed that is highest in organometallics.

The catalyst deactivation rate is also a function of feedstock properties. Heavier reduced crudes with high viscosities, molecular weights, and asphaltene contents tend to be more susceptible to coke formation. Hence, higher pressures and lower space velocities are required for processing these materials. Correlations developed on the basis of commercial data and pilot-plant evaluation of many different reduced crudes can predict the relationship between the hydrodesulfurization reaction rate and the deactivation rate and reduced-crude properties.

The UOP-CCIC catalyst portfolio has been developed to maximize both the removal of sulfur, metals, and other impurities such as nitrogen and Conradson carbon and the life of the catalyst. The CDS-DM series catalysts are typically loaded downstream of the CDS-NP catalysts and are designed for maximum metal-holding capacity. Although their metals-removal activity is high, they maximize the removal of the resin-phase metals and minimize the removal of asphaltene-phase metals. (For an explanation of the terms resin phase and asphaltene phase, see the following "Process

Chemistry" section.) The removal of asphaltene-phase metals can lead to excessive formation of coke precursors, which ultimately reduce the life of downstream catalysts. The CDS-R9 series catalysts are typically located downstream of the CDS-DM catalysts. These transition catalysts have intermediate activity for demetallization and desulfurization and are used to gradually move from maximum demetallization to maximum desulfurization. Once again, the gradual transition helps to minimize the formation of coke precursors, which could lead to shortened catalyst life. The final catalysts are the CDS-R25/R55 series, which have maximum desulfurization activity. By the time the residual oil reaches this series of catalysts, the metals level is sufficiently low to prevent metals deactivation.

In addition to this portfolio of conventional desulfurization and demetallization catalysts, several custom catalysts are available for the RCD Unionfining process. The R-HAC1 catalyst is a residual, mild-hydrocracking catalyst intended for use with lighter feedstocks. Although it has the same hydrodesulfurization activity as conventional HDS catalysts, it produces 3 to 4 vol per cent more diesel fuel without an increase in naphtha or gas yields. The CAT-X catalyst is designed as an FCC feed pretreatment catalyst. The FCC microactivity testing (MAT) of feeds processed over the CAT-X catalyst has shown an increase in the gasoline yield of as much as 5 per cent.

UOP CATALYTIC DEWAXING PROCESS

Introduction

The UOP Catalytic Dewaxing process, formerly known as the Unicracking /DW process, is a fixed-bed process for improving the cold flow properties of various hydrocarbon feedstocks. The process is applied for improving the pour point of lube oil base stocks (LOBSs) and middle distillates, cloud point of diesel fuel, and freeze point of jet fuel. These properties are critical for the low-temperature performance of these products.

The cold flow properties are strongly related to the normal and near-normal (minimally branched) paraffins present in these LOBSs and fuels. As the concentration of the normal and near-normal paraffins increases in these hydrocarbon feedstocks, their pour point, cloud point, and freeze point temperatures increase. The cold flow property temperatures also increase with the molecular weight (chain length) of the paraffins. The UOP Catalytic Dewaxing process improves the cold flow properties by selectively cracking the long-chain normal and near-normal paraffins from the hydrocarbon steams.

The Catalytic Dewaxing process operates across a rather narrow range of design parameters. The primary roles are LOBS dewaxing and middle-distillate flow property improvement. At the same time, the process deep-hydrotreats kerosene or diesel fuel to remove sulfur and nitrogen and also saturates aromatic compounds. Key process features of the Catalytic Dewaxing process include:

- Excellent product stability,
- Excellent product color,
- Constant product quality throughout a catalyst cycle,
- Minimum viscosity reduction compared to other dewaxing processes,
- Long catalyst cycle life,
- Flexibility to produce lube stocks and process distillates in the same unit.

Process Description

The Catalytic Dewaxing process uses a dual-function, non-noble-metal zeolite catalyst to selectively hydrocrack the long-chain paraffinic components in the feedstock. Typically, the first stage of the process involves hydrotreatment of the incoming feedstock through olefin saturation, desulfurization, and denitrification reactions. Pretreating protects the dewaxing catalyst and provides a feed with a low organic sulfur and nitrogen content, which improves the hydrocracking performance. Pretreatment of the feed may not be necessary if the feed is relatively free of organic sulfur and nitrogen.

The process uses two kinds of catalysts. The first is a high-activity desulfurization and dentrification catalyst, which gives an optimum balance between process objectives and cost. The second, a proprietary dewaxing catalyst, selectively cracks straight-chain paraffins. The zeolite support used for the dewaxing catalyst has a pore size that selectively allows the normal and near-normal paraffins to enter the pore structure at the expense of highly branched paraffins. As a result, the rate of cracking for the normal and near-normal paraffins is much higher than that for the branched paraffins. The very selective reduction of the long-chain paraffins thus achieved improves the cold flow properties of the hydrocarbon feedstock. The flexible catalyst system of pretreat and dewaxing catalyst enables a refiner to vary the feedstocks and contaminants without affecting product quality or run length.

UOP has several highly active, long-lived Catalytic Dewaxing catalysts. Process objectives determine the type of catalyst used in a particular unit. Catalytic Dewaxing catalysts typically last 6 to 8 years. During that time, they are regenerated as needed. Typical cycles last 2 to 4 years between regenerations.

Process Flow

Figure shows a simplified process flow for a typical Catalytic Dewaxing unit. Fresh feed is preheated and combined with hot recycle gas. The mixture enters the first reactor or treating by a high-activity denitrogenation-desulfurization catalyst, which converts organic nitrogen and sulfur to ammonia and hydrogen sulfide. The reactions are exothermic and cause a temperature rise in the reactor. The reactions are maintained at as low a temperature as possible to maximize catalyst life. Figure shows two reactors for simplicity. In an actual design, both the pretreat and the dewaxing catalysts an be loaded in a single reactor. The choice of one versus two reactors depends on the feed rate to the unit, reactor size limitations (if any), operational flexibility desired, etc.

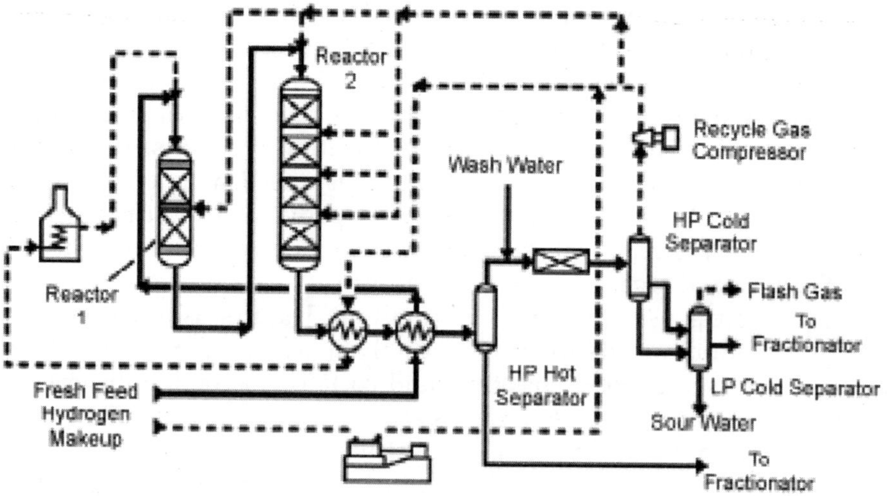

Figure 4.39: Catalytic Dewaxing Process

The effluent from the pretreat section is cooled with cold quench gas before entering the dewaxing section, which contains one of UOP's highly selective Catalytic Dewaxing catalysts. These active catalysts function well in the presence of ammonia and hydrogen sulfide. As the feed flows over the dewaxing catalyst, long-chain normal paraffins are selectively cracked into smaller molecules, thereby improving the desired cold flow property of the feed. The average temperature in the dewaxing section is adjusted to obtain the targeted improvement in the cold flow property.

Dewaxing reactions are exothermic and must be closely controlled because the dewaxing catalyst is sensitive to temperature. As in the pretreater section, reactor temperatures are maintained as low as possible. In the dewaxing section, this low temperature not only prolongs catalyst life, but also maximizes liquid yields and helps maintain control.

In both reactors, temperature is controlled by the injection of cold, hydrogen-rich recycle gas at predetermined points. A unique combination of patented internals allows for sufficient mixing of recycle gas with the hot reactants emerging from the previous catalyst bed and the effective distribution of the quenched mixture to the top of the next catalyst bed.

Effluent from the dewaxing section is cooled by exchange with several process streams, including feed and recycle gas. The effluent is then flashed into a hot high-pressure separator, where liquid products are separated from hydrogen-rich vapors. The liquid fraction from this separator is directed to the fractionation section, and vapors are sent to a cold high-pressure separator after being cooled in reactor effluent coolers. Steam condensate or deaerated boiler feedwater is injected upstream of the reactor effluent air condensers to minimize corrosion and prevent deposition of ammonia salts. The cold-separator vapor is joined by hydrogen makeup gas to become recycle gas. Liquid hydrocarbon flows into a low-pressure separator.

Flash gas from the low-pressure separator is routed to a light-ends recovery unit or to sour fuel gas. Liquid hydrocarbon from the low-pressure separator exchanges heat with the reactor effluent before flowing into the stripper in the fractionation section.

Yield Patterns

The Catalytic Dewaxing process can be applied to a wide range of feedstocks for dewaxing purposes.

Investment and Operating Expenses

The capital investment associated with a Catalytic Dewaxing unit is closely related to the feedstock type and quality as well as the desired level of dewaxing for final products.

UOP UNISAR PROCESS FOR SATURATION OF AROMATICS

Introduction

The UOP Unisar process saturates the aromatics in naphtha, kerosene, and diesel feedstocks. The use of highly active noble-metal catalysts permits the reactions to take place at mild conditions. Because of the mild conditions and the very selective catalyst, the yields are high, and hydrogen consumption is largely limited to just the desired reactions. A total of 20 Unisar units have been licensed worldwide.

Among the applications of the Unisar process are smoke-point improvement in aircraft turbine fuels, reduction of the aromatic content of solvent stocks to meet requirements for air pollution control, production of cyclohexane from benzene, and cetane number improvement in diesel fuels. The Unisar process also produces low-aromatics diesel with excellent color and color stability.

This process was first applied to upgrading solvent naphthas and turbine fuel. The first commercial Unisar plant, a unit processing 250,000 metric tons per year (MTA) [6000 barrels per stream day (BPSD)] was built in Beaumont, Texas, and went on-stream early in 1969. It was designed to saturate the aromatics in untreated straight-run solvent naphtha containing 100 wt ppm sulfur. The aromatics were reduced from 15 to 1.0 vol per cent. The first catalyst cycle lasted more than 8 years. Another of the early plants was started up at Unocal's San Francisco Refinery in 1971. This unit, which processes 600,000 MTA (14,500 BPSD), reduces the aromatics in hydrocracked turbine stock from 30 wt per cent to less than 4 wt per cent. During a test at the latter unit, the aromatics were reduced from 29 wt per cent to less than 0.1 wt per cent.

Application to Diesel Fuels

In the 1990s, the need to increase the cetane number of diesel fuel has grown. Increasing the cetane number improves engine performance and decreases emissions. Cetane number is a strong function of hydrocarbon type and number of carbon atoms. Figure plots cetane number versus number of carbon atoms for compounds in the diesel boiling range. The graph shows that normal paraffins have the highest cetane number, which increases with chain length.

Isoparaffins and mononaphthenes with side chains are intermediate in cetane number, and polynaphthenes and polyaromatics have the lowest cetane numbers.

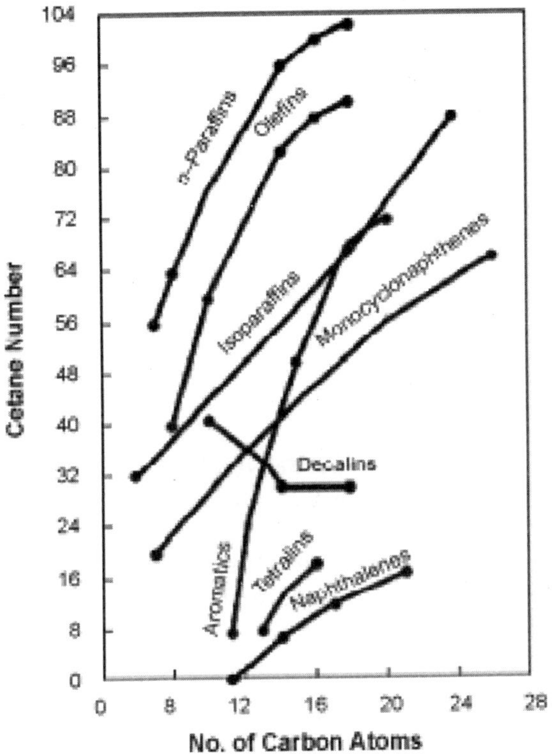

Figure 4.40: Cetane Number versus Hydrocarbon Type

The saturation of aromatics in diesel-range feeds leads to an increase in cetane number (Figure 4.41). However, whether this reaction alone is sufficient to reach cetane numbers near 50, as required in some European countries, depends on the overall compound distribution in each feed. One version of the advanced Unisar catalyst described later in this chapter has built into it some hydrocracking activity to promote naphthenic ring opening and upgrade these low-cetane feedstocks.

These diesel-range feeds typically have substantially higher boiling points and much higher levels of nitrogen and sulfur than the lighter kerosene and solvent feedstocks for which the original Unisar process was developed.

The original noble-metal catalysts used for the Unisar process have limited tolerance to the nitrogen and sulfur contaminants in diesel-range feeds. Thus, new catalysts have been developed to effectively treat these more difficult feeds. In addition, these feeds must be substantially hydrotreated to remove sulfur and nitrogen before they can be treated with the noble-metal Unisar catalysts.

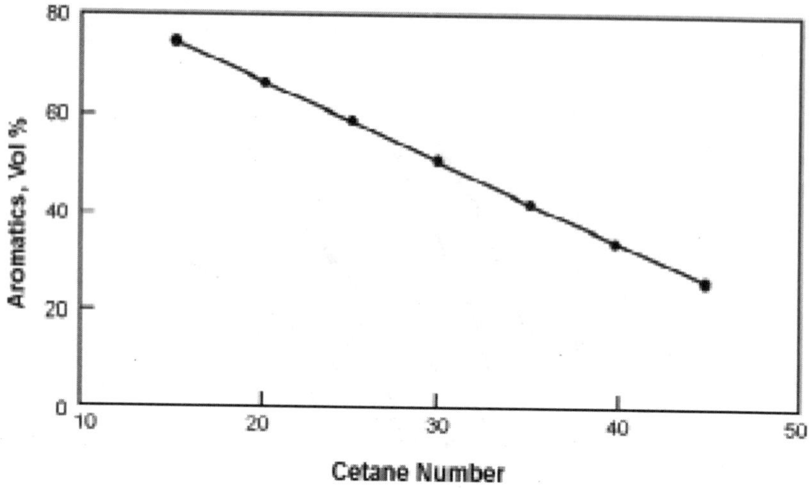

Figure 4.41: Typical Aromatics-cetane Relationship

A flow scheme that integrates the hydrotreating and aromatics-saturation stages has been developed so that low-sulfur, high-cetane diesel fuels can be efficiently produced.

Process Description

The UOP Unisar process is carried out at moderate temperatures and pressures over a fixed catalyst bed in which aromatics are saturated in a hydrogen atmosphere. Exact process conditions vary, depending on the feedstock properties and the level of aromatics desired in the product.

Chemistry

The primary reaction in the Unisar process is the hydrogenation of aromatics. Other reactions that occur are hydrogenation of olefins, naphthenic ring opening, and removal of sulfur and nitrogen. At conditions that result in significant aromatics hydrogenation, olefins in the feed are completely hydrogenated. When the concentrations of sulfur and nitrogen in the feed are relatively high, the hydrogenation of aromatics is severely limited until the concentrations of heterocompounds have been greatly reduced.

The overall aromatics-saturation reaction rate increases with increases in aromatics concentration, hydrogen partial pressure, and temperature. The reaction rate decreases with increases in the concentration of sulfur and nitrogen compounds and the approach to equilibrium. At low temperatures, aromatics in the product are reduced by increasing the temperature. At these low temperatures, reaction kinetics control the aromatics conversion. However, as temperatures are further increased, a point is reached at which additional temperature increases actually increase aromatics in the product. Above this temperature, the reverse dehydrogenation reaction has become dominant, and aromatics conversion is controlled by chemical reaction equilibrium. Thus, using highly active catalysts is important so that lower temperatures—that is, temperatures that are farther from the equilibrium limitation—can be used.

Naphthalene and tetralin have been used as a model to study the reaction mechanism for saturation of diaromatics with the Unisar noble-metal catalyst. The saturation of naphthalene to tetralin and of tetralin to decalin both fit first-order kinetics quite well. However, when these rate constants were used together in a sequential mechanism in which naphthalene is saturated to tetralin and the tetralin is then saturated to decalin, the resultant calculated product distribution did not fit the distribution observed experimentally.

CHEVRON LUMMUS GLOBAL EBULLATED BED BOTTOM-OF-THE-BARREL HYDROCONVERSION (LC-FINING) PROCESS

Introduction

Ongoing trends in the petroleum refining industry have resulted in the need to upgrade bottom-of-the-barrel heavy oils that otherwise are difficult to transport and market due to their high viscosity and high levels of contaminants, such as sulfur, metals, asphaltenes, carbon residues, and solid particles. Petroleum refiners find it necessary to process heavier crudes that require deep residual conversions to produce clean, high-quality finished products. The LC-Fining residual hydroconversion process was developed to specifically target hydrocracking the world's most difficult, heavy, lower-value hydrocarbon streams (petroleum residuals, heavy oils from tar sands, shale oils, solvent-refined coal extracts, etc.) at conversion levels of 80 per cent and higher.

Increasing demand for light and middle distillates, as well as changing environmental regulations and specifications for fuel oil production, has further increased the need for more efficient residuum upgrading processes. The LC-Fining process, when coupled with an integrated, fixed-bed, wide-cut hydrotreater/hydrocracker, produces high-quality finished products without significant quantities of undesirable by-products.

Earlier heavy vacuum residual technologies (carbon rejection or hydrogen addition type) were generally limited to distillate yields of 40 to 60 per cent. The remaining unconverted bottoms were used as coke, low-BTU gas, or residual fuel oil.

A major process route for coping with these challenges is residue hydrocracking. This process is characterized by both thermal cracking and hydrogenation reactions whereby the heavy, hydrogen-deficient components in the feed are converted to lighter products. The LC-Fining process is a commercially proven hydrocracking process for the upgrading of residues

Residual upgrading process requirements should include the ability to (1) handle high heats of reaction without wasting reactor volume, (2) handle extraneous material without plugging, (3) provide uniform distribution of reactants and efficient contacting, and (4) operate over extended periods without shutdown. The nature of the LC-Fining process makes it ideally suited for the conversion of residues to lighter, more valuable products. The process can be tailored for the feedstocks, the degree of conversion, and the product qualities required, especially the production of high-quality residual fuel oil with low sulfur content and good pipeline stability, or high-quality synthetic crude oils.

The LC-Fining process is based on technology initially developed and commercially demonstrated by Cities Service, and subsequently improved and refined by ABB Lummus Global, BP (formerly Amoco Oil Company), and ChevronTexaco Corp. (formerly Chevron). With this process, heavy oil feeds—including gas oils, petroleum atmospheric and vacuum residue, coal liquids, asphalt, bitumen from tar sands, and shale oil—are hydrogenated and converted to a wide spectrum of lighter, more valuable products such as naphtha, light and middle distillates, and atmospheric and vacuum gas oils. Residual products can be used as fuel oil, synthetic crude, or feedstock for a coker, visbreaker, solvent deasphalter, or residual catalytic cracker. Operating conditions and catalyst type and activity can be varied to achieve the desired conversion, Conradson carbon reduction (CCR), desulfurization, and demetallization of residual oil feeds.

Development and Commercial History

From 1957 to 1975, Cities Service Research and Development Company participated with Hydrocarbon Research Institute (HRI) to pilot-test and develop an ebullated bed hydroconversion process (H-Oil). During this period, research and development programs were continually carried out in several pilot units. Based on the pilot tests, the first commercial unit was designed and operated at Lake Charles, Louisiana, in 1963.

In 1975, ABB Lummus Global (Lummus) joined together with Cities Service to license, market, design, and generally improve upon the technology from Cities Service. Pilot-plant facilities for this technology, called LC-Fining, were built in New Jersey. Lummus carried out comprehensive process pilot-plant studies and mechanical design developments and offered initial operation and process simulation services.

The first license was sold to Amoco Oil Company in 1981. Amoco operated the commercial plant in Texas City, conducted extensive pilot-plant and catalyst development work, and eventually became a joint licensor with Lummus in 1984. The Amoco unit started up quickly and performed well right from the beginning: All design targets were met or exceeded. To maximize profits, throughout its operation, the Amoco unit processed the optimum-priced crudes available based on the characteristics of the residual bottoms to be processed in the LC-Fining unit (i.e., high sulfur/high metals content feedstocks, including blends with over 40 per cent Mexican Maya). Amoco installed its own LC-Fining pilot-plant facilities in 1980. Lummus was given access to much of the information from these pilot units as well as that from Syncrude Canada's pilot unit, which operated from 1988 to 1998.

Over many years, Lummus developed a large pilot-plant and commercial units database on various residual feeds from different geographic locations with a wide range of contaminant levels (metals, sulfur, nitrogen, CCR, asphaltenes, etc.). Some of these residual feeds included the world's most difficult, very heavy, lower-value hydrocarbon streams.

Lummus's pilot units were used to conduct many programs for design data development for various potential clients and the U.S. government. At one time, there were three ebullated bed pilot plants operating continuously in addition to mini-stirred autoclave test facilities. Many of the Middle Eastern, Mexican, Venezuelan, western Canadian, southeastern Asia, Russian, and U.S. reduced crudes and vacuum bottoms, solvent-refined coal extracts, heavy oils from tar sands, and shale oils were processed in Lummus's pilot-plant facilities.

When the H-Oil process was developed, the initial goal was to process Athabasca bitumen for Cities Service Athabasca (now Syncrude Canada). Many of the pilot programs conducted from 1957 to 1972 were devoted to Athabasca bitumen. In the 1970s and 1980s Lummus also conducted numerous pilot runs on bitumen feeds for Syncrude.

In the early 1990s, an extensive pilot program was performed for Alberta Oil Sands Technology Research Administration (AOSTRA) in order to demonstrate Lummus's high conversion technology (LC-Fining) with Athabasca bitumen and other Alberta heavy crudes. During this work, the integrated hydrotreater was closely studied and piloted at conditions that could be applied commercially.

In 1990, a 70-day pilot-plant run was conducted on Arabian Heavy vacuum residue in Lummus's research facilities to support AGIP Petroli Raffineria di Milazzo's (RAM) commercial LC-Fining unit design. In this run, a series of test programs were conducted at varying operating conditions and using various feed diluents and cutter stocks. The primary objective of the run was to establish the reactor operating conditions and proper feed diluent blends that would permit the maximum level of conversion to be attained, consistent with meeting RAM's low-sulfur fuel oil requirements. The commercial unit, commissioned in September 1998, has been running with an on-stream time in excess of 96 per cent while producing 1 wt per cent sulfur stable fuel oil.

In 1995, Russian vacuum residual supplied by Slovnaft for its refinery in Bratislava, Slovakia, was processed in Lummus's pilot-plant facility at conversion levels ranging from 60 to 88 vol per cent. For operations at higher conversion levels, an aromatic solvent (heavy cycle oil) was used with the vacuum residual, using high-HDS and low-sediment activity catalysts. This catalyst was superior to the standard LC-Fining catalyst tested in earlier runs in terms of sediment control and hydrodesulfurization (HDS) and CCR activity. The objective of this pilot run was to establish the reaction yields and product qualities to be used in the design and guarantees for Slovnaft's commercial LC-Fining unit. The Slovnaft plant is similar to the RAM unit and has been successfully operating since 2000.

In 2000, Lummus and Chevron joined forces to jointly develop and market the residual upgrading technologies of both companies—including the LC-Fining process—under a single entity, Chevron Lummus Global LLC (CLG).

In 1999 and 2000, two LC-Fining pilot plant runs were conducted at CLG's facilities to support Shell Canada's and Petro-Canada's commercial LC-Fining units design efforts. During the Shell Canada run, a close-coupled, online, integrated, wide cut distillate, fixed-bed hydrotreater was also tested. In addition, off-line hydrotreating tests were performed to replicate the inhibition effects of H2S and NH3 expected in commercial operation.

Under a joint cooperative agreement, CLG and AGIP Petroli conducted several pilot-plant runs in the LC-Fining pilot-plant facility at AGIP's research center related to process optimization/ development and catalyst screening programs for RAM, Shell Canada, and Petro-Canada. The long-term goals of this joint effort are to extend the database, further enhance the correlations and models, test new process designs (e.g., interstage separator/stripper, optimize quantity and interstage injection location of aromatic diluents), and continue to screen and evaluate new residual conversion catalysts.

CLG has built and is building several small pilot units at Chevron's research facilities in Richmond, California to support its residual upgrading technologies: ARDS, VRDS, OCR, upflow reactor (UFR), and LC-Fining.

UOP Bensat Process

The introduction of reformulated gasoline with mandated limits on benzene content has caused many refiners to take steps to reduce the benzene in their gasoline products. The major source of benzene in most refineries is the catalytic reformer. Reformate typically contributes 50 to 75 per cent of the benzene in the gasoline pool.

The two basic approaches to benzene reduction involve prefractionation of the benzene and benzene precursors in a naphtha splitter before reforming, postfractionation in a reformate splitter of the benzene after it is formed, or a combination of the two (Figure 4.42a). The benzene-rich stream must then be treated to eliminate the benzene by using extraction, alkylation, isomerization, or saturation (Figures. 4.42 b and c).

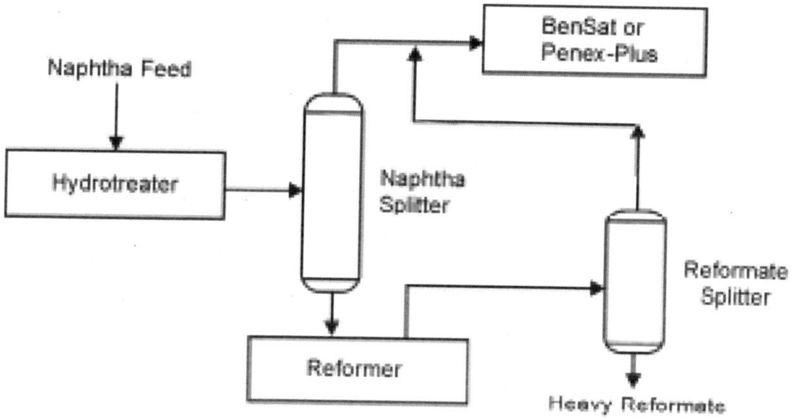

Figure 4.42(a): Fractionation for benzene reduction.

If the refiner has an available benzene market, the benzene-rich stream can be sent to an extraction unit to produce petrochemical-grade benzene. Alkylation of the benzene may also be an attractive option if propylene is available, as in a fluid catalytic cracking (FCC) refinery. An isomerization unit saturates the benzene and also increases the octane of the stream by isomerizing the paraffins to a higher-octane mixture. Saturation in a stand-alone unit is a simple, low-cost option.

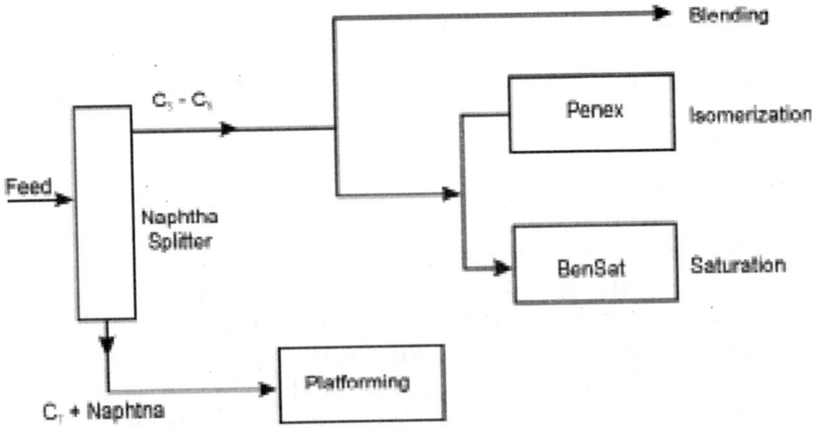

Figure 4.42(b): Prefractionation Options

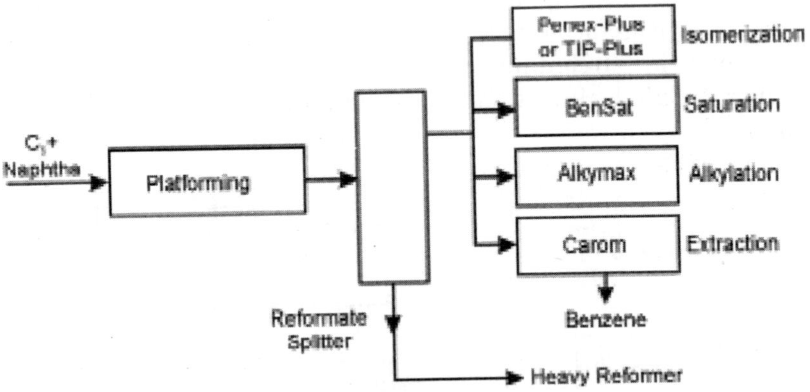

Figure 4.42(c): Postfractionation Options

Process Discussion

The UOP BenSat process was developed to treat C_5-C_6 feedstocks with high benzene levels. Because almost all the benzene is saturated to cyclohexane over a fixed bed of catalyst, no measurable side reactions take place. Process conditions are moderate, and only a slight excess of hydrogen above the stoichiometric level is required. The high heat of reaction associated with benzene saturation is carefully managed to control the temperature rise across the reactor. Product yield is greater than 100 liquid volume per cent (LV %), given the volumetric expansion associated with saturating benzene and the lack of any yield losses from cracking to light ends.

The product has a lower octane than the feed as a result of the conversion of the high-octane benzene into lower-octane cyclohexane. However, the octane can be increased by further processing the BenSat product in an isomerization unit, such as a UOP Penex unit.

Process Flow

The BenSat process flow is shown in Figure The liquid feed stream is pumped to the feed-effluent exchanger and to a preheater, which is used only during start-up. Once the unit is on-line, the heat of reaction provides the required heat input to the feed via the feed-effluent exchanger. Makeup hydrogen is combined with the liquid feed, and flow continues into the reactor. The reactor effluent is exchanged against fresh feed and then sent to a stabilizer for removal of light ends.

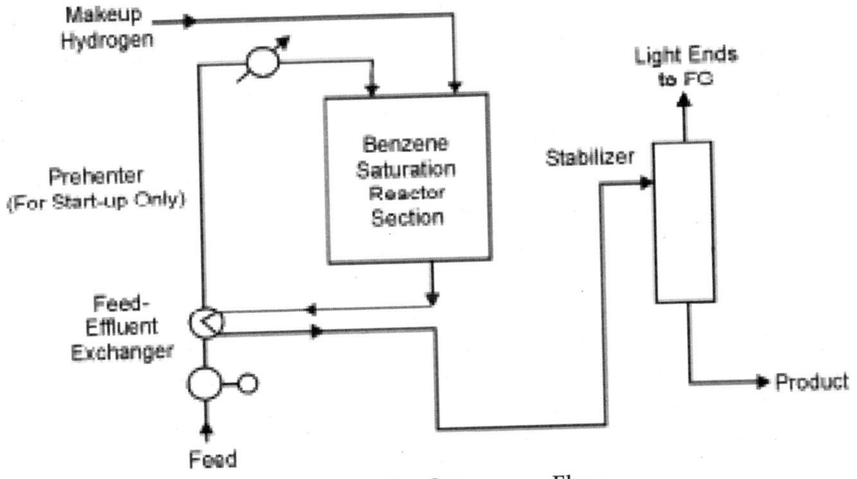

Figure 4.43: BenSat process Flow

Catalyst and Chemistry

Saturating benzene with hydrogen is a common practice in the chemical industry for the production of cyclohexane. Three moles of hydrogen are required for each mole of benzene saturated. The saturation reaction is highly exothermic: the heat of reaction is 1100 Btu per pound of benzene saturated. Because the benzene-cyclohexane equilibrium is strongly influenced by temperature and pressure, reaction conditions must be chosen carefully.

The UOP BenSat process uses a commercially proven noble metal catalyst, which has been used for many years for the production of petrochemical-grade cyclohexane. The catalyst is selective and has no measurable side reactions. Because no cracking occurs, no appreciable coke forms on the catalyst to reduce activity. Sulfur contamination in the feed reduces catalyst activity, but the effect is not permanent. Catalyst activity recovers when the sulfur is removed from the system.

UOP BUTAMER PROCESS

Introduction

The first successful efforts in the research and development of catalytic systems for the isomerization of normal paraffins came in the early 1930s. The requirement for high-octane aviation gasoline during World War II accelerated the application of early isomerization research. Light olefinic hydrocarbons were available from the newly developed fluid catalytic cracking (FCC) process and from other mainly thermal operations. These olefinic hydrocarbons could be alkylated with isobutane ($/C_4$) to produce a high-octane gasoline blending component. However, the supply of isobutane from straight-run sources and other refinery processing was insufficient, and a new source of supply had to be found. Isobutane produced from the new normal paraffin isomerization process met that need.

The first commercial butane isomerization unit went on-stream in late 1941. By the end of the war, 38 plants were in operation in the United States and 5 in allied countries for a total capacity of approximately 50,000 barrels per stream day (BPSD). Five principal isomerization processes, including one developed by UOP, were used in the United States. All were based on Friedel-Crafts chemistry and used aluminum chloride in some form.

The wartime units fulfilled the needs of the time. However, despite many improvements, the units remained difficult and costly to operate. Corrosion rates were excessive, plugging of catalyst beds and equipment was common, and catalyst consumption was high. The units were characterized by high maintenance and operating costs and low onstream efficiency.

The introduction of UOP's Platforming process in 1949 and the rapid spread of such catalytic reforming over dual-functional catalysts in the 1950s served to focus attention on the development of similar catalysts for paraffin isomerization. The term dual functional refers to the hydrogenation and controlled-acidity components of a catalyst. Isomerization was known to be one of several reactions that occurred during catalytic reforming, and so isolating this reaction for use with feeds that did not require any of the other reactions was a natural next step.

Although the earlier of these dual-functional catalysts eliminated many of the shortcomings of the wartime aluminum chloride catalyst systems, they required relatively high operating temperatures. At these temperatures, unfavorable equilibriums limited per-pass conversion. Further research was conducted, and in 1959, UOP made available to the industry a butane isomerization process, UOP's Butamer process, that used a highly active, low-temperature hydroisomerization catalyst capable of achieving butane conversion at temperature levels equivalent to the wartime Friedel-Crafts systems without the attendant corrosion or sludge formations. Industry acceptance of the UOP process was rapid, and in late 1959, the first Butamer unit, the first commercial butane isomerization unit to use a low-temperature, dual-functional catalyst system, was placed on-stream on the United States West Coast.

Process Description

The Butamer process is a fixed-bed, vapor-phase process promoted by the injection of trace amounts of organic chloride. The reaction is conducted in the presence of a minor amount of hydrogen, which suppresses the polymerization of olefins formed as intermediates in the isomerization reaction. Even though the chloride is converted to hydrogen chloride, carbon steel construction is used successfully because of the dry environment. The process uses a high-activity, selective catalyst that promotes the desired conversion of normal butane (nC_4) to isobutane at low temperature and, hence, at favorable equilibrium conditions.

Regardless of the $/C_4$ content of the feed, the butane fraction leaving the unit contains approximately 60 per cent by volume of $/C_4$. Therefore, to obtain optimum plant performance, the refiner wants to charge a butane cut containing the highest practical content of nC_4.

The catalytic reaction is highly selective and efficient and results in a minimum of hydrocracking to light ends or the formation of heavy coproduct. Volumetric yield of $/C_4$ product based on an nC_4 feed typically approximates slightly more than 100 per cent.

Process Chemistry

Isomerization by dual-functional catalysts is thought to operate through an olefin intermediate. The formation of this intermediate is catalyzed by the metallic component, which is assumed for this discussion to be platinum:

$$CH_3—CH_2—CH_2—CH_3 \overset{Pt}{\rightleftharpoons} CH_3—CH_2—CH = CH_2 + H_2 \quad \text{...........(1)}$$

This reaction is, of course, reversible, and because these catalysts are used under substantial hydrogen pressure, the equilibrium is far to the left. However, the acid function (H^+A^-) of the catalyst consumes the olefin to form a carbonium ion and thus permits more olefin to form despite the unfavorable equilibrium:

$$CH_3—CH_2—CH = CH2 + H^+A^- ® \; CH_3—CH_2—CH—CH_3 + A^- \quad \text{...........(2)}$$

The usual rearrangement ensues:

$$CH_3—CH_2—\underset{+}{C}H_2—CH_3 ® \; CH_3—CH_3—\underset{+}{\overset{CH_3}{\overset{|}{C}}}—CH_3 \quad \text{...........(3)}$$

The isoolefin is then formed by the reverse analog of Eq. (2):

$$CH_3—\overset{CH_3}{\overset{|}{C}}—CH_3—A^- ® \; CH_3—\underset{+}{\overset{CH_3}{\overset{|}{C}}} = CH_2 + H^+ A^- \quad \text{...........(4)}$$

The isoparaffin is finally created by hydrogenation:

$$CH_3—\overset{CH_3}{\overset{|}{C}}—CH_2 + H_2 \overset{Pt}{\rightleftharpoons} CH_3— \overset{CH_3}{\overset{|}{CH}} = CH_3 \quad \text{...........(5)}$$

Process Variables

The degree of isomerization that occurs in the Butamer process is influenced by the following process variables.

Reactor Temperature

The reactor temperature is the main process control for the Butamer unit. An increase in temperature increases the /C4 content of the product toward its equilibrium value and slightly increases cracking of the feed to propane and lighter.

Liquid Hourly Space Velocity (LHSV)

An increase in LHSV tends to decrease the $/C_4$ in the product at a constant temperature when other conditions remain the same.

Hydrogen-to-Hydrocarbon Ratio (H_2/HC)

The conversion of nC_4 to $/C_4$ is increased by reducing the H_2/HC ratio; however, the hydrogen effect is slight over the usual operating range. Significant capital savings do result when the H_2/HC ratio is low enough to eliminate the recycle hydrogen compressor and product separator. UOP's standard (and patented) design calls for a H_2/HC ratio of 0.03 molar and allows operation with once-through hydrogen.

Pressure

Pressure has no effect on equilibrium and only a minor influence on the conversion of normal butane to isobutane.

Process Contaminants

Water poisons the Butamer catalyst. A simple but effective molecular-sieve drying system is used on unit hydrocarbon and gas feeds. Sulfur is a temporary poison that inhibits catalyst

UOP PENEX PROCESS

Introduction

A component of refinery gasoline pools that frequently offers the best opportunity for quality improvement is the pentane-hexane fraction, or light straight-run (LSR) naphtha. The LSR is characterized by a low octane number, ordinarily 60 to 70 research octane number (RON), clear. Historically, this fraction, which constitutes approximately 10 per cent of a typical gasoline pool in the United States and often a higher per centage in Europe, has been blended directly into gasoline without additional processing except perhaps treating for mercaptan removal. The low octane rating could be increased by approximately 16 to 18 numbers because of its excellent lead susceptibility. The low octane placed the C_5-C_6 straight run in the position of being that segment of the gasoline pool helped most by the addition of lead and least in need of upgrading by processing.

As the petroleum industry moved toward the marketing of motor fuels with reduced or zero lead levels, accommodating the LSR in the gasoline pool became increasingly difficult. The conversion of these C5 and C6 paraffins to the corresponding branched isomers to increase their RON, clear, octane number was recognized as a logical and necessary step. One option that UOP offers to accomplish this upgrading is the Penex process, which uses a highly active, low-temperature hydroisomerization catalyst. The reliability of this catalyst has been commercially demonstrated since 1959 in butane isomerization (UOP's Butamer process) and since 1969 in C_5-C_6 isomerization.

As a result of U.S. reformulated gasoline legislation for benzene reduction during the 1990s, a variation of UOP's Penex process is being used to saturate all the benzene in the LSR cut and boost the octane of this gasoline fraction.

Process Discussion

The UOP's Penex process is specifically designed for the catalytic isomerization of pentane, hexanes, and mixtures thereof. The reactions take place in the presence of hydrogen, over a fixed bed of catalyst, and at operating conditions that promote isomerization and minimize hydrocracking. Operating conditions are not severe, as reflected by moderate operating pressure, low temperature, and low hydrogen partial pressure requirements.

Ideally, this isomerization catalyst would convert all the feed paraffins to the high-octane-number branched structures: normal pentane (nC_5) to isopentane ($\backslash C_5$) and normal hexane (nC_6) to 2,2- and 2,3-dimethylbutane. The reaction is controlled by a thermodynamic equilibrium that is more favorable at low temperature.

Table 4.12: Typical C_5-C_6 Chargestock and Product Compositions

	Percent of total	*Chargestock*	*Product*
C_5 paraffins, wt %:	*47.5*		
Isopentane		42.0	77.0
n-C_5		58.0	23.0
C_6 paraffins, wt %:	*45.2*		
2,2-dimethylbutane		0.9	31.6
2,3-dimethylbutane		5.0	10.4
Methylpentanes		48.2	46.9
n-C_6		45.9	11.1
C_6 cyclic, wt %:	*7.3*		
Methylcyclopentane		57.0	52.0
Cyclohexane		17.0	48.0
Benzene		26.0	0
Unleaded octane numbers:			
Research		70.1	83.8
Motor		66.8	81.1

Table shows typical charge and product compositions for a C_5-C_6 Penex unit. The compositions of both the C_5 and C_6 fractions correspond to a close approach to equilibrium at the operating temperature.

With C_5 paraffins, interconversion of normal pentane and isopentane occurs. The C_6-paraffin isomerization is somewhat more complex. Because the formation of 2- and 3-methylpentane and 2,3-dimethylbutane is limited by equilibrium, the net reaction involves mainly the conversion of normal hexane to 2,2-dimethylbutane. All the feed benzene is hydrogenated to cyclohexane, and a thermodynamic equilibrium is established between methylcyclopentane and cyclohexane. The octane rating shows an appreciation of some 14 numbers.

Process Flow

As shown in Figure, light naphtha feed is charged to one of the two dryer vessels. These vessels are filled with molecular sieves, which remove water and protect the catalyst. After mixing with makeup hydrogen, the feed is heat-exchanged against reactor effluent. It then enters a charge heater before entering the reactors. Two reactors normally operate in series.

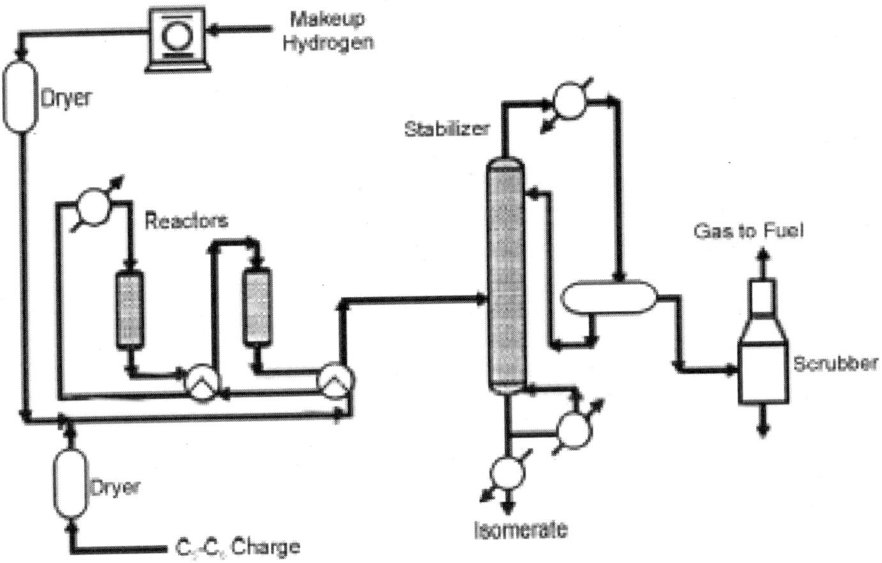

Figure 4.44: UOP Penex Process

The reactor effluent is cooled before entering the product stabilizer. In new Penex designs, both the recycle gas compressor and the product separator have been eliminated. Only a slight excess of hydrogen above chemical consumption is used. The makeup hydrogen, which can be of any reasonable purity, is typically provided by a catalytic reformer. The stabilizer overhead vapors are caustic scrubbed for removal of the HCl formed from organic chloride added to the reactor feed to maintain catalyst activity. After scrubbing, the overhead gas then flows to fuel. The stabilized, isomerized liquid product from the bottom of the column then passes to gasoline blending.

Alternatively, the stabilizer bottoms can be separated into normal and isoparaffin components by fractionation or molecular-sieve separation or a combination of the two methods to obtain recycle of the normal paraffins and low-octane methylpentanes (MeC_5). Product octanes in the range of 87 to 92 RON, clear, can be achieved by selecting one of the various possible schemes.

The least capital-intensive recycle flow scheme is achieved by combining the Penex process with a deisohexanizer column. The deisohexanizer column concentrates the low-octane methylpentanes into the sidecut stream. This sidecut stream combines with the fresh feed before entering the Penex reactor. The deisohexanizer column overhead, which is primarily isopentane, 2,2-dimethylbutane, and 2,3-dimethylbutane, is recovered for gasoline blending. A small bottoms drag stream, consisting of C_6 naphthenes and C_7's, is also removed from the deisohexanizer column and used for gasoline blending or as reformer feed.

An efficient recycle operation is obtained by combining the Penex process with UOP's Molex process, which uses molecular sieves to separate the stabilized Penex product into a high-octane isoparaffin stream and a low-octane normal paraffin stream. In this system, fresh feed together with the recovered low-octane normal paraffin stream is charged to the Penex unit. The isomerized product is denormalized in the Molex unit and recovered for gasoline blending.

Many configurations of separation equipment are possible. The optimum arrangement depends on the specific chargestock composition and the required product octane number.

In addition to increasing octane, another benefit of all Penex-based flow schemes is the saturation of all benzene to cyclohexane. This aspect is particularly important to refiners who want to reduce the level of benzene in their gasoline pool.

Some feedstocks, such as light reformate, can contain high levels of benzene. The performance of the Penex process can be compromised when processing these feedstocks because benzene hydrogenation is a highly exothermic reaction. The heat generated by the benzene hydrogenation reaction can cause the reactors to operate at conditions that are less favorable for octane upgrading. For these applications, UOP offers the Penex-Plus process, which includes two reactor sections. The first section is designed to saturate the benzene to cyclohexane. The second section is designed to isomerize the feed for an overall octane increase. Each reactor is operated at conditions that favor the intended reactions for maximum conversion.

UOP TIP AND ONCE-THROUGH ZEOLITIC ISOMERIZATION PROCESSES

Introduction

Light straight-run (LSR) naphtha fractions made in the refinery are predominantly C_5's and C_6's. Some C7's are also present. They are highly paraffinic and have clear research octane numbers (RONC) usually in the 60s. The nonnormal components have higher octanes than normal paraffins (Table) and are excellent gasoline-blending feedstocks.

Table 4.13: Properties of Common Gasoline Components

	Molecular weight	*Boiling point,* °F*	*Density, * lb/gal*	*RONC*
Isobutane	58.1	10.9	4.69	100+
n-butane	58.1	31.1	4.86	93.6
Neopentane	72.1	49.0	4.97	116
Isopentane	72.1	82.2	5.20	92.3
n-pentane	72.1	96.9	5.25	61.7
Cyclopentane	70.0	120.7	6.25	100
2,2-dimethylbutane	86.2	121.5	5.54	91.8
2,3-dimethylbutane	86.2	136.4	5.54	101.7
2-methylpentane	86.2	140.5	5.57	73.4
3-methylpentane	86.2	145.9	5.44	74.5
n-hexane	86.2	155.7	5.48	94.8
Methylcyclopentane	84.2	161.3	6.28	91.3
2,2-dimethylpentane	100.2	174.6	5.64	92.8
Benzene	78.1	176.2	7.36	100+
2,4-dimethylpentane	100.2	176.9	5.64	83.1
Cyclohexane	84.2	177.3	6.53	83
2,2,3-trimethylbutane	100.2	177.6	5.78	112
3,3-dimethylpentane	100.2	186.9	5.81	98
2,3-dimethylpentane	100.2	193.6	5.83	88.5
2,4-dimethylpentane	100.2	194.1	5.68	55
3-methylhexane	100.2	197.5	5.76	65
Toluene	92.1	231.1	7.26	100+

For the refiner who wants to upgrade the octane of a gasoline pool and has use for a high-purity normal paraffin product, UOP's IsoSiv separation technology is a good fit. However, if octane improvement is of primary importance, isomerization technology is the best choice.

Paraffin isomerization to upgrade the octane of light-naphtha streams has been known to the refining industry for many years and has gained importance since the onset of the worldwide reduction in the use of lead antiknock compounds. This technology continues to be important in view of current U.S. legislation on reformulated gasoline.

However, refiners with idle hydroprocessing equipment, such as old catalytic reformers or hydrodesulfurization units, can consider converting this equipment to a UOP Once-Through (O-T) Zeolitic Isomerization process (formerly known as the Shell Hysomer process). The process scheme is similar to that of a simple hydrotreater, as shown in Figure, and conversions can be accomplished quickly and at low cost. With O-T Zeolitic Isomerization, a 10 to 12 octane-number increase for the C_5–71°C (160°F) light naphtha can be achieved.

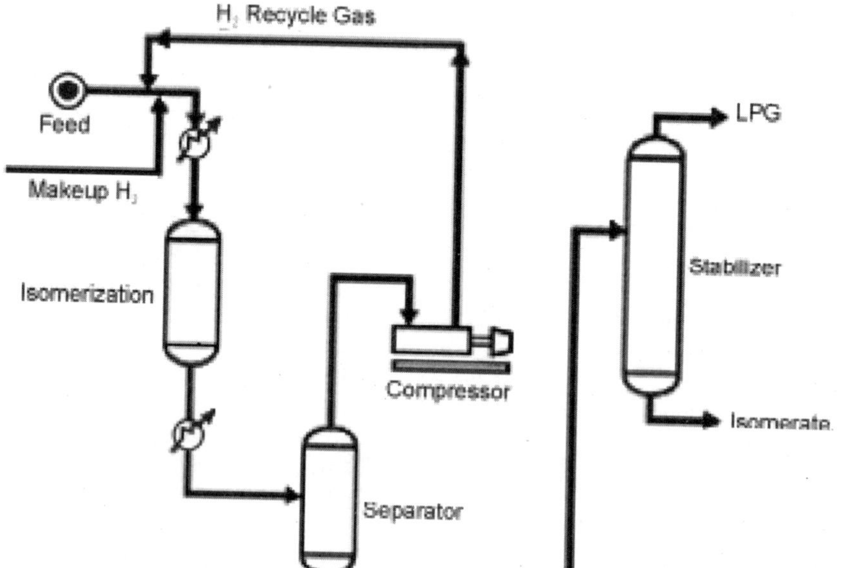

Figure 4.45: UOP Once-through Zeolitic Isomerization Process

For those refiners who need more octane than can be achieved from the once-through operation, an additional 8 to 10 RONC can be gained by adding molecular sieve adsorption to the O-T Zeolitic Isomerization process. Molecular sieve adsorption is used to extract the unreacted normal paraffins so they can be recycled to extinction. This approach of complete isomerization is referred to as UOP's TIP process. Because O-T Zeolitic Isomerization is an integral part of the TIP process, the ensuing discussion begins with the once-through operation and concludes with a discussion of TIP.

O-T Zeolitic Isomerization Process

Process Chemistry

Thermodynamically, low temperatures are preferred for obtaining maximum amounts of branched paraffins in the reaction product. Operation below 150°C (302°F) for maximum activity requires a catalyst that uses a halide activator. For these catalysts, feed drying is required to eliminate any corrosion or catalyst stability concerns.

The catalyst used in the O-T Zeolitic Isomerization process, however, is based on a strongly acidic zeolite with a recoverable noble-metal component. No external acid activators are used and the catalyst does not produce a corrosive environment. Therefore, feed drying is not necessary.

The catalyst base behaves as an acid of the Brönsted type because it has a high activity for normal-pentane isomerization in the absence of a metal component. At a relatively low hydrogen partial pressure, the carbonium ion concentration generated by the activated low-sodium zeolite is apparently higher than it would have been if the paraffin-olefin equilibrium had been established. This excessive carbonium ion concentration leads to not only high initial conversion but also unstable operation and low selectivity under preferred operating conditions (Figure). This figure also shows that incorporation of the metal function stabilizes the conversion and lowers the initial activity. These results are to be ascribed to the lower olefin and carbonium ion concentration in the presence of the dual-function catalyst as a result of the paraffin-olefin equilibrium.

Carbonium ions and isoparaffins are generated from normal paraffins by a combination of hydride-ion abstraction and hydride-ion transfer reactions. In the adsorbed state, skeletal rearrangement reactions occur. Alternatively, while the normal pentane is in the carbonium ion state (nP^+ or \P$^+$), it may surrender a proton to form an olefin, which in turn is hydrogenated to form a paraffin (these two paths are vertical).

Even a minute amount of the noble metal stabilizes the conversion to isopentane, provided that the noble metal is well dispersed and distributed throughout the zeolite. However, in commercial applications, more than the minimum amount of noble metal is required. Normally the catalyst contains a few tenths of a per cent of precious metal. Proper catalyst preparation methods and start-up procedures are essential for optimal results.

Process Description

The O-T Zeolitic Isomerization process is a fixed-bed, vapor-phase process for the catalytic isomerization of low-octane normal pentane or normal hexane or both to high-octane isoparaffins. The isomerization reaction is carried out at 245 to 270°C (470 to 520°F) and 21 to 35 kg/cm2 (300 to 500 lb/in^2 gage) in the presence of hydrogen. Equipment requirements are a reactor vessel, heater, recycle hydrogen compressor, feed-product heat exchanger, product cooler, phase separator drum, and product stabilizer section.

UOP Par-Isom Process

Light straight-run (LSR) naphtha fractions are predominantly C5's and C6's. Some C7's are also present. They are highly paraffinic and have clear research octane numbers (RONC) usually in the 60s. This fraction, which constitutes 10 per cent of a typical gasoline pool in the United States, is usually upgraded with paraffin isomerization technology.

The use of paraffin isomerization technology to upgrade the octane of light naphtha streams has been known to the refining industry for many years and has gained importance since the onset of the worldwide reduction in the use of lead antiknock compounds and benzene. This technology continues to be important in view of current U.S. and European legislation on reformulated gasoline.

The most cost-effective means to upgrade an LSR feedstock in a grassroots situation is the UOP Penex process. This process relies on a highly active chlorided alumina catalyst to produce an isomerate product with a RONC of 82 to 85. However, the catalyst is sensitive to contaminants and is not regenerable.

Alternatively, refiners with idle processing equipment such as old catalytic reformers or hydrodesulfurization units can consider converting this equipment to the UOP Once-Through (O-T) Zeolitic Isomerization process (formerly known as the Shell Hysomer† process). These conversions can be accomplished quickly and at low cost to provide a 10 to 12 octane number increase for the light naphtha. Zeolitic catalysts are tolerant of contaminants and are regenerable, but operate at relatively high temperatures that limit the maximum octane that can be achieved.

With the commercialization of the UOP Par-Isom process, the refiner has another option for light paraffin isomerization. The key to this new process is the LPI-100 catalyst, an innovative, high-performance sulfated metal oxide catalyst with activity approaching that of chlorided alumina catalysts, but with the benefit of being both robust and regenerable. The basic formulation for LPI-100 catalyst was originally developed by Cosmo Research Institute and Mitsubishi Heavy Industries in Japan. With the UOP Par-Isom process, an LSR feedstock can be upgraded to 79 to 82 RONC

Process Discussion

The UOP Par-Isom process is specifically designed for the catalytic isomerization of pentanes, hexanes, and mixtures thereof. The reactions take place in the presence of hydrogen, over a fixed bed of catalyst, and at operating conditions that promote isomerization and minimize hydrocracking. The unit operates at moderate temperature and pressure.

Table 4.14: Typical Estimated Performance, Par-Isom Isomerization Unit 10,000 BPD

Hydrogen consumption		1,602,000 SCF/day
Light-gas yield		
C_1		137,000 SCF/day
C_2		67,000 SCF/day
C_3		246,000 SCF/day
Component	*Fresh feed to reactor*	*Product*
Flow rate, BPD		
IC_4	10	278
NC_4	58	135
IC_5	1,684	3,185
NC_5	2,907	1,368
Cyclo-C_5	169	169
2,2-Dimethylbutane	51	984
2,3-Dimethylbutane	193	461
2-Methylpentane	1,208	1,555
3-Methylpentante	880	903
NC_6	1,935	572
Methylcyclopentane	195	216
Cyclo-C_6	341	121
Benzene	175	0
C_7	194	86
Total	*10,000*	*10,033*
C_4+ properties:		
Specific gravity	0.659	0.647

Ideally, an isomerization catalyst would convert all the feed paraffins to the high-octane-number branched structures: nC_5 to isopentane and nC_6 to 2,2- and 2,3-dimethylbutane. These reactions are controlled by a thermodynamic equilibrium that is more favorable at low temperature. The Penex process operates at a lower temperature than the Par-Isom process, which in turn operates at a lower temperature than the Once-

Through Zeolitic Isomerization process. Consequently, the Penex process produces the highest product octane, followed by the Par-Isom process, with the Once-Through Zeolitic Isomerization process offering the lowest product octane.

Table 4.14 shows typical charge and product compositions for a C_5-C_6 Par-Isom unit. With C_5 paraffins, interconversion of normal pentane and isopentane occurs. The C_6 paraffin isomerization is somewhat more complex. Since the formation of 2- and 3-methylpentane and 2,3-dimethylbutane is limited by equilibrium, the net reaction involves mainly the conversion of normal hexane to 2,2-dimethylbutane. All the feed benzene is hydrogenerated to cyclohexane, and a thermodynamic equilibrium is established between methylcyclopentane and cyclohexane. The octane rating shows an increase of some 13.7 numbers.

Process Flow Scheme

The Par-Isom process flow scheme is shown in Figure and is identical to the O-T Zeolitic Isomerization process flow scheme. In fact, since the two processes operate over the same pressure range, LPI-100 catalyst is a drop-in replacement for HS-10 catalyst that results in a 2 to 3 octane number improvement.

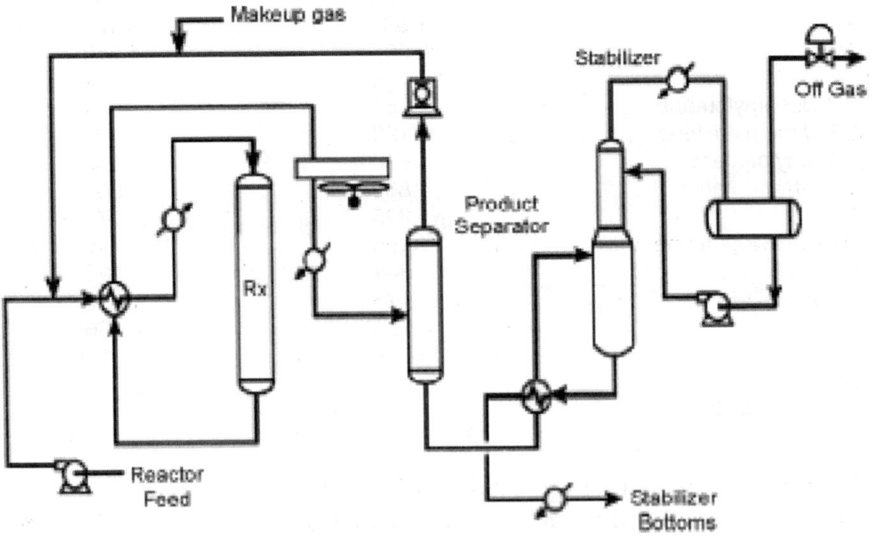

Figure 4.46: Par-Isom Process Flow Scheme

The light naphtha feed is combined with makeup and recycle hydrogen before being directed to the charge heater where the reactants are heated to reaction temperature. A fired heater is not required in the Par-Isom process, due to the much lower reaction temperature needed for LPI-100 catalyst than for zeolitic catalysts. Hot oil or high-pressure steam can be used as the heat source in this exchanger. The heated combined feed is then sent to the isomerization reactor.

The reactor effluent is cooled and then sent to a product separator where the recycle hydrogen is separated from the other products. Recovered recycle hydrogen is directed to the recycle compressor and then returned to the reaction section. The liquid product is sent to a stabilizer column where the light ends and dissolved hydrogen are removed. The stabilized isomerate product can be sent directly to gasoline blending.

Alternatively, the stabilizer bottoms can be fractionated in a deisohexanizer column to concentrate the normal hexane and low-octane methylpentanes into a sidecut stream. This sidecut stream combines with the fresh feed before entering the Par-Isom reactor. The deisohexanizer column overhead, which is primarily isopentane, 2, 2-dimethylbutane, and 2, 3-dimethylbutane, is recovered for gasoline blending. A small bottoms drag stream, consisting of C_6 naphthenes and C_7's, is also removed from the deisohexanizer column and used for gasoline blending or as reformer feed. Product octanes in the range of 85 to 87 RONC can be achieved with this flow scheme.

CHEVRON LUMMUS GLOBAL ON-STREAM CATALYST REPLACEMENT TECHNOLOGY FOR PROCESSING HIGH-METAL FEEDS

Introduction

In processing less expensive, high-metal feeds, the need for frequent catalyst change-outs can make conventional fixed-bed residuum hydrotreating technology uneconomical. Chevron Lummus Global (CLG) developed on-stream catalyst replacement (OCR) to remove metals from feed before it is hydrotreated in fixed-bed residuum desulfurization (RDS) units. The ability to add and withdraw catalyst from the high-pressure, moving-bed OCR reactor while it is onstream gives refiners the opportunity to process heavier, high-metal feeds or to achieve deeper desulfurization while maintaining fixed-bed run lengths and improving product properties.

Development History

Development of the OCR process started in 1979 as part of the research on new reactor concepts which might be applied to synthetic fuels and heavy oil upgrading. Most of these alternative fuels are difficult to upgrade to transportation fuels. Typically high in nitrogen, sulfur, and metals, they tend to deactivate catalyst very rapidly. Consequently, conventional fixed-bed hydrotreating processes can not upgrade these feedstocks economically. CLG determined that if fresh catalyst could be continually moved through a reactor, then catalytic activity could be maintained without shutting down the unit. Theoretically, the metal capacity of the catalyst would be fully utilized in the OCR unit, thus reducing the necessary size of the downstream RDS unit and lowering total operating costs

Using cold model and catalytic testing, CLG showed that with the proper equipment design these process improvements could be obtained. The feasibility of the process was proved during the operation of a 200 barrel per day (BPD) demonstration unit a ChevronTexaco's Richmond refinery in 1985. Critical to the design's success was the proof that the valves could operate reliably at the high temperatures and pressures required for residuum upgrading.

Process Description

CLG's OCR process is a countercurrent, moving-bed technology that removes metals and other contaminants from feedstocks prior to processing in fixed-bed residuum hydrotreating reactors.

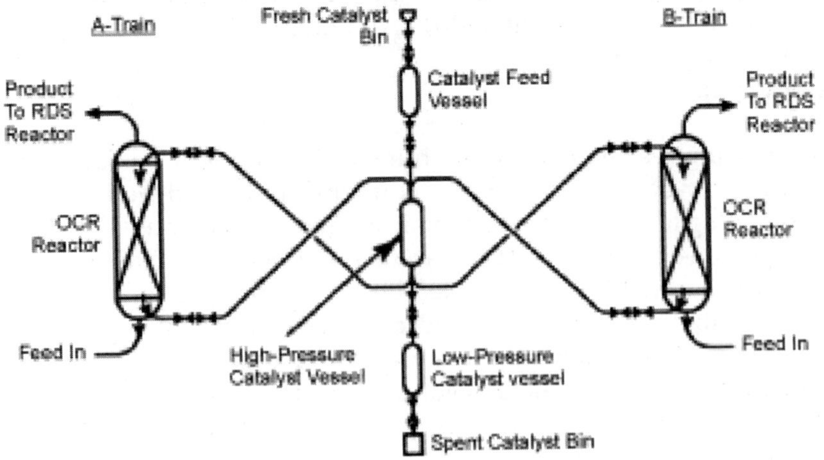

Figure 4.47: OCR Reactor System CLG's OCR technology adds and removes catalysts from a high-pressure reactor while it is operating, thus providing refiners the opportunity to process less expensive, high-metal feeds.

In the OCR rector, residuum and hydrogen flow upward through the reactor, and the catalyst flows downward. This process removes the metals and carbon residue that cause plugging and catalyst deactivation in conventional fixed-bed RDS units. Figure shows the OCR reactor system, including the equipment used to transfer catalyst into and out of the reactor. OCR reactors can be paired and serve as pretreatment beds for two parallel trains of fixed-bed reactors. In this case, only one set of catalyst transfer vessels is needed to move catalyst for two OCR reactors.

Catalyst Transfer System

In a parallel train system, catalyst transfer is done, on average, once a week to and from each OCR reactor. The amount of catalyst transferred varies from 1.5 to 8 per cent of the OCR reactor catalyst capacity. The quantity removed is dictated by the nickel and vanadium content of the feed and the metals concentration on the removed catalyst. The transfer rate is adjusted to allow for changes in operating requirements. Once the requirements are defined, the catalyst is added and withdrawn batchwise on a regular schedule to maintain the required OCR activity.

All the steps required to transfer the catalyst are controlled by a computer-driven, automatic sequencer. Use of the automatic sequencer minimizes the need for operator attention and ensures consistent adherence to all necessary procedures. Operation of the OCR catalyst transfer system is easily monitored by the existing RDS board operator.

Catalyst transfer in and out of the OCR reactor is accomplished in a series of steps which do not interrupt the operation of the unit:

1. Fresh catalyst is transferred by gravity into the low-pressure catalyst feed vessel.
2. There, flush oil (usually a heavy gas oil) is added, and the mixture is transferred as a slurry to the high-pressure catalyst vessel (HPCV).
3. The low-pressure catalyst feed vessel is then isolated, and the pressure in the HPCV is equalized with the top of the OCR reactor.
4. The fresh catalyst is then transferred as a slurry to the top of the OCR reactor.
5. Once the transfer is complete, as indicated by the level in the HPCV, the double isolation valves are flushed to remove catalyst, and the HPCV is isolated from the system.

Spent catalyst is removed from the bottom of the reactor in a similar manner:

1. The HPCV is pressure-equalized with the bottom of the OCR reactor.
2. The spent catalyst is moved as a slurry in the feed residuum from the bottom of the reactor. Once the desired amount of catalyst has been transferred, as indicated by the level in the HPCV, the transfer is stopped and the valves and lines are flushed with oil.
3. The double isolation valves are closed, and the HPCV is isolated from the OCR reactor and depressurized. The spent catalyst is washed of residuum and cooled.
4. The catalyst is transferred as a slurry to the low-pressure catalyst vessel, where the flush oil is drained.
5. The spent catalyst flows by gravity into the spent catalyst bin for disposal.

Since the catalyst is transferred in a low-velocity oil slurry, catalyst attrition is prevented and the system's lines and valves are protected from erosion. OCR lines are smaller than main process lines, and special full port valves are used in the catalyst transfer lines. These valves are flushed clear of catalyst before closing, to minimize valve wear.

OCR Reactor

The catalyst bed in the OCR unit is essentially a fixed bed, which intermittently moves down the reactor. The catalyst level in the OCR reactor is monitored by a level detector at the top of the reactor. As fresh catalyst is added at the top of the reactor, residuum is fed into the bottom. Both move through the reactor in a countercurrent flow, causing the dirtiest, most reactive residuum to contact the oldest catalyst first. The upflow of the residuum through the OCR reactor slightly expands the catalyst bed. This slight expansion enhances residuum/catalyst contact, minimizes reactor plugging, and creates a consistent pressure drop, thus providing for optimum flow patterns through the reactor. Meanwhile, the fully spent OCR catalyst is removed at the bottom of the reactor.

The specially designed cone at the bottom of the reactor allows for plug flow of the catalyst to the removal port at the bottom of the reactor. This plug flow ensures that the most metal-loaded, least active catalyst is removed from the reactor. Consequently catalyst activity is maximized and cost is minimized.

The Rose Process

Background

The Residuum Oil Supercritical Extraction process is the premier deasphalting technology available in industry today. This state-of-the-art process extracts high-quality deasphalted oil (DAO) from atmospheric or vacuum residues and other feedstocks. Depending on solvent selection, the DAO can be an excellent feedstock for catalytic cracking, hydrocracking, or lube oil blending. The asphaltene product from the ROSE process is often blended to fuel oil, but can also be used in the production of asphalt blending components, solid fuels, or fuel emulsions. Other possible options for the asphaltenes include use as feedstock to conversion processes such as partial oxidation, coking, or visbreaking.

The ROSE process was originally developed and commercialized by Kerr-McGee Corporation and first licensed by the company in 1979. In 1995, KBR (Kellogg Brown & Root, Inc.) acquired the ROSE process from Kerr-McGee. To date, 33 ROSE units with a total capacity of over 600,000 BPSD have been licensed and/or designed. All these units utilize supercritical fluid technology. KBR is responsible for the design or revamp of more than 400,000 BPSD of this total capacity, including the conversion of the world's largest solvent deasphalting facility for Chevron in Richmond, California, to a 50,000 BPSD ROSE unit.

Advantages

Processing residues in a ROSE unit merits serious consideration for today's refiner. A processing scheme utilizing a ROSE unit offers several operational and economic advantages over competing schemes. These advantages include.

- Increased yield and improved quality of valuable DAO product compared to other deasphalting processes.
- Significantly reduced fuel oil production for refineries blending vacuum residue to fuel.
- Flexibility to process atmospheric/vacuum residues from varying crude sources with little difficulty.
- State-of-the-art supercritical solvent recovery that significantly reduces operating costs compared to other solvent deasphalting processes
- Significantly lower capital and operating costs compared to other upgrading processes

Rose Dao Yield and Quality Advantage

ROSE technology offers significant DAO yield and quality advantages compared to other technologies. Superior process performance is ensured by utilizing state-of-the-art asphaltene and DAO separator internals (ROSEMAX).

Achieving the maximum yield and quality benefits from countercurrent extraction requires that limits to mass transfer be minimized. The capacity of the separator vessels must be maximized for a given size for economical design. These issues are addressed by the new generation of ROSE separator internals. A brief discussion of the ROSEMAX internals is provided in the next few paragraphs.

A commitment to enhance performance of the asphaltene and DAO separators for two ROSE licensees prompted KBR to consider significant design improvements to the previous Kerr-McGee internals design. KBR and Koch Engineering formed a team to identify design improvements and to quantify potential benefits. A significant amount of engineering analysis, pilot-plant testing, and computer flow modeling was done to support design changes that would significantly improve performance. A major advance resulting from these efforts was the development of our new proprietary ROSEMAX separator internals that are now available to all ROSE licensees.

New packing capacity correlations were developed based on laboratory and pilot-plant test work done by Koch and KBR for both liquid-liquid and supercritical service. These correlations can be used for both structured and random dumped packing. The correlations were verified for the conditions found in the ROSE separators, i.e., very high phase rates, low interfacial tension, and near-critical and supercritical conditions. These correlations provide improved understanding of how the packing crimp size, crimp angle, and surface treatment affect extraction capacity and efficiency and coalescing capacity and efficiency. A complete understanding of how to vary packing parameters to achieve desired performance is required for proper selection of packing size and arrangement.

The use of ROSEMAX internals allows the ROSE separators to operate at about twice the phase rates of conventional separators and provides about twice the mass-transfer efficiency of conventional extraction contacting devices.

Rose Operating Cost Savings

ROSE utility costs (steam, power, fuel, and cooling water) are typically 40 to 70 per cent of the costs associated with a conventional solvent deasphalting process. These savings are primarily a result of recovering over 90 per cent of the extraction solvent as a supercritical fluid. Other processes remove the solvent from the DAO by flashing at low pressure. The solvent is then compressed and condensed before being reused in the process.

These utility savings can play a significant role in minimizing total project costs associated with conversion of an existing solvent deasphalting unit to ROSE technology or for a grassroots installation.

Conventional versus Supercritical Solvent Recovery

Figure 4.48 illustrates the energy requirements to recover the solvent in the DAO for conventional solvent recovery processes. All the solvent exits the extractor as a solution of DAO and solvent at a relatively low-temperature and high pressure (point A). The stream is heated and flashed at some higher temperature and reduced pressure (point E). At this condition the majority of the solvent flashes from the solution and is condensed. The DAO and the remaining solvent are further heated and enter the product stripper (point F) at a greatly reduced pressure, where the remaining solvent is recovered. In this scheme, all the solvent is vaporized and condensed prior to being recycled to the extraction conditions. The energy requirements for this path are proportional to the quantity of solvent that follows each course.

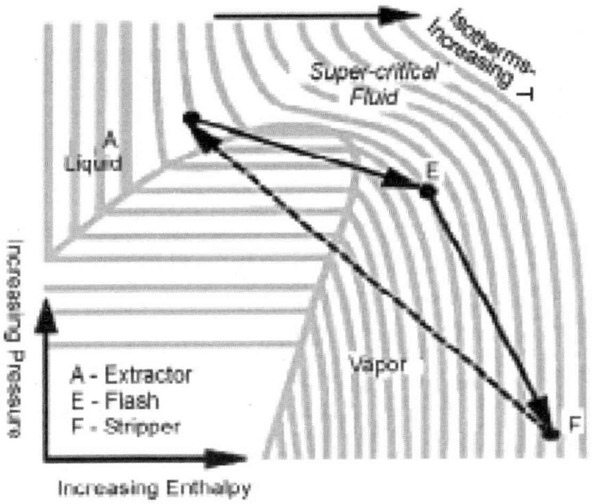

Figure 4.48: Conventional Solvent Recovery

Figure 4.49 illustrates the energy requirements to recover the solvent in the DAO for a supercritical solvent recovery processes. All the solvent exits the extractor as a solution of DAO and solvent at approximately the same relatively low temperature and high-pressure conditions as the conventional scheme (point A). The DAO solvent solution flows through the ROSE exchanger, gaining heat from the recycled supercritical solvent (point B). The solution is further heated by gaining heat from the stripped DAO product.

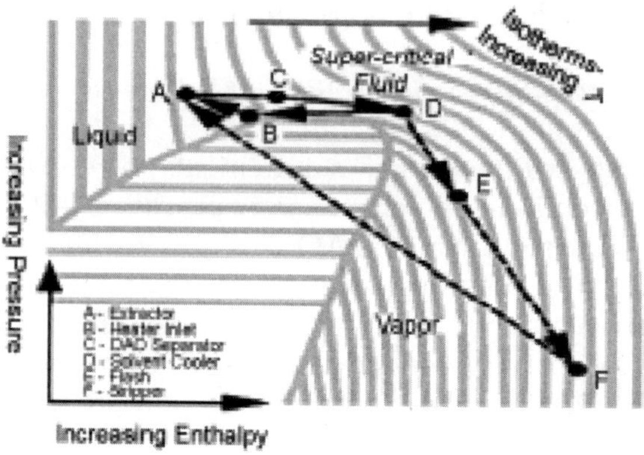

Figure 4.49: Supercritical Solvent Recovery

UOP SORBEX FAMILY OF TECHNOLOGIES

Introduction

The Sorbex name is applied to a technique, developed by UOP, that is used to separate a component or group of components from a mixture by selective adsorption on a solid adsorbent. The Sorbex technology is a continuous process in which feed and products enter and leave the adsorbent bed at substantially constant composition. This technology simulates the countercurrent flow of a liquid feed over a solid bed of adsorbent without physically moving the solid. The principles of Sorbex technology are the same regardless of the type of separation being conducted. The following are examples of commercially proven UOP technologies based on the Sorbex principle; each makes use of a specific adsorbent-desorbent combination uniquely tailored to the specific separation:

- *Parex*: separation of para-xylene from mixed C_8 aromatic isomers,
- MX Sorbex: meta-xylene from mixed C8 aromatic isomers,
- *Molex*: linear paraffins from branched and cyclic hydrocarbons,
- *Olex*: olefins from paraffins,

- *Cresex*: para-cresol or meta-cresol from other cresol isomers,
- *Cymex*: para-cymene or meta-cymene from other cymene isomers,
- *Sarex*: fructose from mixed sugars,

In addition to these applications, numerous other commercially interesting separations have been identified and demonstrated using the Sorbex process. These applications include monomethyl paraffins, 2,6-dimethyl naphthalene, ethylbenzene, 1-butene, ethyl toluenes, toluidines, terpenes, chloro and nitro aromatics, alpha and beta naphthols, alkyl naphthalenes, alpha olefins, and tall oil. Some of these separations have been commercialized under tolling agreements at a large-scale Sorbex plant that UOP has operated in Shreveport, Louisiana.

Principles of Adsorptive Separation

Adsorbents can be visualized as porous solids. When the adsorbent is immersed in a liquid mixture, the pores fill with liquid, but the equilibrium distribution of components inside the pore is different from the distribution in the surrounding bulk liquid. The component distributions inside and outside the pores can be related to one another by enrichment factors analogous to relative volatilities in distillation. The adsorbent is said to be selective for any components that are more concentrated inside the pores than in the surrounding bulk liquid.

Adsorption has long been used for the removal of contaminants present at low concentrations in process streams. In some instances, the objective is removal of specific compounds. In other cases, the objective is improvement of general properties, such as color, taste, odor, or storage stability. Common adsorbents are generally classified as polar or nonpolar. Polar, or hydrophilic, adsorbents include silica gel, activated alumina, molecular sieves, and various clays. Nonpolar adsorbents include activated carbons and other types of coal-derived carbons. Polar adsorbents are used when the components to be removed are more polar than the bulk process liquid; nonpolar adsorbents are used when the target components are less polar. Particularly useful are those adsorbents based on synthetic crystalline zeolites, which are generically referred to as molecular sieves. A wide variety of selectivities can be obtained in molecular sieves by varying the silica/alumina ratio, crystalline structure, and nature of the replaceable cations in the crystal lattice.

In one commercial separation, linear paraffins are separated from branched-chain and cyclic hydrocarbons by adsorption on 5A molecular sieves. The diameter of the pores is such that only the linear molecules may enter, and branched or cyclic molecules are completely excluded. In this case, the selectivity for linear hydrocarbons is infinite, and the adsorbent acts as a true molecular sieve. Adsorbents that completely exclude unwanted components are rare. In most applications, the pores are large enough to admit molecules of all the components present, and selectivity is the result of electronic interactions between the surface of the adsorbent pores and the individual components.

Adsorption is more efficient than conventional techniques such as liquid-liquid extraction or extractive distillation for many commercially important separations. Considerable development work has identified many adsorbents that are much more selective for specific components than any known solvents. In addition, adsorptive separation exhibits much higher mass-transfer efficiency than conventional extraction or extractive distillation. For example, laboratory chromatographs commonly achieve separation efficiencies equivalent to many thousands of theoretical equilibrium stages in columns of modest length. Such high mass-transfer efficiency stems from the use of small particles of adsorbent with high interfacial area and the absence of significant axial mixing.

In contrast, the trays of conventional liquid-liquid extractors and distillation columns are designed to obtain almost complete axial mixing in each physical stage. Thus, the number of theoretical equilibrium stages is essentially limited to the number of physical stages installed. In theory, this limitation can be partly overcome by the use of packed columns. However, if the packing is small enough to provide interfacial area comparable to that of an adsorbent, maintaining uniform countercurrent flow of the vapor and liquid phases becomes difficult. This flow limitation is less troublesome in an adsorptive system because only one fluid phase is involved.

The Sorbex Concept

In spite of the potential advantages of adsorptive separation, it did not achieve wide commercial acceptance until the introduction of the UOP Sorbex process in the early 1960s. Prior to the Sorbex process, adsorptive processes were designed much as laboratory chromatographs. Feed was introduced in pulses, and the composition of products varied with time. Integrating such an intermittent process with continuous processes operating both upstream and downstream was difficult. The Sorbex process, for the first time, offered a truly continuous adsorptive separation process that produced products with essentially constant compositions.

The easiest way to understand the Sorbex process is to think of it as a countercurrent flow of liquid feed and solid adsorbent (Figure). For simplicity, assume the feed is a binary mixture of components A and B, and the adsorbent has a selective attraction for component A. In practice, the feed to a Sorbex unit may contain a multitude of components from which one or more components would be selectively recovered.

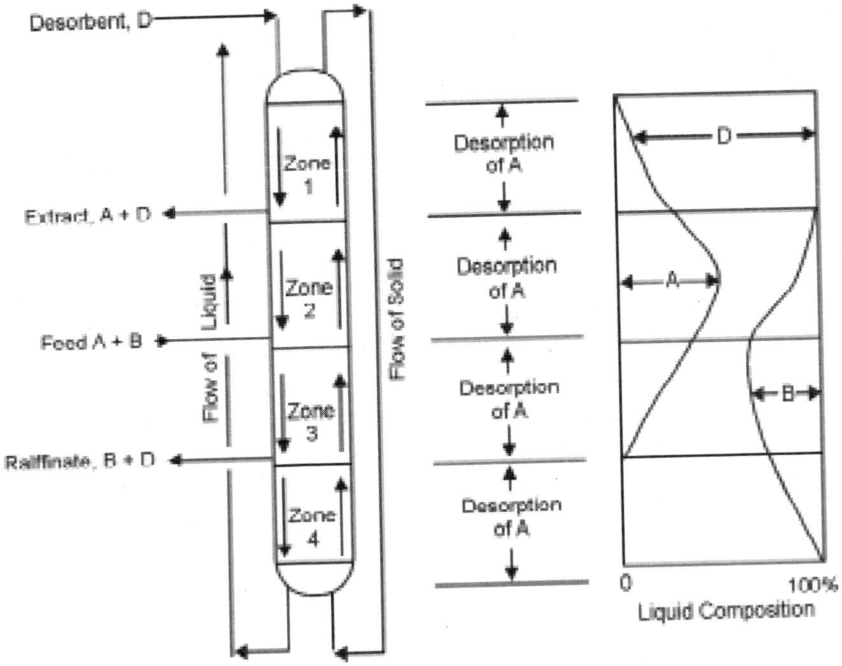

Figure 4.50: Moving-bed Analogy

The positions of injection and withdrawal of the four net streams divide the adsorbent bed in four zones:

- *Zone 1*: adsorption of component A. This zone is between the point of feed injection and raffinate withdrawal. As the feed flows down through zone 1, countercurrent to the solid adsorbent flowing upward, component A is selectively adsorbed from the feed into the pores of the adsorbent. At the same time, the desorbent (component D) is desorbed from the pores of the adsorbent to the liquid stream to make room for A in the pores.
- *Zone 2*: desorption of component B. This zone is between the point of feed injection and extract withdrawal. At the fresh-feed point, the upward-flowing solid adsorbent contains the quantity of component A that was adsorbed in zone 1. However, the pores will also contain a large amount of component B, because the adsorbent has just been in contact with fresh feed. The liquid

entering the top of zone 2 contains no B, only A and D.

UOP/FW USA SOLVENT DEASPHALTING PROCESS

Introduction

The UOP/FWUSA Solvent Deasphalting (UOP/FWUSA SDA) process is a solvent extraction process developed and jointly offered by UOP and Foster Wheeler USA Corporation (FW) for the processing of vacuum residues (VR) or atmosphere residues (AR) feedstock. The UOP/FWUSA process contains process features from both UOP's Demex solvent extraction process and FW's LEDA solvent deasphalting process. This combination of features has resulted in an advanced solvent deasphalting technology (UOP/FWUSA Solvent Deasphalting process) that is capable of achieving the highest product qualities with the lowest operating costs. The UOP/FWUSA SDA process employs a unique combination of features to separate VR into components whose uses range from incremental feedstock for downstream conversion units to the production of lube base stock and asphalts. Because the UOP/FWUSA process provides the refiner with increased flexibility regarding future processing decisions, including crude section, refinery debottlenecking, and the potential to reduce crude runs and fuel oil yields, it represents an important element in the refiner's overall bottom-of-the-barrel processing strategy.

Process Description

The UOP/FWUSA SDA process typically divides VR into two components: a relatively contaminant-free nondistillable deasphalted oil (DAO) and a highly viscous pitch. Like propane deasphalting, the UOP/FWUSA SDA process is based on the ability of light paraffin hydrocarbons to separate the residue's heavier asphaltenic components. Associated with these heavier materials is the majority of the crude's contaminants. Consequently, the lower contaminant content of the recovered DAO allows this material to be used in many refining applications, probably the most important of which is as increment feedstock to catalytic processes such as fluid catalytic cracking (FCC) or hydrocracking for conversion into transportation fuel products.

Because the pitch recorded from the UOP/FWUSA SDA unit contains most of the contaminants present in the crude, it typically has a high viscosity and a relatively low penetration value. Commercially, UOP/FWUSA SDA pitch has been used in the manufacturing of asphalts and cement and as a blending component in refinery fuel oil pools. Other potential uses include the production of hydrogen, synthesis gas, or low-Btu fuel gas and as a solid-fuel blending component.

Unlike conventional propane deasphalting, the UOP/FWUSA SDA process uses a unique combination of heavier solvents, supercritical solvent techniques, and patented extractor internals to efficiently recover high-quality DAO at high yield. A schematic flow scheme of a modern UOP/FWUSA SDA design is shown in Figure 4.51. This design, which has evolved from experience gained from both pilot-plant and commercial operations as well as detailed engineering analyses of its various components, minimizes operating and capital costs and efficiently recovers the desired product yields at the required product qualities.

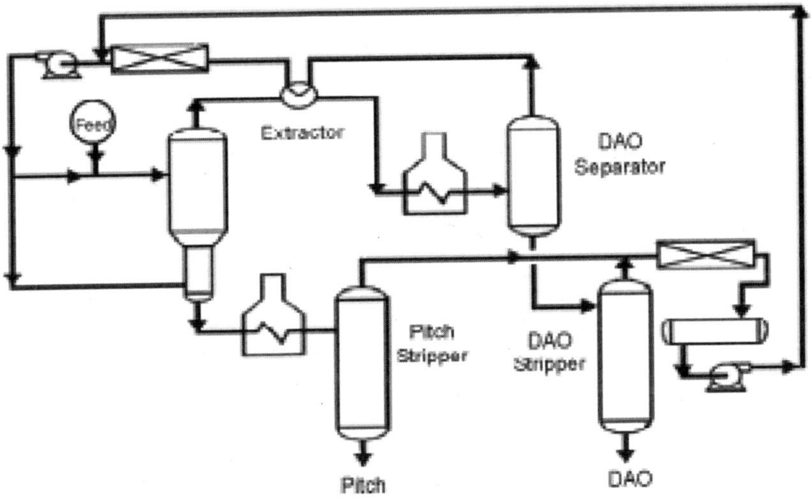

Figure 4.51: Schematic Flow Diagram of UOP/FW USA SDA Process

Incoming VR is mixed with solvent and fed to the vertical extractor vessel. At the appropriate extractor conditions, the VR-solvent blend is separated into its DAO and pitch components. The yield and quality of these components are dependent on the amount of contaminants in the feedstock, the composition and quantity of solvent used, and the operating conditions of the extractor.

With the extractor, the downflowing asphaltene-rich pitch component and the upflowing DAO solvent mixture are separated by patented extractor internals. The extractor design also includes a unique liquid flow distribution system to minimize the possibility of fouling the internals. Compared to previous designs, the increased separation efficiency achieved by these two features significantly reduces the size of the extractor vessel and the overall cost of the UOP/FWUSA SDA unit.

The combination of heat exchange with recovered solvent and a direct-fired heater or a hot oil heating fluid heats the DAO solvent mixture leaving the top of the extractor to its critical temperature. The separation of the DAO and solvent components of this mixture is accomplished at supercritical conditions within the DAO separator. Recovered solvent is recycled back to the extractor. Because most of the solvent is recovered supercritically, this material can be effectively used for process heat exchange. Consequently, compared to earlier subcritical solvent-recovery designs, supercritical solvent recovery can reduce utilities requirements by more than one-third.

To minimize solvent loss, any traces of solvent remaining in both the DAO exiting the DAO separator and the pitch from the extractor are recovered in the DAO and pitch strippers, respectively. This recovered solvent is also recycled to the extractor. If the recovery of an intermediate-quality resin stream is desired—for instance, when specialty asphalts are produced or when independent control of DAO and pitch quality is desired—a resin settler may be added between the unit's extractor and DAO separator.

Typical Feedstocks

The SDA process (normally using propane or a propane-butane mixture as the solvent) has been in commercial use for the preparation of lubricant-bright-stock feeds from asphalt-bearing crude residue for many years. Many commercial SDA units have also been used for preparing paving and specialty asphalts from suitable vacuum residues.

The increasing use of the fluid catalytic cracking (FCC) process together with the increasing price of crude oil resulted in the need to maximize the quantity of FCC feed-stock obtained from each barrel of crude. These conditions led to the extension of the SDA process to the preparation of cracking feedstocks from vacuum residues. The current trend for maximizing distillate oil production has also prompted the increased use of the SDA process to prepare hydrocracking feedstocks from vacuum residues.

SDA supplements vacuum distillation by recovering additional high-quality paraffinic oil from vacuum residues beyond the range of practical distillation. Although atmospheric residues have been commercially solvent-deasphalted, typical SDA feedstocks are 570°C + (1060°F$^+$) TBP cut-point vacuum residues. These vacuum residues often contain high levels of metals (primarily nickel and vanadium), carbon residue, nitrogen, sulfur, and asphaltenes. Table gives three examples of vacuum residue feedstocks, covering a wide range of properties, that can be processed in an SDA unit.

Table 4.15: Typical SDA Feedstocks

	Vacuum residue TBP cut point, °C	*Gravity, °API*	*Conradson carbon residue, wt %*	*Sulfur, wt %*	*Ni + V, wt ppm*
Heavy Arabian	570	3.6	25.1	5.5	193
Heavy Canadian	570	8.1	17.4	2.7	110
Canadian	570	11.7			

UOP ISOSIV PROCESS

Introduction

Light straight-run (LSR) naphtha fractions made in the refinery are predominantly C_5's through C_7's, with traces of C_8's. They are highly paraffinic and contain moderate amounts of naphthenes, low aromatics, and no olefins. The average clear research octane number (RONC) is usually in the 60s.

The paraffinicity of light naphtha is what makes it a desirable petrochemical cracking stock. The aromatic rings are too thermally stable for cracking, and the naphthenes produce more liquid products. The straight-chain normal paraffins produce more ethylene and less pyrolysis gasoline than the branched-chain paraffins.

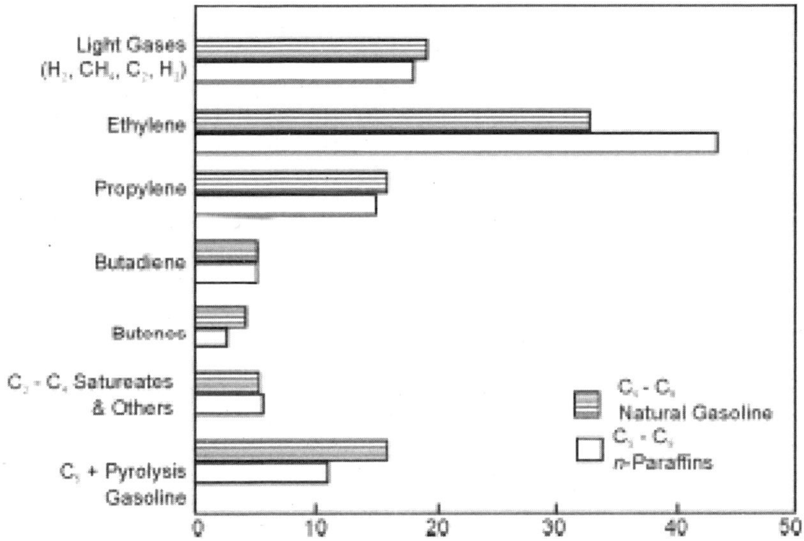

Figure 4.52: Pyrolysis Yield Data

Figure compares pyrolysis unit yields from a normal paraffin feed with yields from a mixed natural gasoline feed. The yields are based on a single-pass pyrolysis operation at equivalent high furnace severities for both feeds. The normal paraffin feed was extracted from a C_5 through C_9 natural gasoline stream. The natural gasoline feed contained 54.4 per cent straight-chain paraffins and 45.6 per cent branched and cyclic hydrocarbons. The ethylene yield is about 30 per cent higher for the all-normals fractions. Propylene, butene, and light-gas yields decrease slightly. The pyrolysis gasoline yield is considerably reduced.

As the endpoint of naphtha is decreased, the paraffinicity of the stream increases; as a result, ethylene production increases and the production of pyrolysis gasoline and fuel oil decreases. The LSR naphtha—especially the 70°C (C_5–160°F) portion, which is about 95 per cent paraffinic—is therefore a prime substitute for natural gas liquids as an ethylene plant feed. The nonnormal components of the LSR naphtha fraction have higher octanes than the normal paraffins (Table) and are excellent gasoline blending components.

Table 4.16: Properties of Common Gasoline Components

Component	*Molecular weight*	*Boiling point, °C (°F)*	*Density, kg/m3(lb/gal)*	*RONC*
Isobutane	58.1	–11.7 (10.9)	562 (4.69)	100+
n-butane	58.1	–0.5 (31.1)	582 (4.86)	93.6
Neopentane	72.1	9.4 (49.0)	596 (4.97)	116
Isopentane	72.1	27.9 (82.2)	623 (5.20)	92.3
n-pentane	72.1	36.1 (96.9)	629 (5.25)	61.7
Cyclopentane	70.0	49.3 (120.7)	749 (6.25)	100
2,2-dimethlybutane	86.2	49.7 (121.5)	664 (5.54)	91.8
2,3-dimethylbutane	86.2	58.0 (136.4)	664 (5.54)	101.7
2-methylpentane	86.2	60.3 (140.5)	667 (5.57)	73.4
3-methylpentane	86.2	66.3 (145.9)	652 (5.44)	74.5
n-hexane	86.2	68.7 (155.7)	657 (5.48)	24.8
Methylcyclopentane	84.2	71.8 (161.3)	753 (6.28)	91.3
2,2-dimethylpentane	100.2	79.2 (174.6)	676 (5.64)	92.8
Benzene	78.1	80.1 (176.2)	882 (7.36)	100+
2,4-dimethylpentane	100.2	80.5 (176.9)	676 (5.64)	83.1
Cyclohexane	84.2	80.7 (177.3)	782 (6.53)	83
2,2,3-trimethylbutane	100.2	80.9 (177.6)	693 (5.78)	112
3,3-dimethylpentane	100.2	86.1 (186.9)	696 (5.81)	98
2,3-dimethylpentane	100.2	89.8 (193.6)	699 (5.83)	88.5
2,4-dimethylpentane	100.2	90.1 (194.1)	681 (5.68)	55
3-methylhexane	100.2	91.9 (197.5)	690 (5.76)	65
Toluene	92.1	110.6 (231.1)	870 (7.26)	100+

UOP's IsoSiv process uses molecular sieves to physically remove normal paraffins from the LSR feedstock. In the past, gasoline-range IsoSiv units were primarily used to produce specialty chemicals. The normal paraffin product having a 95 to 98 per cent purity was cut into single-carbon-number fractions for special solvents. The normal-paraffin–free fraction was usually sent to the gasoline pool as an octane booster. The more recent IsoSiv units were built to produce high-octane gasoline components; the normal paraffin by-product was sold as petrochemical feedstock or sent to an isomerization reactor.

General Process Description

The LSR naphtha fractions usually contain 40 to 50 per cent normal paraffins. The IsoSiv process (Figure) separates the normal paraffins from a hydrocarbon mixture by selective adsorption on a molecular sieve material. This material is a crystalline zeolite having uniform pore dimensions of the same order of magnitude as the size of individual hydrocarbon molecules. The molecular sieve used for normal paraffin separation has pore openings in the crystalline structure that are sized to allow molecules of normal paraffin to pass through the pore openings into the internal crystal cavity, where they are retained. Nonnormal hydrocarbons, such as isoparaffins, naphthenes, and aromatics, have larger molecular diameters and are, therefore, excluded from entering the crystal cavity through the pore opening.

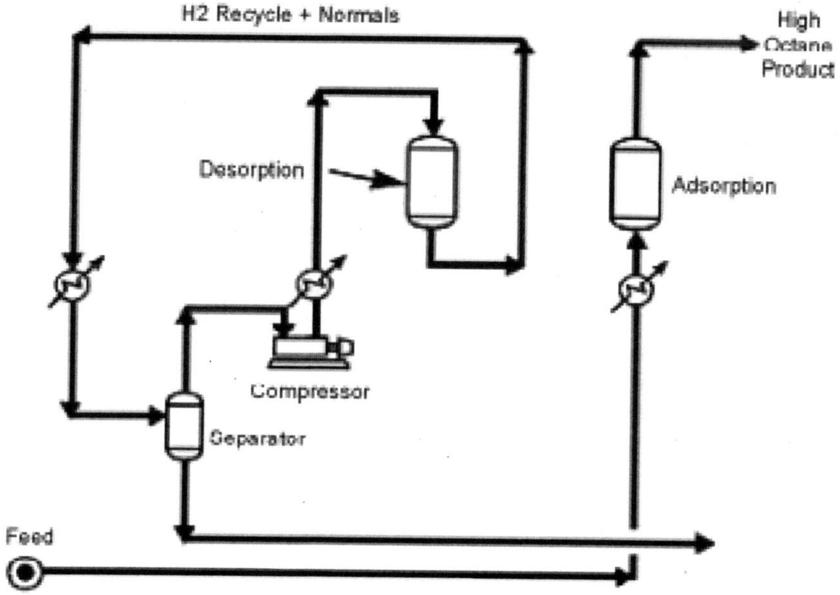

Figure 4.53: UOP IsoSiv Process

The heart of the IsoSiv process is the adsorber section, which consists of vessels filled with molecular sieve adsorbent. The LSR feedstock is fed into one end of an adsorber vessel. The normal paraffins in the feedstock remain in the vessel by being adsorbed into the molecular sieve, and the remainder of the feedstock passes out the other end of the vessel as a nonnormal product. In a subsequent process step, the normal paraffins are recovered from the adsorber vessel as a separate product by use of a purge material. All process hardware in an IsoSiv unit is conventional refinery equipment, such as pumps, furnaces, heat exchangers, and compressors, that is designed to deliver the feedstock and the purge material to the adsorber section.

Typical performance results in an isomer product that is 98 to 99 per cent free of normal paraffins and a normal paraffin product of 95 to 98 per cent purity. The high-octane isomer product can have a RON approximately 15 numbers higher than the feed, depending on feed composition. The IsoSiv-grade molecular sieve adsorbent is fully regenerable and has an expected life of 10 to 15 years.

Kerosene Isosiv Process for Production of Normal Paraffins

The straight-chain normal paraffins in the kerosene range (C_{10} to C_{18}) have their principal uses in detergent manufacture, chlorinated fire retardants, plasticizers, alcohols, fatty acids, and synthetic proteins. The separation of these straight-chain normal paraffins from other classes of hydrocarbons, such as branched-chain isoparaffins, naphthenes, and aromatics, was a virtual impossibility prior to the advent of the synthetic zeolites known as molecular sieves.

Table 4.17: Kerosene ISOSIV Commercial Applications

Unit	*Feed type*	*Start-up*	*Location*	*Normal paraffin capacity, BPSD*
1.	Kerosene	1964	United States	2300
2.	Kerosene	1971	West Germany	650
3.	Kerosene-gas oil	1972	Italy	2600
4.	Kerosene	1973	Italy	2600
5.	Kerosene-gas oil	1974	Italy	5800
6.	Kerosene-gas oil	1976	Italy	5800
7.	Kerosene	1983	Brazil	2600
8.	Kerosene	1992	People's Republic of China	950

These uniform, molecular-pore-sized adsorbents, developed by Union Carbide in the early 1950s, opened the way for refiners and petrochemical producers to add adsorption as a means of separating hydrocarbon classes to those already known, such as distillation and liquid-liquid extraction. To date, eight kerosene IsoSiv systems have been started up (Table). The IsoSiv process is licensed by UOP* subsequent to the joint venture ownership of UOP by Union Carbide and Allied Signal in 1988. Currently as a result of the merger of UCC with Dow, and Allied Signal with Honeywell, UOP is now owned jointly by Dow and Honeywell.

General Process Description

The IsoSiv process separates normal paraffins from a hydrocarbon mixture, such as kerosene or gas oil, by selective adsorption on a molecular-sieve adsorbent material. This material is a crystalline zeolite having uniform pore dimensions of the same order of magnitude as the size of individual hydrocarbon molecules. The molecular sieve used for normal paraffin separation has openings in the crystalline structure that are sized to allow normal paraffin molecules to pass through the pore openings into the internal crystal cavity, where they are retained. Nonnormal hydrocarbons, such as isoparaffins, naphthenes, and aromatics, have larger molecular diameters and are therefore excluded from entering the crystal cavity through the pore opening.

The heart of the IsoSiv process is the adsorber section, consisting of vessels filled with molecular-sieve adsorbent. The kerosene or gas oil feedstock is fed into one end of an adsorber vessel, the normal paraffins in the feedstock remain in the vessel by being adsorbed in the molecular sieve, and the remainder of the feedstock passes out the other end of the vessel as a denormalized kerosene gas oil. The normal paraffins are recovered from the adsorber vessel as a separate product by using a purged material. All process hardware in an IsoSiv unit is conventional refinery equipment, such as pumps, furnaces, heat exchangers, and compressors, designed to deliver the feedstock and the purge material to the adsorber section and to remove the products from the adsorber section. The kerosene IsoSiv process typically recovers 95 wt per cent of the normal paraffins in the feedstock and produces a normal paraffin product of 98.5 wt per cent purity.

Process Perspective

During the early 1960s, the appeal of molecular-sieve adsorption led to widespread efforts at innovating new adsorption technology. Many of these efforts were successful, in that they resulted in molecular-sieve processes capable of separating long-chain normal paraffins from kerosene-range feedstocks at just the time when the detergent industry decided to switch to linear alkylbenzene sulfonates as a basis for its formulations of "soft" detergents. The consequent demand for long-chain normal paraffins led to a worldwide wave of construction: at least 12 adsorption plants were built to process kerosene-range feedstocks and use processes developed by Union Carbide, UOP, Esso, British Petroleum, Shell, and Texaco. Among the first units was the South Hampton Company's naphtha IsoSiv unit, which was converted to the kerosene range in 1961. In 1964 Union Carbide Corporation installed at its Texas City, Texas, petrochemical complex an IsoSiv unit producing 100,000 metric tons/year (MTA) (220,000 lb/yr) of normal paraffins from kerosene. This unit was to remain the world's largest normal paraffin-producing plant for almost 10 years.

At the beginning of the 1970s, a further extension of adsorption technology was required. The normal paraffins used as substrates for protein production extend into the gas oil feedstock range. Suitable modifications can and have been made to existing adsorption technology to allow successful application to the new requirements. In 1972, Liquichimica S.p.A., now Condea Augusta S.p.A. but then a subsidiary of the Liquigas Group of Italy, installed and started up in Augusta, Sicily, a modified IsoSiv unit to produce 110,000 MTA (242,000 lb/yr) of normal paraffins from both kerosene and gas oil feedstocks. Plant expansions put on-stream in 1973 brought normal paraffin production capacity at Augusta up to approximately 250,000 MTA (551,000 lb/yr), making it by a wide margin the largest single normal paraffin-producing installation in the world. A second unit that came on-stream in December 1974 almost doubled previous capacity. A third IsoSiv unit of more than 200,000 MTA (440,000 lb/yr) came on-stream in 1976. Total installed capacity is more than 650,000 MTA (4,862,000 lb/yr) of normal paraffin production. These units have used feedstocks ranging from kerosene to gas oil and intermediate mixtures of both.

A seventh kerosene IsoSiv unit came on-line in Brazil in 1983. An eighth came on-line in China in 1992.

Detailed Process Description

The kerosene IsoSiv process employs the highly selective adsorption capability of molecular sieves. The simplified process flow scheme is shown in Figure. The basic cycle consists of three steps: adsorption, copurge, and desorption. This section describes each in detail.

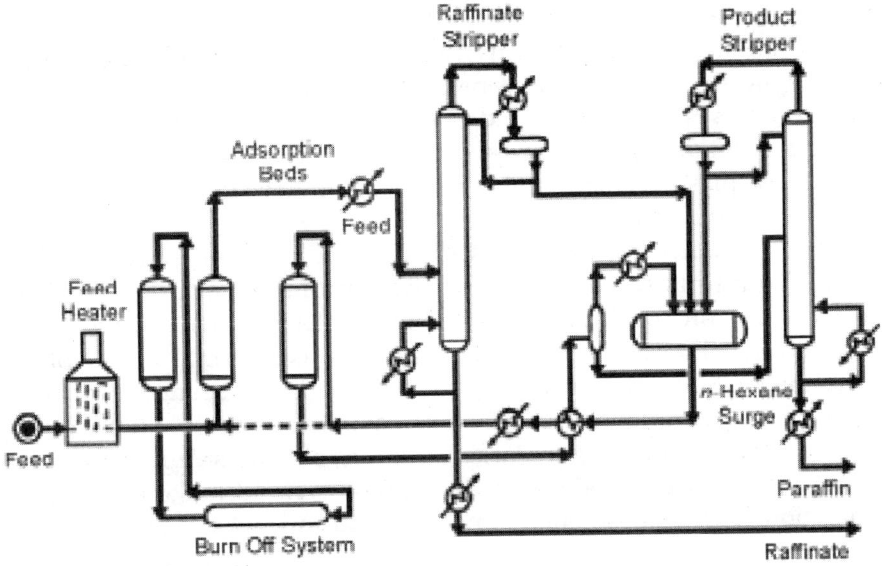

Figure 4.54: Kerosene IsoSiv Process

Adsorption Step

Hydrocarbon feed at elevated temperature and slightly above atmospheric pressure is passed upward through an adsorber vessel, where the normal paraffins are selectively adsorbed in the bed. In processing gas oil feedstock, hexane is added to the gas oil feed to dilute it and prevent capillary condensation from occurring on the adsorbent bed. As the normal paraffins are adsorbed, the liberated heat of adsorption creates a temperature front that travels through the bed. This front closely coincides with the mass-transfer front and gives an indication of when the adsorption step should be terminated to prevent the normal paraffins from breaking through the effluent end of the bed.

UOP Molex Process for Production of Normal Paraffins

The separation of normal paraffins from isoparaffins is done commercially for a number of reasons. In the lighter hydrocarbon range, isoparaffins are often more desirable because of their higher octane values and their superior gasoline alkylation characteristics. In the heavier range, normal paraffins are typically the desired product because of the benefits derived from their linearity in the production of plasticizers, linear alkylbenzene sul-fonates, detergent alcohols, and ethoxylates.

We will discusses the specific application of the UOP Molex process to the separation of normal paraffins from isoparaffins. Although not limited in its application to a particular processing mode or carbon number, the Molex process is most often used in the recovery of normal paraffins for plasticizer and detergent-range applications. Typical carbon numbers are C_6 to C_{10} for plasticizers, C_{10} to C_{14} for linear alkylbenzenes, and C_{10} to C_{18} + for detergent alcohols.

The UOP Molex process is an established, commercially proven method for the liquid-phase adsorptive separation of normal paraffins from isoparaffins and cycloparaffins using the UOP Sorbex separations technology, which uses zeolitic adsorbents. Isothermal liquid-phase operation facilitates the processing of heavy and broad-range feedstocks. Vapor-phase operations, in addition to having considerable heating and cooling requirements, require large variations of temperature or pressure or both through the adsorption-desorption cycle to make an effective separation. Vapor-phase operations also tend to leave a certain residual level of coke on the adsorbent, which must then be regenerated on a cyclic basis. Operation in the liquid phase allows for uninterrupted continuous operation over many years without regenerations.

Figure illustrates the general design characteristics of such units.

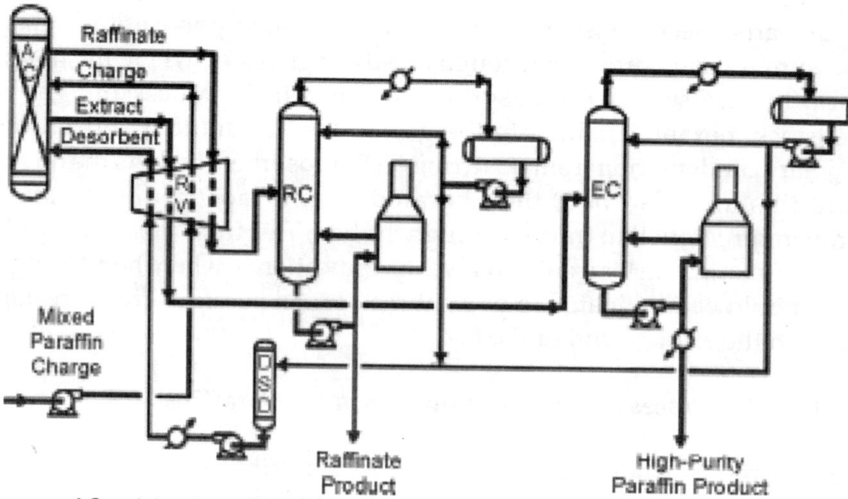

AC = Adsorbent Chamber
DSS = Desorbent Surge Drum
EC = Extract Column
RC = Raffinate Column
RV = Rotary Valve

Figure 4.55: UOP Molex Process Design Characteristics

Yield Structure

Typically, a UOP Molex process unit produces normal paraffins at about 99 wt per cent purity and at about 98 wt per cent recovery, depending on the amount of adsorbent used relative to the volume of feed.

Economics

To a certain extent, the economics of the UOP Molex unit is dependent on the feed quality, because some prefractionation and hydrotreating may be required to control the level of contaminants that might otherwise affect unit performance or adsorbent life. If the feed is assumed to have been properly treated, the estimated erected cost of a UOP Molex unit, feeding 383,000 metric tons per year (MTA) (844 million lb/year) of a paraffinic kerosene in the C_{10} to C_{15} range with about 34 per cent normal paraffins, was about $25 million in 1995. This unit was designed for the recovery of 96,000 MTA (211 million lb/year) of normal paraffins at 99 per cent purity. This cost represents the fully erected cost within battery limits for a particular UOP Molex unit.

The utility requirements for such a unit per metric ton of product are as follows:

Electric power, kWh	54.3
Hot-oil heat, 10^3J/h (Btu/h)	120 (114)
Cooling water circulated [15°C (27°F]) rise], m^3 (gal)	5.1 (180)

Commercial Experience

As of early 2002, a total of 26 UOP Molex process units had been commissioned. Another three were in various stages of design or construction. Product capacities ranged from 2500 MTA (5.5 million lb/year) to 155,000 MTA (340 million lb/year).

UOP OLEX PROCESS FOR OLEFIN RECOVERY

Discussion

The recovery of olefins from mixtures of olefins and paraffins is desirable in a number of areas: recovery of propylene from propane-propylene streams, recovery of C_4 olefins, and especially recovery of heavier olefins for the manufacture of oxoalcohols for plasticizer and detergent applications.

The UOP Olex process, a method of separating olefins from paraffins, is another of the many applications of the UOP Sorbex separations technologies. This technique involves the selective adsorption of a desired component from a liquid-phase mixture by continuous contacting with a fixed bed of adsorbent. Other commercial methods of separation include extraction and extractive distillation. These methods, however, are less efficient in terms of product recovery and purity, much more energy-intensive, limited in application to lighter molecular weights, and limited as to the range of carbon numbers and of the molecular weights over which they apply.

In contrast, adsorbents have been developed that demonstrate the desirable characteristic of preferential relative adsorptivity for olefins as compared with paraffins. This characteristic permits ready separation of olefins and paraffins even with feedstocks that have a wide boiling range.

Much higher mass-transfer efficiency can be achieved in adsorptive operations than with the conventional equipment used for extraction and extractive distillation. As an example, laboratory-scale chromatographic columns commonly show separative efficiencies equivalent to many thousands of theoretical trays in columns of modest length. This high efficiency results from the use of small particles to give high interfacial areas and form the absence of significant axial mixing of either phase.

In contrast, trayed fractionating columns and liquid-liquid extractors are designed to provide practically complete axial mixing in each physical element to create interfacial areas for mass transfer. The number of theoretical equilibrium stages is thus limited substantially to the number of physical mixing stages installed. This limitation could be avoided, in theory, by the use of packaged columns. However, if the packing-particle size is small enough to provide interfacial areas comparable with those obtained in adsorptive beds, the capacity to accommodate the counterflow of two fluid phases becomes low. Thus, great difficulty is encountered in obtaining uniform unchanneled flow of both fluid phases. In an adsorptive bed, these limitations are much less severe because only one fluid phase is involved.

In the past, processes employing solid adsorbents for treatment of liquids have not gained wide acceptance except where the quantity of material to be removed was small and frequent regeneration of the adsorbent was therefore not required. One reason for this slow acceptance was the absence of a design that would permit continuous operation. In the usual fixed-bed adsorptive process, the feed stream is discontinuous, and the product streams vary in composition. Thus, integrating the operation of any intermittent process with continuous processes operating upstream and downstream from it is difficult.

The unique process configuration used in Sorbex units eliminates these problems and facilitates continuous adsorptive separation. The Sorbex flow scheme simulates the continuous countercurrent flow of adsorbent and liquid without actual movement of the adsorbent. This system design makes adsorptive separation a continuous process and eliminates the inherent problems of moving-bed operation.

Essentially, the UOP Olex process is based on the selective adsorptive separation of olefins from paraffins in a liquid-phase operation. The adsorbed olefins are recovered from the adsorbent by displacement with a desorbent liquid of a different boiling point.The flow diagram for the Olex process is shown in Figure.

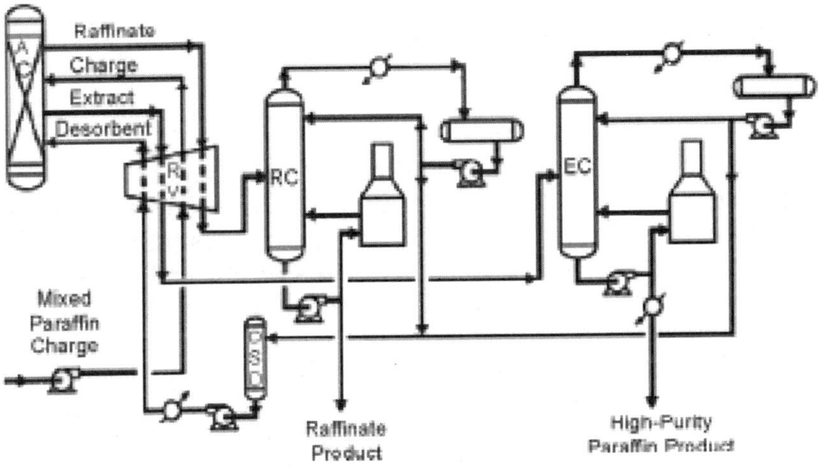

AC = Adsorbent Chamber
DSS = Desorbent Surge Drum
EC = Extract Column
RC = Raffinate Column
RV = Rotary Valve

Figure 4.56: UOP Olex Process

Commercial Experience

Six UOP Olex process units have been commissioned since the first one came on stream in 1972. Five commercial units process heavy (C_{10-13} up to C_{15-18}) olefin feeds and one unit processes a light (C_4) olefin feed. The heavy-feed olefin content ranges from 10 to 13 wt per cent, and the light feed is approximately 80 wt per cent olefins.

Economics

The estimated erected cost of a UOP Olex unit for the production of 52,000 metric tons/year (115 million lb/yr) of olefins in the C_{11} to C_{14} carbon range from a feed stream containing 10 wt per cent olefins was about $25 million. This amount was the fully erected cost within battery limits of a unit.

The utility requirements for such a unit per metric ton of product would be approximately as follows:

Electric power, kWh	110
Hot-oil heat, 10^3J/h (Btu/h)	266 (253)
Cooling water circulated [15°C (27°F]) rise], m^3 (gal)	10.5 (371)

KBR Refinery Sulfur Management

Introduction

Raw crude oil contains sulfur and nitrogen. During processing, the sulfur and nitrogen are converted principally to H2S and NH3 and, to a lesser degree, organic sulfur (COS and CS_2) and mercaptans (RSH).

More stringent environmental standards on the emissions of sulfur and nitrogen compounds, together with the low sulfur specifications for petroleum products, have resulted in making sulfur management critical within today's refinery. The importance of sulfur management cannot be overstressed. Today's refineries are processing crudes with higher sulfur contents and are doing more bottom-of-the-barrel conversion. The need for new or revamped sulfur management facilities is expected to grow as demands for cleaner fuels and environment increase and crude oil slates change.

As illustrated in Figure, sulfur management within a refinery consists of four basic processes. Amine treating units (ATUs) remove H_2S from recycle gas streams in hydroprocessing operations and from fuel gas/liquefied petroleum gas (LPG) recovery units. The amine is regenerated in one or more amine regeneration units (ARUs). Sour water strippers (SWSs) remove H_2S and NH_3 from the sour water streams. Sour water is the result of refinery operations using steam in distillation or steam as a means to reduce hydrocarbon partial pressure or where water injection is used to combat potential corrosion or salt buildup. The sulfur in the acid gas from the ARU and the SWS is removed first by a Claus sulfur recovery unit (SRU) that achieves 92 to 96 per cent of the overall sulfur recovery and then by a tail gas cleanup unit (TGCU) that can boost overall sulfur recovery to 99.9 per cent. Most refineries now degas the molten sulfur produced. The amine, SWS, SRU, and TGCU processes are discussed in the following chapters.

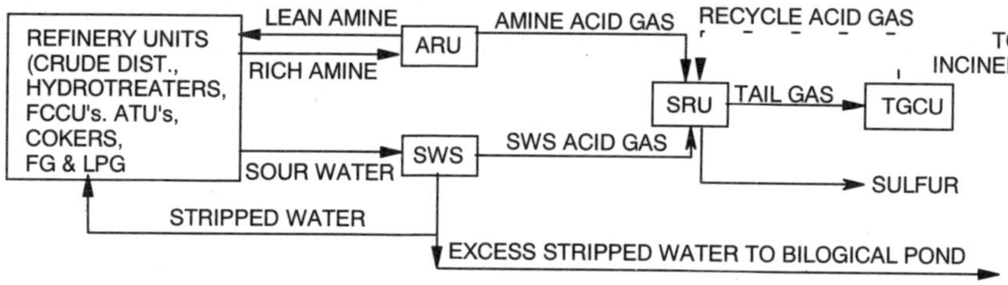

Figure 4.57: Sulfur Removal/Recovery

Amine plant

Introduction

The sulfur in crude oil that is converted to H_2S during processing typically is removed by a suitable amine. Two broad classifications of refinery amine treating applications are recycle gas treating and fuel gas/LPG recovery. In recycle gas treating, shown schematically in Figure, the product oil from a hydroprocessing unit has an upper specification limit on its sulfur content. The sulfur in the feed oil reacts with H_2 at elevated pressure (typically 35 to 150 bar gage) to form H_2S. The reactor product stream is flashed, and a recycle gas stream, containing H_2, H_2S, and some hydrocarbons, is sent to an amine absorber, where the H_2S is removed by the circulating amine stream.

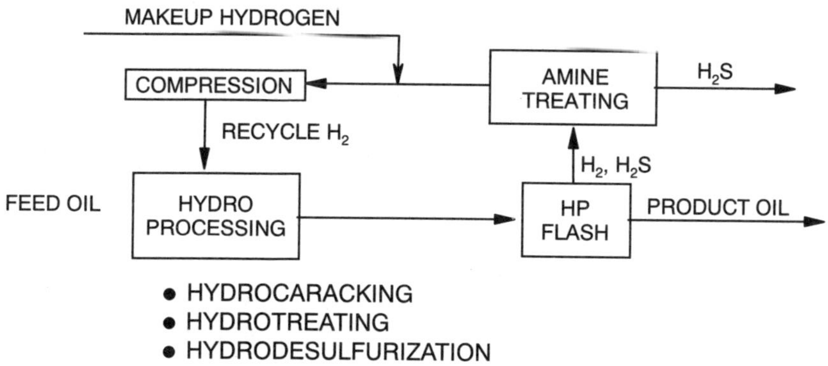

Figure 4.58: Recycle Gas Amine Treating

In fuel gas and LPG recovery units, the off-gases and stabilizer overheads from cracking, coking, and reforming units are sent to gas recovery units. The sour fuel gas has H2S removed at low pressure (3.5 to 14 bar gage typically) by the circulating amine. The LPG stream has the bulk H_2S removed by amine at 14 to 21 bar gage, then the remaining H_2S plus mercaptans is treated by a caustic solution and a proprietary solvent wash that converts the mercaptans to mercapticides. These washes typically achieve a Copper Strip 1A specification. Figure shows a typical block flow design illustrating the processing steps.

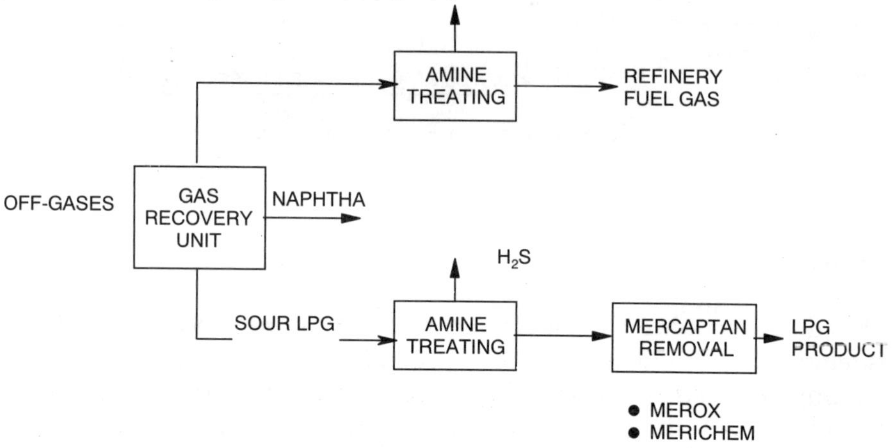

Figure 4.59: Fuel Gas/Lpg Amine Treating

Process Description

Many refineries have multiple amine absorbers served by a common amine regeneration unit. Other refineries have two separate amine regeneration systems, with one system typically dedicated to "clean" users (such as hydrotreaters) and the other dedicated to "dirty" users (such as FCC units or cokers). A dual amine regeneration system is illustrated in Figure show the rich amine streams from the absorbers being combined and sent to the rich amine flash drum to flash off light hydrocarbons and to separate entrained hydrocarbons from the amine. This is necessary to minimize hydrocarbon carryover to the Claus SRUs. The flashed gas is treated with a slipstream of lean amine in the flash gas scrubber prior to routing to the fuel gas system. The flash drum is most often operated at 50 to 75 lb/in^2 gage so that the flashed amine can be delivered to the top of the regenerator without a pump. Steam from the reboilers using 50-lb/in^2 gage saturated steam strips the acid gas (H_2S and CO_2) from the amine. The overhead is cooled to 38 to 49°C to minimize water carryover to the SRU. Provision is made at the reflux accumulator and at the bottom of the regenerator to skim off light and heavy hydrocarbons respectively. The regenerated amine is filtered and cooled, then distributed to the various absorbers.

BELCO EDV WET SCRUBBING SYSTEM: BEST AVAILABLE CONTROL TECHNOLOGY (BACT) FOR FCCU EMISSION CONTROL

The FCCU—A Unique Process for Emissions Control

The control of particulate and SO_2 emissions with wet scrubbing systems is not uncommon. However, the control of these emissions, combined with the special needs and requirements of the fluid catalytic cracking unit (FCCU) process, indeed makes this a special process for wet scrubbing systems.

Uncontrolled particulate (catalyst) emissions from this source vary depending on the number of stages of internal and external cyclones. Although cyclones are effective in collecting the greater constituent of catalyst recirculated in the FCCU regenerator, the attrition of catalyst causes a significant amount of finer catalyst to escape the cyclone system with relative ease. Typically, emissions will range from 3.0 to 8.0 lb per 1000 lb of coke burn-off.

Sulfur emissions in the form of SO_x (SO_2 and SO_3) from the regenerator vary significantly depending on the feed sulfur content and the FCCU design. In the FCCU reactor, 70 to 95 per cent of the incoming feed sulfur is transferred to the acid gas and product side in the form of H_2S. The remaining 5 to 30 per cent of the incoming feed sulfur is attached to the coke and is oxidized into SO_x which is emitted with the regenerator flue gas. The sulfur distribution is dependent on the sulfur species contained in the feed, and in particular the amount of thiophenic sulfur. SO_2 can range from 200 to 3000 parts per million dry volume basis (ppmdv), whereas SO3 typically varies from an insignificant value to a maximum of 10 per cent of the SO2 content.

The FCCU application presents the additional requirement that in order to match the reliability of the FCCU, the air pollution control equipment must operate on-line for 3 to 5 years without interruption. It must be able to tolerate significant fluctuations in operating conditions, withstand the severe abrasion from catalyst fines, and maintain operation through system upsets. The robust design of the wet scrubbing system must tolerate all operations without requiring a shutdown. It is paramount that the operability of the air pollution control system be no less than that of the FCCU process.

Controlled Emissions—A Trend toward Lower Levels

By examining the trends of emissions regulations in the United States, a trend for better control of emissions from FCCUs can be established. The United States Environmental Protection Agency (USEPA) established New Source Performance Standards (NSPS) for emissions from FCCUs for new or significantly modified units. A summary of this standard is provided in Table. Additionally, a maximum achievable control technology (MACT) standard is in the final stages of promulgation. This standard, which is intended to regulate the amount of hazardous air pollutants (HAPs) from the FCCU, essentially established the particulate emission level at the same level as NSPS, or 1.0 lb/1000 lb of coke burned.

Table 4.18: New Source Performance Standards for FCCU Regenerator Emissions

Pollutant	*FCCUs affected*	*Emission regulation*
Particulate	All	1.0 lb/1000 lb Coke burn-off and 30% opacity
SO_2	With add-on SO_2 control device	50 ppm SO_2 or 90% reduction, whichever is least stringent
	Without add-on SO_2 control device	9.8 lb SO_2/1000 lb coke burn-off
		Or 0.3% Sulfur in feed (% by weight)

The USEPA also has been aggressive in pursing enforcement actions against refiners who, in its opinion, have significantly modified their facilities but avoided the NSPS regulations. This has resulted in several consent decrees where a refiner has agreed to install pollution controls to mitigate the impact of any past modifications made to its facility. Refiners who have reached consent decrees with the USEPA include Koch Refining, British Petroleum, Motiva/Equilon/Shell, Marathon Ashland LLC, Holly Corporation, Premcor Refining, Conoco, and Murphy Oil. In many cases, the agreed-to emissions levels (25 ppm SO_2 and 1.0 lb/1000 lb of coke burned) are more restrictive than the NSPS regulations. Wet scrubbing systems are mandated for many of the facilities in the consent decrees.

A Proven Wet Scrubber Design for the FCCU Process

The worldwide leading technology to control emissions from this process is Belco Technologies Corporation's EDV wet scrubbing system. This wet scrubbing system controls particulate (catalyst dust), SO_2 (sulfur dioxide), and SO_3 (sulfuric acid mist) all in one system.

Figure 4.60: EDV Wet Scrubbing System in Texas

Removal of relatively coarse particulate, which constitutes the majority of the particulate from the FCCU, is accomplished in the absorber vessel where caustic soda (NaOH) or other reagents are utilized to absorb SO_2 and discharge it in the form of a soluble salt. Fine particulate control and significant reduction of SO_3 in the form of sulfuric acid mist are accomplished in devices known as filtering modules. Excess water droplets are removed in highly efficient droplet separators. An EDV wet scrubbing installation in Texas is shown in Figure 4.60.

The flue gas from the FCCU enters the spray tower, where it is immediately quenched to saturation temperature. Although the flue gas normally enters the wet scrubber after passing through a heat recovery device, the system is designed so that it can accept flue gas directly from the FCCU regenerator at the temperature at which it exits the FCCU regenerator. The spray tower itself is an open tower with multiple levels of spray nozzles. Each level of nozzles can have one or multiple nozzles depending on the diameter of the absorber vessel. Since it is an open tower, there is nothing to clog or plug in the event of a process upset. In fact, this design has handled numerous process upsets where more than 100 tons of catalyst has been sent to the wet scrubber in a very short period of time.

UOP MEROX PROCESS

Introduction

The UOP Merox process is an efficient and economical catalytic process developed for the chemical treatment of petroleum fractions to remove sulfur present as mercaptans (Merox extraction) or to directly convert mercaptan sulfur to less-objectionable disulfides (Merox sweetening). This process is used for liquid-phase treating of liquefied petroleum gases (LPG), natural-gas liquids (NGL), naphthas, gasolines, kerosenes, jet fuels, and heating oils. It also can be used to sweeten natural gas, refinery gas, and synthetic gas in conjunction with conventional pretreatment and posttreatment processes.

Merox treatment can, in general, be used in the following ways:

- To improve lead susceptibility of light gasolines (extraction)
- To improve the response of gasoline stocks to oxidation inhibitors added to prevent gum formation during storage (extraction and sweetening)
- To improve odor on all stocks (extraction or sweetening or both)

- To reduce the mercaptan content to meet product specifications requiring a negative doctor test or low mercaptan content (sweetening)
- To reduce the sulfur content of LPG and light naphtha products to meet specifications (extraction)
- To reduce the sulfur content of coker or fluid catalytic cracking (FCC) C_3-C_4 olefins to save on acid consumption in alkylation operations using these materials as feedstocks or to meet the low-sulfur requirements of sensitive catalysts used in various chemical synthesis processes (extraction)

Process Description

The UOP Merox process accomplishes mercaptan extraction and mercaptan conversion at normal refinery rundown temperatures and pressures. Depending on the application extraction and sweetening can be used either singly or in combination. The process is based on the ability of an organometallic catalyst to promote the oxidation of mercaptans to disulfides in an alkaline environment by using air as the source of oxygen. For light hydrocarbons, operating pressure is controlled slightly above the bubble point to ensure liquid-phase operation; for heavier stocks, operating pressure is normally set to keep air dissolved in the reaction section. Gases are usually treated at their prevailing system pressures.

Merox Extraction

Low-molecular-weight mercaptans are soluble in caustic soda solution. Therefore, when treating gases, LPG, or light-gasoline fractions, the Merox process can be used to extract mercaptans, thus reducing the sulfur content of the treated product. In the extraction unit (Figure 4.61a), the sulfur reduction attainable is directly related to the extractable-mercaptan content of the fresh feed.

In mercaptan-extraction units, fresh feed is charged to an extraction column, where mercaptans are extracted by a countercurrent caustic stream. The treated product passes overhead to storage or downstream processing.

The mercaptan-rich caustic solution containing Merox catalyst flows from the bottom of the extraction column to the regeneration section through a steam heater, which is used to maintain a suitable temperature in the oxidizer. Air is injected into this stream, and the mixture flows upward through the oxidizer, where the caustic is regenerated by converting mercaptans to disulfides.

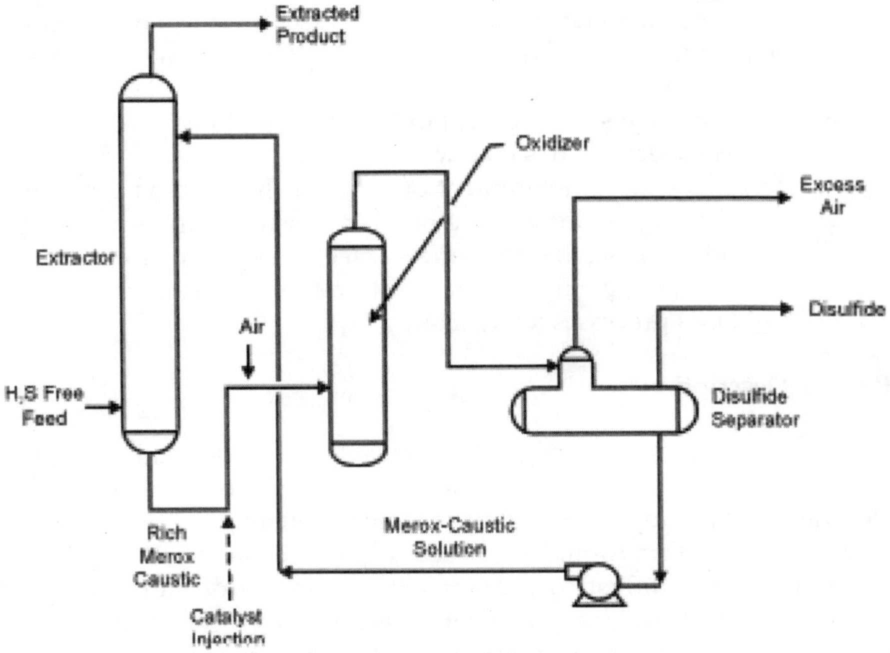

Figure 4.61(a): Merox Mercaptan-Extraction Unit

The oxidizer effluent flows into the disulfide separator, where spent air, disulfide oil, and the regenerated caustic solution are separated. Spent air is vented to a safe place, and disulfide oil is decanted and sent to appropriate disposal. For example, the disulfide oil can be injected into the charge to a hydrotreating unit or sold as a specialty product. The regenerated-caustic stream is returned to the extraction column. A small amount of Merox catalyst is added periodically to maintain the required activity.

Merox Sweetening

In sweetening units, the mercaptans are converted directly to disulfides, which remain in the product; the total sulfur content of the treated stock is not reduced.

Merox sweetening can be accomplished in four ways:

- Fixed-bed processing with intermittent circulation of caustic solution (Figure 4.61b)

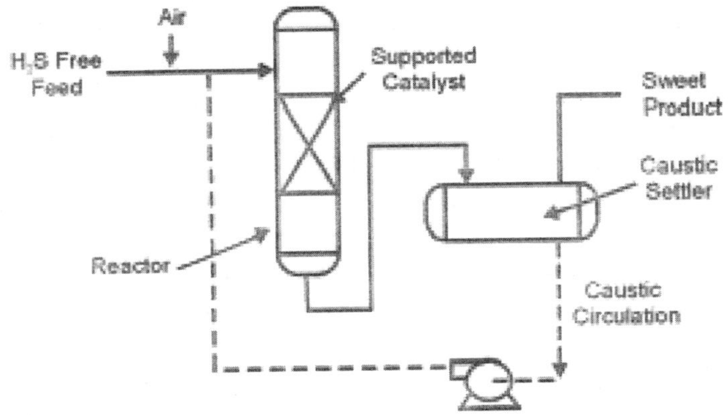

Figure 4.61(b): Fixed-Bed Merox Sweetening Unit

- ➢ Minimum-alkali fixed-bed (Minalk) processing, which uses small amounts of caustic solution injected continuously.
- ➢ Caustic-Free Merox treatment for gasoline and kerosene.
- ➢ Liquid-liquid sweetening.

Fixed-Bed Sweetening (Conventional). Fixed-bed sweetening is normally employed for virgin or thermally cracked chargestocks having endpoints above about 120°C (248°F). The higher-molecular-weight and more branched mercaptan types associated with these higher-endpoint feedstocks are only slightly soluble in caustic solution and are more difficult to sweeten. The use of a fixed-bed reactor facilitates the conversion of these types of mercaptans to disulfides.

Fixed-bed sweetening uses a reactor that contains a bed of specially selected activated charcoal impregnated with nondispersible Merox catalyst and wetted with caustic solution. Air is injected into the feed hydrocarbon steam ahead of the reactor, and in passing through the catalyst bed, the mercaptans in the feed are oxidized to disulfides. The reactor is followed by a settler for separation of caustic and treated hydrocarbon. The settler also serves as a caustic reservoir. Separated caustic is circulated intermittently to keep the catalyst bed wet. The frequency of caustic circulation over the bed depends on the difficulty of the feed-stock being treated and the activity of the catalyst.

An important application of this fixed-bed Merox sweetening is the production of jet fuels and kerosenes. As a result of the development of the Merox fixed-bed system, jet fuels and kerosenes (also diesel and heating oils) can be sweetened at costs that are incomparably lower than those of the simplest hydrotreater. The same basic process flow just described is used. However, because of other particular jet-fuel quality requirements, some pretreatment and posttreatment are needed whenever any chemical sweetening process is used.

Fixed-Bed Sweetening (Minalk). This Merox sweetening version is applied to feedstocks that are relatively easy to sweeten, such as catalytically cracked naphthas and light virgin naphthas.

THE S ZORB SULFUR REMOVAL TECHNOLOGY APPLIED TO GASOLINE

Introduction

The need for improving the environment has resulted in a continuing worldwide effort to produce cleaner-burning fuels. The reduction of sulfur in gasoline is considered to be a means for improving air quality because of the impact on performance of sulfur-sensitive items such as automotive catalytic converters. The presence of oxides of sulfur in auto-motive engine exhaust inhibits and may irreversibly poison noble metal catalysts in the converter. Emissions from an inefficient or poisoned converter contain levels of noncom-busted, nonmethane hydrocarbons, oxides of nitrogen, and carbon monoxide. Such emissions are catalyzed by sunlight to form ground-level ozone, more commonly referred to as smog.

As the concern over the impact of automotive air pollution continues, it is clear that further effort to reduce the sulfur level in automotive fuels will be required. Governments around the world are moving to mandate lower allowable sulfur levels in transportation fuels. In the United States, the U.S. Environmental Protection Agency has promulgated regulations that require refiners to achieve an average pool target of 30 ppm sulfur in gasoline with phase-in beginning in 2004. In Canada, refiners will have to lower sulfur levels in two steps: The first was to 150-ppm level by the year 2002, and then down to 30-ppm level by the year 2005. In Europe, gasoline sulfur levels are required to be below 50 ppm by 2005 and to less than 10 ppm between 2005 and 2008.

In conjunction with the need to be able to produce low-sulfur-content gasoline, there is a need for a process that will not negatively impact the important properties of the gasoline. Any acceptable sulfur reduction process must have a minimal effect on the olefin content of the feed since saturation of olefins greatly affects the octane number (both research and motor octane number). Such adverse effect on the olefin content is generally due to the severe conditions normally employed, such as during hydrodesulfurization, to remove thiophenic compounds, which are some of the most difficult sulfur-containing compounds to remove from cracked gasoline. In addition, there is a need to avoid a system wherein the conditions are such that the aromatic content of the cracked gasoline is lost through saturation. ConocoPhillips Petroleum has developed such a process with its S Zorb sulfur removal technology (SRT), which is a commercialized, economically competitive process to remove sulfur from hydrocarbon streams.

Technology Development

Affected as all other refiners by the new regulations, ConocoPhillips began searching early for the solution, but was unsatisfied with existing and developing technologies. During an intensive research and development effort initiated in early 1998, ConocoPhillips researchers, aided by extensive experience in the areas of sulfur chemistry, hydrotreating, olefin processes, and sorption, were successful in developing a superior process, S Zorb SRT. The S Zorb process can economically reduce the sulfur content of gasoline to ultralow levels with minimal octane loss, minimal hydrogen consumption, and near-zero volume loss. These results are obtained through the use of a novel sorbent that selectively removes sulfur from hydrocarbon molecules. In conjunction with the sorbent development, a reliable plant design was developed that allows continuous operation paralleling major refining units such as the fluidized catalytic cracker (FCC), thus reducing operation and maintenance costs.

To develop and prove the new technology within the time frame mandated by upcoming regulations, ConocoPhillips decided to fast-track the design and construction of the first commercial S Zorb SRT unit at its Borger, Texas, refinery. Implementation of the S Zorb technology at the Borger refinery was conducted by an integrated group, which included research and development, process operations, maintenance, safety, environmental, and construction teams. Through the application of creative engineering approaches, a clear focus on schedule and deliverables, and full support of upper management, the technology was moved from conceptual flow sheet to a fully operational plant in less than 18 months.

The vertical start-up (feed in to on-specification production within 72 hours) and high throughput of the S Zorb unit at the Borger refinery are testimony to the efficacy of the S Zorb sorbent and engineering design. Within 72 hours of unit start-up, product gasoline with only 10 ppm sulfur was produced, and within a week after start-up, the plant was processing full-range FCC naphtha at 110 per cent of the design capacity. The plant continues to produce low-sulfur, full-range FCC naphtha with minimal octane loss exceeding the gasoline sulfur removal requirements for the Borger refinery.

Process

S Zorb SRT is not a modified hydrogenation technology. It uses a revolutionary new sorbent that operates in a fluidized-bed reactor (sorber), allowing optimal conditions to attain the required sulfur removal while minimizing undesired reactions and results. In the sorber, a stream of vaporized sulfur-containing feed and hydrogen is passed over the sorbent, yielding a low-sulfur product with little octane loss and no volume loss. The sorbent circulates between the sorber and a regeneration section with on-line regeneration, allowing for steady-state performance with very selective, deep sulfur removal and extended run lengths.

The ConocoPhillips S Zorb process is a continuous operation, with the sorbent moving between three vessels in a pseudocontinuous process. Sulfur is chemically removed from the gasoline stream in the sorber and stored on the sorbent. The effluent stream from the sorber is a low-sulfur gasoline blendstock with minimal octane loss, minimal hydrogen consumption, and little to no volume loss. The sorbent is removed from the sorber in a continuous process and regenerated in an oxygen-containing atmosphere to produce sulfur dioxide, a small amount of carbon dioxide, and a fresh sorbent.

The sulfur dioxide produced in this process can be separated from this regenerator effluent stream in a number of ways depending on the configuration of the refinery. The primary benefit of the absorptive process is the elimination of hydrogen sulfide from the reactor. The lack of hydrogen sulfide allows the S Zorb process to remove as much as 99+ per cent of the sulfur from a gasoline stream without the need for an additional reactor to remove mercaptans formed from olefin recombination.

After oxidative regeneration, the sorbent is transferred to a small holding vessel, where it is reduced by using a hydrogen stream that can have as little as 50 per cent purity. The S Zorb sorbent reduction is highly invariant to reduction temperature and pressure. Therefore, the reduction vessel is primarily treated as a surge vessel and requires minimal oversight by the operators.

The S Zorb SRT process has many control parameters that allow for treating various streams and conditions by changing the operating conditions while controlling octane loss. A few of these parameters include adjustments in sorber temperature or pressure, sorbent circulation rate, and sorbent sulfur loading. One of the unique features of the S Zorb process is that selectivity toward sulfur removal increases with an increase in temperature.

The result of this increased selectivity is improved octane retention at a given product stream sulfur level. Octane protection and olefin conversion are closely related for gasoline. Because of the decoupling of the olefin loss from sulfur removal activity, S Zorb SRT produces much higher-octane gasoline for a given sulfur level than hydrotreating.

One key aspect to operation of the S Zorb unit is the ability to shut down the regeneration section of the unit for periods of time while the sorber continues to operate.

CONOCOPHILLIPS S ZORB DIESEL PROCESS

Introduction

The S Zorb sulfur removal technology (SRT) diesel process removes sulfur from distillates to meet the new diesel fuel regulatory requirements around the world. Environmental concerns initiated the S Zorb concept over 20 years ago with sulfur removal from furnace exhaust. During the last several years, the S Zorb-Gasoline process for producing ultralow-sulfur gasoline has been commercialized. Today ultralow-sulfur and "zero"-sulfur diesels are being mandated in order to allow auto manufacturers to meet the emissions standards. Advanced engine emission control technologies, in the forms of particulate matter (PM) traps and NOx adsorbers, are much less effective with high-sulfur fuel because of catalyst poisoning. To ease the impact of these regulations on the refining industry, ConocoPhillips has developed a unique and selective sulfur removal technology that applies to current on-road diesel blends and other higher-sulfur, distillate-range blendstocks. Key advantages offered by the S Zorb SRT diesel process are very low hydrogen consumption, high product yields (>99.5 per cent), and the preservation of the fuel properties such as lubricity.

Chemistry

Sulfur removal in the S Zorb process is influenced by a number of feedstock properties such as boiling point range, nitrogen content, and most importantly the sulfur content. The amount of sulfur and its distribution among different organic molecules vary significantly between different types of diesel feedstocks. In Figure 4.62, the concentration of sulfur is plotted against boiling point (in degrees Celsius) for diesel fuels and blendstocks from different sources.

The vertical line on the plots is at 315°C. Below this temperature, the sulfur molecules are primarily benzothiophenes, dibenzothiophenes, and methyldibenzothiophenes which have generally high removal rates. Above 315°C the peaks represent the higher-boiling-point dimethyldibenzothiophenes which are the most difficult to remove sulfur compounds. The plot for straight-run diesels shows that two of the straight-run diesel feeds have very similar sulfur distributions with high concentrations of lower-boiling sulfur compounds, while the third diesel contains significant amounts of the higher-boiling sulfur compounds. The third diesel is significantly more difficult to desulfurize. The blended diesels represent current on-road diesel fuels. The total sulfur concentration is lower, but previous hydroprocessing has removed all but the higher-boiling sulfur compounds. A similar sulfur distribution is seen in a light cycle oil, where the majority of the sulfur is again in the higher-boiling fraction even though this light cycle had not been previously hydrotreated. All the diesel feedstock with high concentrations of sulfur above the 315°C line has been shown to require more severe process conditions to produce ultralow-sulfur products.

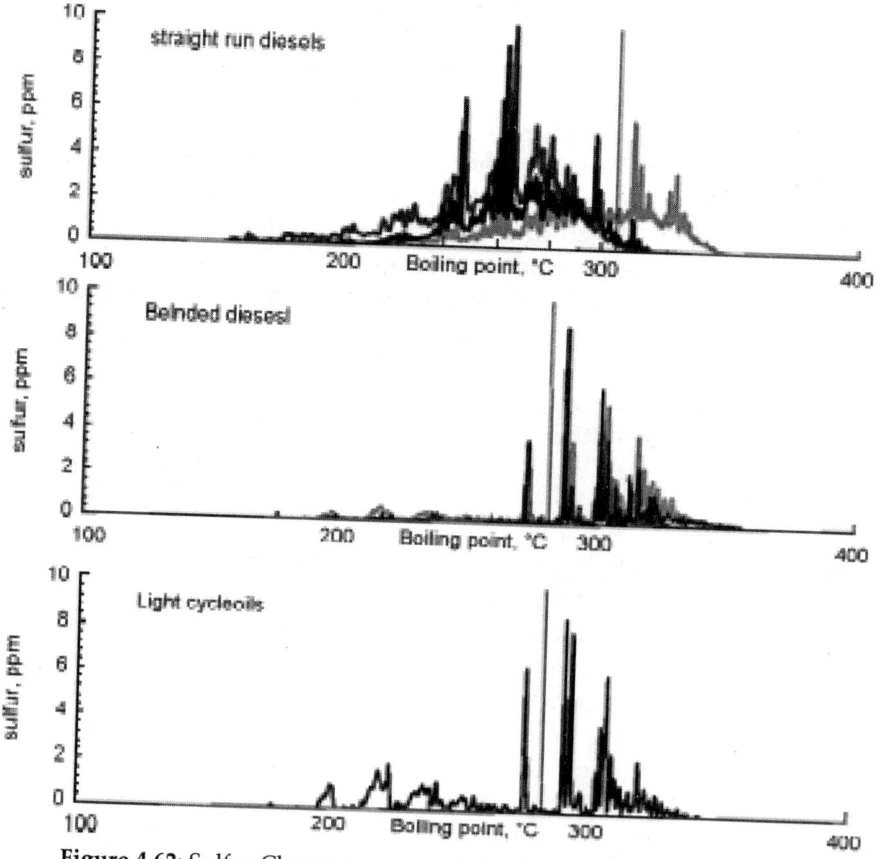

Figure 4.62: Sulfur Chromatograms of Distillate-Range Hydrocarbons

In S Zorb technology, sulfur removal occurs when the entire organic sulfur molecule sorbs onto the S Zorb particle surface in a mechanism similar to hydrodesulfurization. The sulfur atom is then separated from the remainder of the organic molecule, which is released into the bulk fluid. In standard hydrodesulfurization, the absorbed sulfur is reduced to form H2S gas. Also, there is significant hydrogenation of aromatic rings. In S Zorb, the sulfur atom remains bound to the sorbent as a metal sulfide. No H2S is observed in the product, and hydrogenation of aromatic rings is not significant. Consumption of hydrogen is minimal. The production of light hydrocarbons is also extremely low. This produces diesel-range product yields greater than 99.5 per cent. The S Zorb chemistry generates a product that has virtually the same properties as the feed. As shown in Table, even at very low product sulfur levels, there are no significant changes in specific gravity, distillation boiling curve, cetane number, cetane index, ASTM color, cloud point, and pour point. There is also virtually no impact on lubricity, unlike in traditionally hydrodesulfurized diesel products.

Table 4.19: Feed and Product Properties of S Zorb-Treated Diesel

Property	*Feed*	*Product*
Sulfur, wt ppm	523	6
API gravity	33.20	33.22
Hydrogen, wt %	12.72	12.72
Cetane number	43.5	43.5
Cetane index	44.4	44.4
D86 distillation		
IBP	385°F	380°F
10	440°F	438°F
50	513°F	513°F
90	604°F	603°F
ASTM color	0.5	0.5
Cloud point	–10.6°F	–10.6°F
Pour point	–13.5°F	–13.5°F
Lubricity		
SLBOCLE	3700	3600
HFRR	385	315

Sorbent

The S Zorb sorbent is composed of base metals commonly used in conventional refining catalysts. It contains no noble metals. The S Zorb sorbent has a significantly higher active site density than typical HDS catalysts. The higher active site density produces a material that has significantly higher sulfur removal rates and less sensitivity to poisons such as silica and chlorides. Both pilot-plant and commercial data indicate little or no loss in activity even after several hundred regeneration cycles. The sorbent particle size averages 50 to 60 μm and is produced by a spray-drying process similar to the process used to manufacture catalytic cracking catalysts. These small, fluidizable particles have higher external surface areas than conventional 1/16-inch extrudate and exhibit no intraparticle transport limitations. Much effort in the sorbent development process has focused on the production of an attrition-resistant material, commercial manufacturability, and composition optimization. ConocoPhillips licenses certified catalyst vendors to manufacture and market the sorbent in commercial quantities.

Process Description

The process scheme for the S Zorb SRT diesel process is similar to that for a low-pressure hydrodesulfurization process with the exception that a fluidized-bed reactor having continuous regeneration replaces the fixed-bed HDS reactor.

GASOLINE DESULFURIZATION

Introduction

Both government and private industries are actively involved in the search for less polluting or nonpolluting alternatives to the conventional hydrocarbon-fueled internal combustion engine. Unfortunately, issues regarding the economics, infrastructure requirements, and consumer acceptance of these alternatives suggest that for the foreseeable future, automobiles will continue to rely on the internal combustion engine as their primary power source. To reduce automobile emission levels, governments have begun to implement fuel reformulation and emission limits. Automobile manufacturers, meanwhile, argue that current gasoline quality is inadequate for the advanced engine and catalytic converter designs required to meet the new emission standards, citing sulfur as the major deterrent to their ability to further reduce engine emissions.

Governments have responded to these issues by legislating that refiners produce low-and ultralow-sulfur gasoline. A survey of U.S. refiners conducted in the mid-1990s indicated the average refinery gasoline pool contained approximately 350 wt ppm sulfur.1 By January 2005, U.S. tier II regulations will establish a maximum standard gasoline sulfur content of 30 wt ppm—a more than 90 per cent reduction in gasoline sulfur content. Geographic phase-in has been provided for U.S. refiners located in PADD

IV (Colorado, Idaho, Montana, and Utah) where, because of the large number of small refineries, 300 wt ppm sulfur gasoline will be allowed during 2004 to 2006, with an 80 wt ppm cap being implemented by 2006.

As Figure indicates, although the timing may vary, other countries around the world are also imposing gasoline sulfur limits. In Europe, a 150 wt ppm gasoline sulfur specification has been enacted as an intermediate step prior to the implementation of a 50 wt ppm limit by 2005. Other European countries including the United Kingdom and Germany have introduced tax incentives for the production of 50 wt ppm sulfur gasoline, with Germany also intending to introduce incentives for 10 wt ppm sulfur gasoline in 2003. In addition, Japan will implement a 50 wt ppm gasoline sulfur specification in 2005, with the potential for a further reduction to 10 wt ppm by 2008. This legislation will require refiners to make numerous technical and economic decisions. For the decision process to be most effective, refiners must become familiar with the various methods for reducing gasoline sulfur content, the variety of gasoline desulfurization technologies that are available, and the often subtle differences between them. The first step in this process involves a review of the origins of the sulfur present in gasoline.

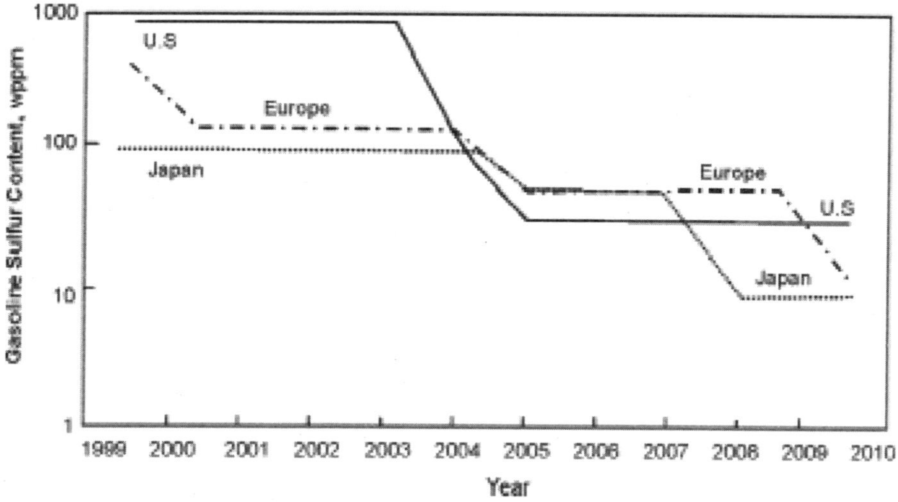

Figure 4.63: Gasoline Sulfur Limits

Why FCC Naphtha?

Most refinery gasoline desulfurization strategies have focused on the FCC naphtha component of the refinery gasoline pool. As illustrated in Figure, a survey of U.S. refiners conducted in the mid-1990s identified

FCC naphtha as the largest single contributor of sulfur to the gasoline pool, accounting for more than 90 per cent of the pool's total sulfur content.

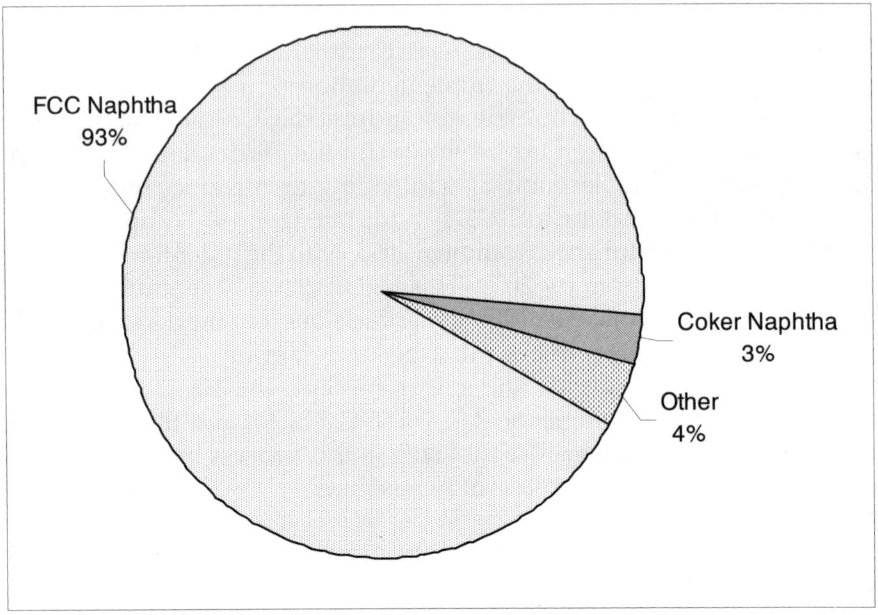

Figure 4.64: Sulfur Sources in U.S. Gasoline Pool

This survey also provided insight into the magnitude of the effort that would be required to meet lower gasoline sulfur targets since, as indicated in Figure 4.65, FCC naphtha was shown to represent almost 40 per cent of the typical refinery gasoline pool. For these reasons, any strategy for reducing gasoline sulfur content clearly must focus on desulfurizing the FCC naphtha component of the gasoline pool.

An effective gasoline desulfurization strategy should begin with an examination of the general features of FCC naphtha. The FCC naphtha's molecular structure, as represented by its GC-PIONA analysis, and the types and amounts of sulfur within FCC naphtha can vary significantly. In addition, the types of sulfur compounds and their distribution within the FCC naphtha vary with boiling range. The naphtha's light fraction contains predominantly mercaptan-type sulfur, while the naphtha's intermediate-range fraction contains primarily thiophenic-type sulfur and its heavy fraction contains mostly benzothiophenic-type sulfur compounds.

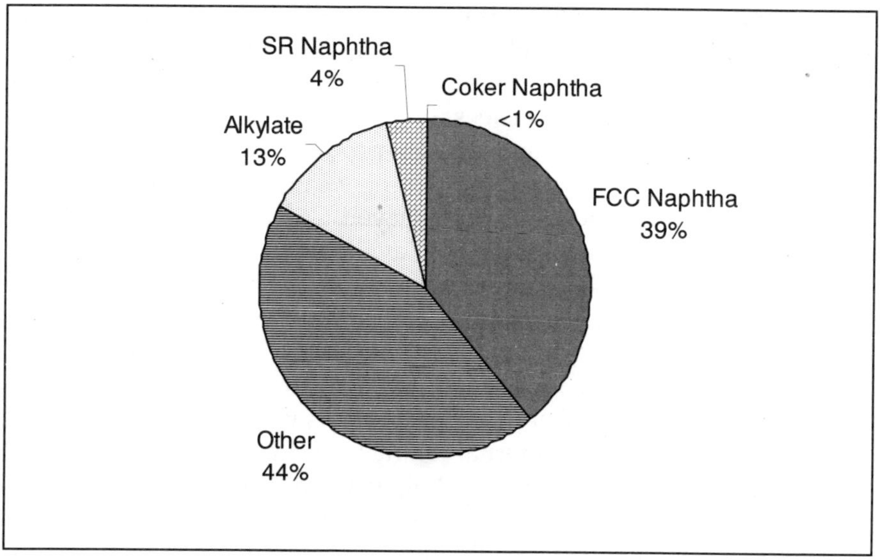

Figure 4.65: U.S. Gasoline Pool Composition

The two major components that influence the naphtha's octane—olefins and aromatics—are not evenly distributed. Rather, similar to sulfur, the naphtha's aromatic components are concentrated in its heavy fraction, while the lighter naphtha contains primarily olefinic compounds. As a result of the octane differences between these components, and their distribution within the FCC naphtha, the Road octane of the intermediate fraction of FCC naphtha typically is lower than that of either the naphtha's light or heavy fractions.

Reducing the Sulfur Content of FCC Naphtha

Before new processing units are added, other options for reducing the sulfur content of the FCC naphtha should be considered. For example, FCC manufacturers have claimed the ability to reduce the sulfur content of FCC naphtha by 20 to 30 per cent. These additives tend to reduce the amount of thiophenes and alkylthiophenes in the FCC naphtha.

CONOCOPHILLIPS DELAYED COKING PROCESS

Introduction

In today's competitive refining environment, delayed coking still remains the industry's leading economic choice in heavy oil upgrading technology.

At most delayed coking sites, it is more profitable to limit coke generation due to its relatively low market value in comparison with the coker's other products. The top coking facilities are continually optimizing operations to increase flexibility for processing a variety of feedstocks, while maximizing higher-valued liquid and gas products. Equally critical in daily operations is the emphasis on maintaining a safe and reliable processing unit. The patented ConocoPhillips Delayed Coking Process is the most effective coking method commercially available today for designing new units and retrofitting existing sites.

The proprietary ConocoPhillips Delayed Coking Process is based upon patented processes and many years of operating experience. The heritage Conoco company first entered the realm of delayed coking with construction of its first coker in the early 1950s in Ponca City, Oklahoma. Conoco has designed all its internal coking facilities constructed after 1960. Then in the early 1980s, Conoco began licensing its delayed coking technology to outside companies. Now refiners throughout the world are utilizing the combined ConocoPhillips company's expertise for residual oil conversion.

Over the last 50 years, ConocoPhillips has gained valuable design and operating experience in all areas of delayed coking. This experience enabled ConocoPhillips to greatly improve the economics, reliability, and flexibility of the delayed coking process. ConocoPhillips is also the industry leader in the premium needle coke market. Although ConocoPhillips does not license needle coke technology, this is a more severe process and much experience in this arena has been applied in the fuel and anode coke licensing technology. This chapter highlights many of the advantages associated with the patented delayed coking technology.

Technology Highlights

The patented ConocoPhillips Delayed Coking Process offers licensees the following benefits:

- Best qualified delayed coking licensor. ConocoPhillips is the only licensor that designs, owns, and operates delayed cokers.
- Most experienced team. Fifteen delayed coking licenses worldwide in the last 10 years alone have been awarded to ConocoPhillips. The experienced engineering team utilizes operating and maintenance personnel from existing ConocoPhillips coking facilities to aid in design, construction, start-up, and optimization efforts.
- Maximum coking capacity. ConocoPhillips' ultralow coke drum

cycle time operation and large coke drum diameter design, combined with other design advantages, help units achieve high unit throughput at a lower capital investment cost.

- Higher liquid yields. ConocoPhillips' patented Distillate Recycle, Zero Natural Recycle Technology, and other processes reduce coke yields while maximizing more valuable liquid products.
- Improved safety. ConocoPhillips is committed to personnel safety in all phases of design, start-up, and operation. Dedication to safety has recently generated several exciting improvements in the arenas of safety rationale and emergency egress strategy.
- Greater operability and reliability. Key information from internal coking facilities and recent licensees helps ConocoPhillips to continually improve unit on-stream factors and operations. Operations and maintenance personnel are integrally involved in the design and start-up phases for new and retrofit coker projects.

FW Delayed-Coking Process

The delayed coker is an important residue conversion process, or so-called "bottom-of-the-barrel" process where residues from heavy, high-sulfur crudes are converted to transportation fuels.

Statistically, in terms of number of units installed and total current operating capacity, it is quite easy to show that delayed coking is the residue-conversion process which is most often used today. In addition, because of its wide commercial acceptance, delayed coking has been referred to as the yardstick against which other, less commercially proven, processes must be measured.

Despite its wide commercial use, only relatively few contractors and refiners are truly knowledgeable in delayed-coking design, so that this process carries with it a "black art" connotation.

The year in which delayed coking was first developed is given in historical listings of petroleum advances as 1928. We know that in early refineries severe thermal cracking of residue would result in the deposit of unwanted coke in the heaters. By evaluation of the art of heater design, methods were found by which it was possible to raise rapidly the temperature of the residue above the coking point without depositing the coke in the heater itself. Provision of an insulated surge drum downstream of the heater so that the coking took place after the heater, but before subsequent processing, resulted in the name "delayed coking."

The next step was to add a second coke drum, which doubled the run length and led to the development of the art of switching coke drums while still maintaining operation.6 In the early 1930s the drums were limited in size to 10 ft in diameter.7 Coke drums as large as 30 ft in diameter have recently been installed. Figure shows the general trend in the growth in coke-drum diameter from 1930 to 1995.

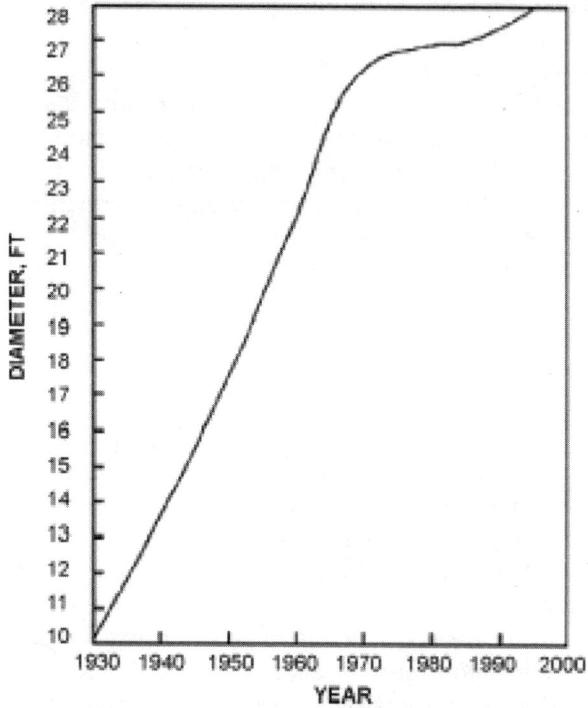

Figure 4.66: Trend in Delayed-Coker Maximum Drum Size

As of Jan. 1, 1995, there were in operation in the United States 45 delayed cokers with the capacity to process 1,598,000 BPSD of fresh feed. Table shows coking-plant-capacity statistics for the United States from 1946 to 1995.

Table 4.20: United States Coking-Plant Statistics

Year	*United States coking capacity, BPSD*
1946	102,000
1950	158,000
1960	475,000
1970	835,000
1973	1,008,000
1982	1,118,100
1995	1,598,000

Coking Section

Reduced-crude or vacuum-residue fresh feed is preheated by exchange against gas oil products before entering the coker-fractionator bottom surge zone. The fresh feed is mixed with recycle condensed in the bottom section of the fractionator and is pumped by the heater charge pump through the coker heater, where the charge is rapidly heated to the desired temperature level for coke formation in the coke drums. Steam is often injected into each of the heater coils to maintain the required minimum velocity and residence time and to suppress the formation of coke in the heater tubes.

The vapor-liquid mixture leaving the furnace enters the coke drum, where the trapped liquid is converted to coke and light-hydrocarbon vapors. The total vapors rise upward through the drum and leave overhead.

A minimum of two drums is required for operation. One drum receives the furnace effluent, which it converts to coke and gas while the other drum is being decoked.

Fractionation Section

The coke-drum overhead vapors flow to the coker fractionator and enter below the shed section. The coke-drum effluent vapors are often "quenched" and "washed" with hot gas oil pumped back to the trayed wash section above the sheds. These operations clean and cool the effluent-product vapors and condense a recycle stream at the same time. This recycle stream, together with the fresh feed, is pumped from the coker fractionator to the coking furnace. The washed vapors pass to the rectifying section of the tower. A circulating heavy gas oil pumparound stream, withdrawn from the pumparound pan, is used to remove heat from the tower, condensing the major portion of heavy gas.

FW/UOP VISBREAKING PROCESS

Introduction

Visbreaking is a well-established noncatalytic thermal process that converts atmospheric or vacuum residues to gas, naphtha, distillates, and tar. Visbreaking reduces the quantity of cutter stock required to meet fuel oil specifications while reducing the overall quantity of fuel oil produced.

The conversion of these residues is accomplished by heating the residue material to high temperatures in a furnace. The material is passed through a soaking zone, located either in the heater or in an external drum, under proper temperature and pressure constraints so as to produce the desired products. The heater effluent is then quenched with a quenching medium to stop the reaction.

With refineries today processing heavier crudes and having a greater demand for distillate products, visbreaking offers a low-cost conversion capability to produce incremental gas and distillate products while simultaneously reducing fuel oil viscosity. Visbreaking can be even more attractive if the refiner has idle equipment available that can be modified for this service.

When a visbreaking unit is considered for the upgrading of residual streams, the following objectives are typically identified:

- Viscosity reduction of residual streams which will reduce the quantity of high-quality distillates necessary to produce a fuel oil meeting commercial viscosity specifications.
- Conversion of a portion of the residual feed to distillate products, especially cracking feedstocks. This is achieved by operating a vacuum flasher downstream of a visbreaker to produce a vacuum gas oil cut.
- Reduction of fuel oil production while at the same time reducing pour point and viscosity. This is achieved by utilizing a thermal cracking heater, in addition to a visbreaker heater, which destroys the high wax content of the feedstock.

Specific refining objectives must be defined before a visbreaker is integrated into a refinery, since the overall processing scheme can be varied, affecting the overall eco-nomics of the project

Coil Versus Soaker Design

Two visbreaking processes are commercially available. The first process is the coil, or furnace, type, which is the type offered through Foster Wheeler and UOP. The coil process achieves conversion by high-temperature cracking within a dedicated soaking coil in the furnace. With conversion primarily achieved as a result of temperature and residence time, coil visbreaking is described as a high-temperature, short-residence-time route. Foster Wheeler has successfully designed many heaters of this type worldwide.

The main advantage of the coil-type design is the two-zone fired heater. This type heater provides for a high degree of flexibility in heat input, resulting in better control of the material being heated. With the coil-type design, decoking of the heater tubes is accomplished more easily by the use of steam-air decoking.

Foster Wheeler's coil-type cracking heater produces a stable fuel oil. A stable visbroken product is particularly important to refiners who do not have many options in blending stocks.

The alternative soaker process achieves some conversion within the heater. However, the majority of the conversion occurs in a reaction vessel or soaker which holds the two-phase effluent at an elevated temperature for a predetermined length of time. Soaker visbreaking is described as a low-temperature, high-residence-time route. The soaker process is licensed by Shell. Foster Wheeler has engineered a number of these types of visbreak-ers as well.

By providing the residence time required to achieve the desired reaction, the soaker drum design allows the heater to operate at a lower outlet temperature. This lower heater outlet temperature results in lower fuel cost. Although there is an apparent fuel savings advantage experienced by the soaker-drum-type design, there are also some disadvantages. The main disadvantage is the decoking operation of the heater and soaker drum. Although decoking requirements of the soaker drum design are not as frequent as those of the coil-type design, the soaker design requires more equipment for coke removal and handling.

The customary practice of removing coke from a drum is to cut it out with high-pressure water. This procedure produces a significant amount of coke-laden water which needs to be handled, filtered, and then recycled for use again. Unlike delayed cokers, visbreakers do not normally include the facilities required to handle coke-laden water. The cost of these facilities can be justified for a coker, where coke cutting occurs every day. However, because of the relatively infrequent decoking operation associated with a visbreaker, this cost cannot be justified.

Product qualities and yields from the coil and soaker drum design are essentially the same at a given severity and are independent of visbreaker configuration.

Feedstocks

Atmospheric and vacuum residues are normal feedstocks to a visbreaker. These residues will typically achieve a conversion to gas, gasoline, and gas oil in the order of 10 to 50 per cent, depending on the severity and feedstock characteristics. This will therefore reduce the requirement for fuel oil cutter stock. The conversion of the residue to distillate and lighter products is commonly used as a measurement of the severity of the visbreaking operation. Per cent conversion is determined as the amount of 650°F+ (343°C+) material present in the atmospheric residue feedstock or 900°F+ (482°C+) material present in the vacuum residue feedstock which is visbroken into lighter boiling components.

The extent of conversion is limited by a number of feedstock characteristics, such as asphaltene, sodium, and Conradson carbon content. A feedstock with a high asphaltene content will result in an overall lower conversion than a normal asphaltene feedstock, while maintaining production of a stable fuel oil from the visbreaker bottoms. Also the presence of sodium, as well as higher levels of feed Conradson carbon, can increase the rate of coking in the heater tubes. Minimizing the sodium content to almost a negligible amount and minimizing the Conradson carbon weight per cent will result in longer cycle run lengths.

Variations in feedstock quality will impact the level of conversion obtained at a specific severity. Pilot plant analyses of a number of different visbreaker feedstocks have shown that, for a given feedstock, as the severity is increased, the viscosity of the 400°F+ (204°C+) visbroken tar initially decreases and then, at higher severity levels, increases dramatically, indicating the formation of coke precursors.

The point at which this viscosity reversal occurs differs from feed to feed but typically coincides with approximately 120 to 140 standard cubic feet (SCF) of C_3–gas production per barrel of feed (20.2 to 23.6 normal m^3/m^3). It is believed that this reversal in viscosity defines the point beyond which fuel oil instability will occur. Fuel oil instability is discussed in the next section of this chapter, "Yields and Product Properties."

The data obtained from these pilot tests have been correlated. The viscosity reversal point can be predicted and is used to establish design parameters for a particular feedstock to avoid the formation of an unstable fuel oil, while maximizing conversion.

Pilot plant work has also been done relating visbreaker heater run length to conversion and feedstock quality. Figure graphically represents the decrease in heater run length with increasing feedstock conversion. This graph has been plotted with data for three atmospheric residues with varying feed Conradson carbon. Figure 4.67 shows that, for a given per cent conversion, as the feed quality diminishes (i.e., as Conradson carbon grows higher), coking of the heater tubes increases, resulting in shorter run lengths.

It has been found that visbreaking susceptibilities bear no firm relationship with API density, which is the usual chargestock property parameter utilized in thermal cracking correlations. However, feedstocks with low n-pentane insolubles and low softening points show good susceptibility to visbreaking.

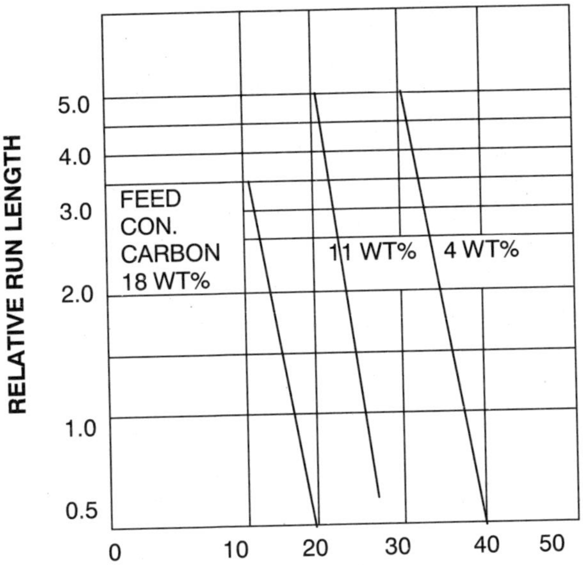

Figure 4.67: Relative Run Length versus Conversion at Various Feed Qualities

HÜLS ETHERS PROCESSES

Introduction

During the early 1990s, the oxygenate portion of the gasoline pool was the fastest-growing gasoline component, and the majority of this growth was in methyl tertiary butyl ether (MTBE). The major reasons for this growth were generally considered to be environmental concerns and octane upgrades. Countries such as the United States, Korea, and Taiwan mandated the use of oxygenates in gasoline to promote cleaner-burning fuels. Lead phasedown programs, the introduction of midgrade and higher-octane premium gasolines, and newer, more sophisticated car engines all contributed to a steadily increasing demand for higher-quality gasoline and thus a continuing need to increase the octane of the refinery gasoline pool.

In the late 1990s regulators in the United States passed legislation requiring the phaseout of MTBE from many markets in the early 2000s, due to numerous instances of groundwater contamination with MTBE. The major cause of the contamination is leaking underground storage tanks and pipelines and limited biodegradability. Some people have a very low taste threshold of ~10 ppb MTBE in H_2O. Very few new ether units for fuels applications have been constructed since the U.S. action. By 2002 many U.S. producers stopped MTBE production and were considering alternative technology to consume isobutene by reusing MTBE plant equipment.

An important source of MTBE, as well as other ethers, is the refinery. In 1994, installed refinery MTBE capacity of more than 5.57 million metric tons per annum (MTA) [129,000 barrels per stream-day (BPSD)] represented about 28 per cent of the worldwide MTBE production. Other major sources of MTBE production are from the dehydrogenation of isobutene, as a by-product from propylene oxide production, and from naphtha cracker C_4s.

Although MTBE is the most common ether, it is not the only ether used in gasoline blending today. Tertiary amyl methyl ether (TAME), ethyl tertiary butyl ether (ETBE), and diisopropyl ether (DIPE) are also used as gasoline blending ethers. Table provides a list of the gasoline blending properties of ethers being used in gasoline pools. In addition to providing a gasoline oxygenate source, these ethers have excellent research and motor blending octanes.

Table 4.21: Refinery Oxygenates

	Blending octane			Blending RVP		
Ethers	*RONC*	*MONC*	*(R + M)/2*	*kg/cm²*	*lb/in²*	*Oxygen, wt %*
MTBE	118	100	109	0.56–0.70	8–10	18.2
DIPE	112	98	105	0.28–0.35	4–5	15.7
TAME	111	98	105	0.21–0.35	3–5	15.7
ETBE	117	102	110	0.21–0.35	3–5	15.7

Note: RVP = Reid vapor pressure; RONC = research octane number clear; MONC = motor octane number clear; (*R* + *M*)/2 (RONC)/2 (sometimes referred to as road octane).

Ethers are generally favored over alcohols in gasoline blending for two reasons: they have a very low solubility in water compared to alcohols, and they have a low blending vapor pressure compared to alcohols.

Hüls Ethers Process for Mtbe, ETBE, and Tame

The Hüls ethers processes for MTBE, TAME, ETBE, and tert-amyl ethyl ether (TAEE), formerly colicensed by UOP and Hüls AG, are now licensed exclusively from UOP in Des Plaines, Illinois, as part of the UOP Ethermax process technology. This process can be used to produce the ethers for gasoline blending from olefin feedstocks available within a refinery. Depending on the type of hydrocarbon and alcohol feed, the following etherification reactions take place:

$$CH_2 = C(CH_3)_2 + CH_3OH \circledR\ (CH_3)_3{-}C{-}O{-}CH_3 \quad \text{.................} \ (1)$$

Isobutylene Methanol MTBE

$$CH_2 = C(CH_3)_2 + CH_3CH_2OH \circledR\ (CH_3)_3{-}C{-}O{-}CH_2CH_3 \quad \text{.................} \ (2)$$

Isobutyleen Ethanol ETBE

$$CH_3CH = C(CH_3)_2 + CH_3OH \circledR\ CH_3CH_2\,(CH_3)_2{-}C{-}O{-}CH_3 \quad \text{.................} \ (3)$$

Isoamylene Methanol TAME

The reactions proceed in the liquid phase at mild conditions in the presence of a solid acidic catalyst. The catalyst typically is a sulfonic ion-exchange resin. The reaction temperature is kept low and can be adjusted over a fairly broad range. Higher temperatures are possible, but excessive temperatures are not recommended because resin fouling from polymers can occur. Around 130°C (266°F), sulfonic ion-exchange resins become unstable. Operation in the lower temperature range ensures stable operation and long catalyst life.

The reaction of an isoolefin with alcohol is conducted in the presence of a small excess of alcohol relative to that required for the stoichiometric reaction of the isoolefin contained in the hydrocarbon feed. Operation with a small excess of alcohol has a number of advantages and practically no drawbacks because any excess alcohol is recovered and recycled. Some of the advantages are:

- ➢ The equilibrium is displaced toward the production of ether to favor higher per-pass conversion.
- ➢ Production of high-octane ether is maximized, and production of lower-octane oligimers is minimized.
- ➢ Process temperature is more efficiently and securely controlled.

In the absence of a small excess of alcohol, isoolefin dimerization, also exothermic, can take place rapidly. This reaction can result in a sharp temperature rise in the resin bed. Such an increase causes irreversible catalyst fouling, and catalyst destruction can occur if the temperature rise is excessive.

Under proper conditions, the etherification reaction is nearly 100 per cent selective except for minor side reactions resulting from the presence of certain feed impurities. Water contained in the feed results in equivalent amounts of tertiary butyl alcohol (TBA) in the MTBE or ETBE product. Water in isoamylene feed yields tertiary amyl alcohol (TAA) in the TAME product. In small quantities, these alcohol by-products are unimportant. They need not be separated from the ether product because they have high octane values and can be used as gasoline blending agents.

Process Flow

The fresh hydrocarbon feed must be treated in a water wash if it comes from a fluid catalytic cracking (FCC) unit. The treatment step is needed to remove basic nitrogen compounds, which are catalyst poisons. This procedure is not necessary if the feed comes from either a steam cracker or a UOP Oleflex unit. In the case of TAME production, diolefins must also be removed in a hydrogenation unit. The clean fresh feed is mixed with fresh and recycled alcohol and charged to the reactor section. The reactor can be a tubular reactor, or more typically, two adiabatic reactors with recycle are used. The majority of the reaction occurs in the first reactor. The second reactor completes the reaction of isoolefins to ether. Cooling between reactors is required to maximize the approach to equilibrium in the second reactor.

The product from the reactor section primarily contains ether, excess alcohol (methanol or ethanol), and unreacted C_4 or C_5 hydrocarbons. This stream is sent to a fractionation column, where high-purity MTBE, ETBE, or TAME is recovered from the bottoms. The unreacted hydrocarbon, typically referred to as the raffinate stream, and alcohol are taken off the top of the fractionator. Before it leaves the unit, the raffinate is water-washed to remove excess alcohol. The water-alcohol mixture from the water wash is fractionated in the alcohol recovery section. The recovered alcohol is recycled back to the reactor, and the water is recycled back to the water wash.

UOP Ethermax Process for MTBE, ETBE, and Tame Production

Process Description

The Ethermax process, licensed exclusively by UOP, can be used to produce methyl tertiary butyl ether (MTBE), tertiary amyl methyl ether (TAME), or ethyl tertiary butyl ether (ETBE). This process combines the Hüls fixed-bed etherification process with advanced RWD catalytic distillation technology from Koch-Glitsch, Inc. The combined technology overcomes reaction equilibrium limitations inherent in a conventional fixed-bed etherification process.

The Ethermax process reacts tertiary olefins, such as isobutylene and isoamylene, over an acid resin in the presence of alcohol to form an ether. The reaction chemistry and unit operating conditions are essentially the same as those of a conventional ether process, such as the Hüls MTBE process, except that KataMax† packing has been added to increase the overall conversion.

KataMax packing represents a unique and proprietary approach to exposing a solid catalyst to a liquid stream inside a distillation column. The reactive distillation zone of the RWD column uses KataMax packing to overcome reaction equilibrium constraints by continuously fractionating the ether product from unreacted feed components. As the ether product is distilled away, the reacting mixture is no longer at equilibrium. Thus, fractionation in the presence of the catalyst promotes additional conversion of the reactants. Isobutylene conversions of 99 and 97 per cent, respectively, for MTBE and ETBE are typical, and isoamylene conversions of up to 94 per cent can be achieved economically with this process. These design specifications are typical for gasoline blending; however, practically any olefin conversion is achievable by designing a unit to accommodate individual refinery needs. For example, the Ethermax process can be designed to convert 99.9+ per cent of the isobutylene when butene-1 production is a design objective.

The flexibility of the Ethermax process provides refiners with many routes to increase oxygenate or octane levels in their gasoline pool. Existing MTBE units can be converted to TAME or ETBE production. Increases in throughput and olefin conversion are possible in an existing ether unit by revamping it to the Ethermax process. The revamp increases the oxygenate level of the gasoline pool, and the resulting octane improvement gives a refiner the flexibility to optimize gasoline production from other refinery processes.

Process Flow

The process flow for the Ethermax process is shown in Figure 4.68. The majority of the reaction is carried out in a simple fixed-bed adiabatic reactor. The effluent from this reactor feeds the RWD column, where the ethers are separated from unreacted feed components. The bottoms from the RWD column are the MTBE, ETBE, or TAME product. The unreacted components move up the column and enter the catalytic section of the fractionator for additional conversion. The catalytic section of the RWD column uses KataMax packing to overcome reaction equilibrium constraints by simultaneously reacting the feed component and fractionating the ether product.

The overhead from the RWD column is routed to the alcohol (either methanol or ethanol) recovery section. In this system, water is used to separate the alcohol from the hydrocarbon in a simple countercurrent extraction column, and a distillation column is used to recover the alcohol. The recovered alcohol is recycled to the reactor section. The hydrocarbon raffinate is generally sent downstream to an alkylation unit in the case of C_4s and some of the C_5s. Depending on the gasoline Reid vapor pressure (RVP) specification, the C_5s may be blended directly into the gasoline pool.

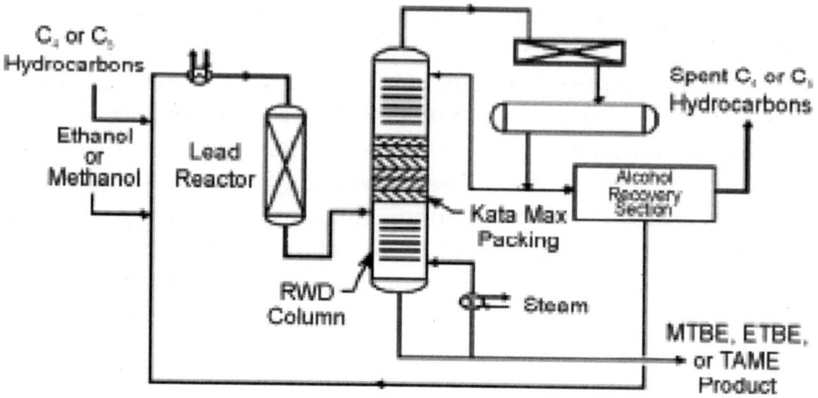

Figure 4.68: Ethermax Process

Yields

The example in Table represents oxygenate production from fluid catalytic cracking (FCC) olefins using the Ethermax process to separately process the C_4 and C_5 cuts. The TAME feed is assumed to be pretreated to saturate dienes.

Table 4.22: Oxygenate Production from Ethermax Process

	MTBE operation, MTA (BPD)	*TAME operation, MTA (BPD)*
Total FCC hydrocarbon feed	174,800 (5632)	200,600 (5960)
Reactive components in the feed:		
Hydrocarbons:		
/C4	33,000 (1052)	—
/C5	—	61,000 (1688)
Alcohols:		
Methanol	18,200 (436)	21,500 (515)
Ethers product:		
MTBE	50,000 (1279)	—
TAME	—	70,000 (1729)

Note: MTA = metric tons per annum; BPD 5 barrels per day; i = iso.

Operating Cost AND Economics

The estimated utilities for a 50,000 metric ton per annum (MTA) (1279 BPD) Ethermax unit producing MTBE and a 70,000 MTA (1729 BPD) Ethermax unit producing TAME, respectively, are given in Table 13.2.2. The 2002 estimated erected cost for an Ethermax unit to produce 50,000 MTA (1279 BPD) of MTBE is $8 million U.S. and $11.5 million U.S. to produce 70,000 MTA (1729 BPD) of TAME including the SHP, based on the inside battery limits of the process unit erected in the U.S. Gulf Coast.

Ethermax Commercial Experience

The first Ethermax process unit was commissioned at the Hüls AG, Marl, Germany, facility in March 1992. As of 2002, another 10 Ethermax process units have been placed onstream. These units process a wide variety of feedstocks from FCC, stream cracking, and dehydrogenation units. The performance of all operating units has exceeded representations. An additional 11 Ethermax units have been licensed. Together these units represent more than 7,925 kMTA (200,000 BPSD) of ethers capacity.

An improved KataMax packing design was commercialized in 1996. The new design further improved the already superior hydraulic capacity of the KataMax packing while maintaining excellent pressure drop and mass-transfer characteristics.

UOP OLEFIN ISOMERIZATION

Introduction

UOP developed olefin isomerization catalysts and process technology primarily for use in a fuel ethers complex where methyl tertiary butyl ether (MTBE) or tertiary amyl methyl ether (TAME) is the desired product. These ethers are reaction products from isoolefins while normal olefins are not reactive.

While presented here integrated in an ethers complex, UOP's olefin isomerization processes can be used in any other application where conversion of normal olefins to branched olefins is needed or desirable.

The idea of skeletal isomerization of normal olefins to isoolefins is not new. During the 1960s, UOP scientists, as well as others, worked on catalyst development based on chlorided alumina. However, these previous catalyst systems were not attractive because they showed poor stability and low selectivity. The increasing demand for oxygenates has resulted in a higher demand for isobutylene and isoamylene to be used in ether production. UOP began active research in 1989 for a skeletal olefin isomerization catalyst. The timing of this demand was fortunate because of the emergence of a number of new catalytic materials. The challenge was to develop a catalyst with high selectivity and stability. With its unique expertise in the development of new materials, UOP developed a proprietary catalyst for skeletal isomerization of light normal olefins. The catalyst was further improved and then successfully produced in a commercial-scale manufacturing trial run..

Description of the Pentesom Process

The UOP Pentesom process isomerizes normal C_5 olefins to reactive isoamylene for conversion to TAME. This unit, coupled with an ethers unit, such as the Ethermax process, maximizes the production of TAME derived from fluid catalytic cracking (FCC) unit C_5 olefins. The high-conversion Pentesom unit normally can increase the TAME production from an FCC unit by 1.7 times that of the stand-alone TAME Ethermax unit. The Pentesom-Ethermax flow scheme consumes more than 80 per cent of the available C_5 olefins in a typical FCC feed stream. This consumption compares to only about 50 per cent C_5 olefin utilization with stand-alone TAME production.

UOP's analysis has shown that, in most cases, operating the Pentesom unit on a oncethrough basis by adding a second Ethermax unit downstream of the Pentesom unit is preferable to recycling the Pentesom effluent to a single Ethermax unit. The block flow diagram of this flow scheme is shown in Figure.

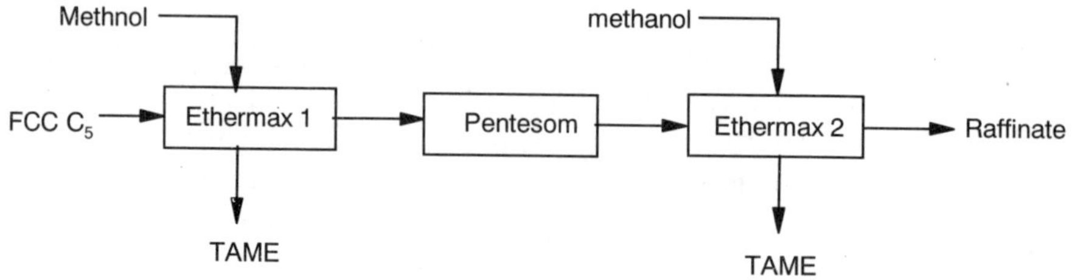

Figure 4.69: Typical FCCU C_5 Processing Scheme

The primary benefit of this flow scheme is a savings in utilities. The FCC unit has a substantial amount of saturate unreactive C_5s contained with the C_5 olefins. When the Ethermax-Pentesom units are operated in a recycle mode, the resulting buildup of normal paraffins consumes both utilities and capacity that are not directed toward TAME production. The normal paraffins must be purged from the recycle loop by a bleed, which also results in a loss of normal pentenes from TAME production. Operation in a once-through flow scheme eliminates these concerns.

Pentesom Process Flow

The Pentesom flow scheme consists of a single reactor containing a high-activity molecu-lar-sieve-based catalyst (Figure). The Ethermax effluent passes through a fired heater and is combined with a small amount of hydrogen before entering the single, fixed-bed Pentesom reactor. The reactor effluent is cooled and condensed before entering a separator. The separator overhead stream, which is rich in hydrogen, is compressed and recycled to the Pentesom reactor. A small amount of makeup hydrogen is added to the recycle stream. Separator bottoms are routed to a stripper. Stripper bottoms are sent to a second Ethermax unit for additional TAME production.

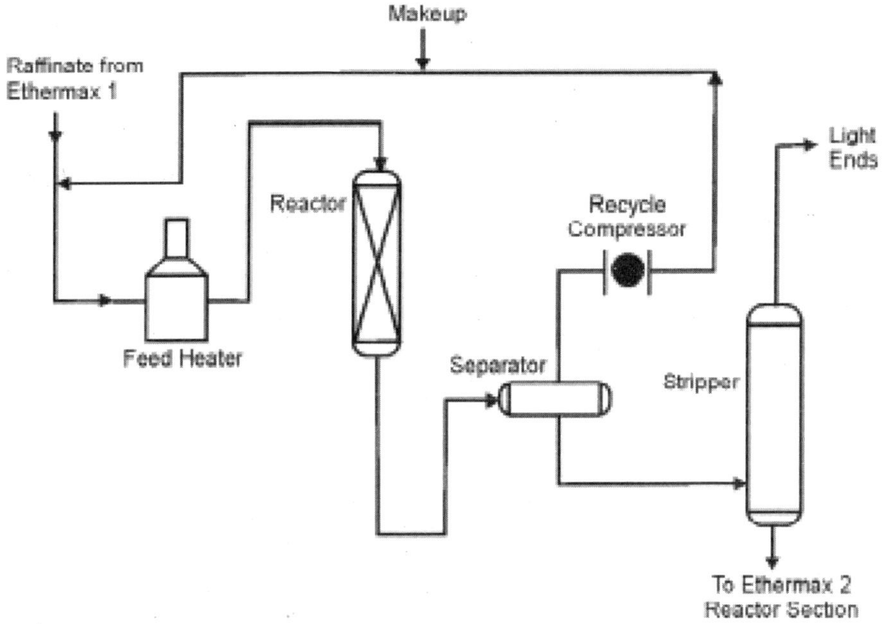

Figure 4.70: UOP Pentesom Process

No feed pretreatment other than that required for the Ethermax unit is required for the Pentesom process. The catalyst operates for 1 year between regenerations. The regeneration is normally conducted in situ using the existing process equipment. No additional regeneration equipment is needed.

Description of the Butesom Process

The Butesomprocess is UOP's C_4 olefin isomerization process. The process isomerizes normal butenes to isobutylene, which can then be further converted to MTBE.

Butesom unit is operated in a recycle mode with an ethers unit, such as UOP's Ethermax process, for MTBE production.

The Butesom-Ethermax flow scheme is targeted for streams rich in C_4 olefins and low in paraffins. Such a stream is that available from naphtha-based steam crackers. In these situations, the raffinate is typically a relatively low-value product because no other conversion processes are available. The olefin utilization for MTBE in these cases is typically less than 30 per cent. The Butesom-Ethermax flow scheme increases olefin utilization to 80 per cent.

The feedstock from the FCC unit contains a high saturate C4 concentration, which is normally not a good fit for a Butesom unit. The primary reason that the Butesom unit is more effective on high-olefin feedstocks is that the conversion of normal butene to isobutylene is equilibrium-limited to 40 to 50 per cent under normal operating conditions. A recycle operation is used to achieve higher overall conversion of n-butene above the equilibrium limit. However, the high paraffin concentrations present in FCC feedstocks require a substantial bleed on the recycle loop to purge the paraffins. The recycle bleed drags butenes out of the flow scheme, thereby reducing the availability of butenes for conversion to MTBE.

Butesom Process Flow

This simple unit uses a molecular-sieve-based catalyst and swing reactors. The Ethermax effluent passes through a combined feed exchanger and fired heater before entering one of the swing fixed-bed Butesom reactors. The reactors are operated in a swing mode with one reactor on-line and the other in regeneration. The reactor effluent is exchanged with the fresh feed in the combined feed exchanger and cooled before being compressed and condensed in the effluent compressor system.

OXYPRO PROCESS

The UOP Oxypro process is a unique, low-cost, refinery-based catalytic process for the production of di-isopropyl ether (DIPE) from propylene and water. The ether DIPE has high octane, low vapor pressure, and excellent gasoline blending properties.

The Oxypro process is especially well suited for processing propylene derived from the fluid catalytic cracking (FCC) unit within the refinery.

After amine and Merox* treating, the FCC-derived propylene is fed directly along with water to the Oxypro process. The propylene and water are converted to DIPE at more than 98 wt per cent selectivity. The Oxypro product has a purity of more than 98 wt per cent DIPE and a research octane number clear (RONC) and motor octane number clear (MONC) that are comparable to other ethers, such as MTBE and TAME.

The Oxypro product shows a clear octane advantage over both catalytic polymerization and alkylation of propylene. DIPE from the Oxypro process generates 112 RONC and 98 MONC compared to only 90 RONC and 89 MONC for C_3 alkylate and 93 RONC and 82 MONC for catalytic polymerization gasoline. The combination of high-octane product and near 100 per cent overall conversion gives the Oxypro process superior performance compared to other refinery C_3 alternatives.

Process Flow Scheme

A simplified flow scheme of the Oxypro process is shown in Figure. The amine- and Merox-treated mixed-C_3 stream from the FCC unit enters the unit and is mixed with makeup water and internal recycle streams of propylene, isopropyl alcohol (IPA), and water. The combined streams are processed downflow in a fixed-bed reactor. The reactor effluent is sent to fractionation, where the light ends and propane are removed. The propane product meets typical liquefied petroleum gas (LPG) specifications of less than 5 wt per cent propylene. The fractionation product is then sent to product recovery, where water, IPA, and DIPE are eparated. The IPA and water are recycled to the reactor to maximize DIPE production. The DIPE product has a purity of more than 98 wt per cent.

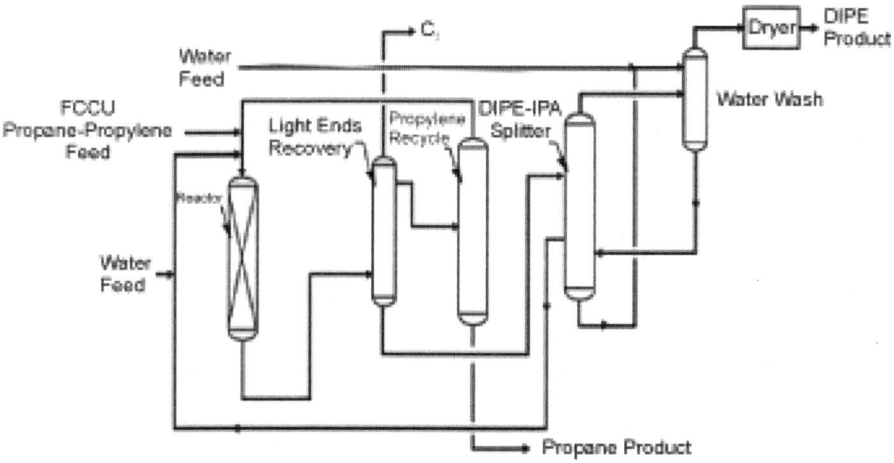

Figure 4.71: Oxypro Flow Scheme

Yields

The yields from an Oxypro unit designed to produce 96,000 metric tons per year (MTA) [2500 barrels per day (BPD)] of DIPE product are shown in Table.

Table 4.23: Oxypro Unit Yield Summary

	BPSD	*kg/h*
Feed:		
Propylene	2,918	10,050
Propane	1,239	4,100
H_2O	328	2,170
Product:		
LPG	1,300	4,310
DIPE	2,500	12,010

Note: BPSD = barrels per stream-day.

The feedstock used to generate these yields is representative of a mixed propane-propylene stream from an FCC after amine and Merox treating. Specifications typical for an Oxypro unit DIPE product are shown in Table.

Table 4.24: Typical Oxypro Unit Product Specifications

Specific gravity	0.73
DIPE, wt %	98
Water, wt ppm	<100
IPA, wt %	<0.5
C_6+,wt %	<2
Octane:	
RONC	112
MONC	98

Operating Costs and Economics

The estimated inside-battery-limits erected cost of an Oxypro unit built on the U.S. Gulf Coast in 2002 with a capacity to produce 88,500 MTA (2300 BPD) of DIPE is about $26 million.

HYDROGEN PROCESSING

Introduction

A Historical Perspective

This chapter provides a historic perspective of the importance of hydrogen processing in the refining industry. The technologies referred to in this chapter are dealt with in greater detail in the respective technology chapters.

The potential for applying hydrogenation reactions within the refining industry has been known since the early years of the twentieth century. In October 1930 the American Chemical Society conducted a symposium in Cincinnati on the subject of "Industrial HighPressure Reactions." In one of the papers, Haslam and Russell discussed the five adaptations of hydrogenation which appeared to be of the most immediate importance.

Horne and McAfee in 1958 noted that of these five, the second, third, and fourth were already being practiced, and they predicted that the first and fifth would also soon be commercialized. They were right. Such reactions are now commonplace in modern refineries, largely because of the plentiful supply of hydrogen produced either in catalytic reformers or from inexpensive sources of natural gas.

The modern version of distillate hydrocracking was introduced in the United States in the 1960s to convert excess fuel oil to predominantly motor gasoline and some jet fuel. Fluid catalytic cracking (FCC) cycle oils were popular feedstocks at first. The process was then used to upgrade the liquids produced in delayed and fluid coking and solvent deasphalting. While these residuum conversion schemes were being implemented, a demand for low-sulfur fuel oil (LSFO) developed and fixed-bed residuum hydrotreating was commercialized. As LSFO demand declined, these hydrotreaters were used to prepare feed for FCC units. On-stream catalyst replacement technology has been commercially successful in extending the capabilities of fixed-bed residuum hydrotreaters. Ebullating-bed hydrocrackers have now found Deceased. Wide commercial application to processing various types of residues containing high levels of contaminants such as sulfur, nitrogen, metals, and asphaltenes.

Since the early distillate hydrocrackers were built, the demand for motor gasoline has not grown as quickly as that for middle distillates. The more recent hydrocracking units have been designed, therefore, to make good-quality kerosene and diesel fuel. Besides this trend, hydrocracking is steadily replacing conventional extraction processes in lube oil base stock manufacture because it can produce much more valuable by-products than the older process. The introduction of Chevron's ISODEWAXING and hydrofinishing process in conjunction with lube ISOCRACKING gave refiners the opportunity to produce highquality base oils with viscosity indexes greater than 120.

Our understanding of the basic reactions, catalyst development and selection, and the subtleties of applying hydroprocessing to various applications has advanced considerably; and hundreds of technical articles have been written on the subject. This chapter summarizes the current understanding. By reviewing the literature references, the reader can explore the background of the subject in greater depth. Emphasis will be placed on the hydroprocessing of middle distillate and heavier feedstocks, since this is the field that has seen most of the advances.

For clarity, throughout the chapter we will use the term hydroprocessing to describe all the different processes in which hydrocarbons react with hydrogen. Hydrotreating will be used to describe those hydroprocesses dealing predominantly with impurity removal from the hydrocarbon feedstock. Hydrocracking will be used to describe those hydroprocesses that accomplish a significant conversion of the hydrocarbon feedstock to lower-boiling products.

Hydro processing Objectives

Hydroprocessing feedstocks—naphthas, atmospheric gas oils, vacuum gas oils (VGOs), and residuum—have widely different boiling character. Within each of these different boiling ranges exist a variety of molecular types. This depends on both the crude oil source and whether the material was produced in a cracking reaction or as a straight-run component of the original crude oil. The vacuum residuum is the lowest-value fraction in the crude oil. Historically it has been blended into heavy fuel oil. The demand for this product, however, has not kept pace with the tremendous increase in demand for transportation fuels. Environment pressures have widened this gap by restricting the use of high-sulfur fuel oil while mandating cleaner light products.

The introduction of residuum hydroprocessing in the 1960s was a response to an increasing demand for LSFO to replace the high-sulfur heavy fuel. Over the years the increased demand for light products has focused attention on converting residuum to higher-value products. Products from residuum hydrotreating are often fed to a fluid catalytic cracking unit to produce good-quality motor gasoline. In this case the removal of nitrogen, Conradson carbon, and metal contaminants in the hydrotreater is just as important as sulfur removal. Hydroprocessing feedstocks which boil in the 650 to 1050°F VGO range can be straight-run stocks or stocks produced in cokers, thermal crackers, visbreakers, or other VGO boiling-range materials produced from residue conversion units. Again these are often processed to produce either LSFO or FCC feed. Sometimes they are hydrocracked to produce diesel, kerosene jet fuel, and/or naphtha. The hydrocracked heavy products are also excellent ethylene plant feedstocks or lube oil base stocks (when produced from the hydroprocessing of straight-run feedstocks) because the process removes the less desirable heavy aromatics.

Olefin Production from Methanol

The conversion of natural gas into syngas is the first step in utilizing natural gas for methanol and for the conversion of gas-to-liquids hydrocarbon products. Methanol as a base chemical offers limited opportunities for natural gas utilization unless linked to other derivative markets. Gas-to-liquids (GTL) technology using Fischer-Tropsch type of catalysts offers large market opportunities for natural gas utilization but is challenged by the economics of high capital costs with relatively low transportation fuel product values.

Syngas and methanol production technologies are achieving greater economies of scale. World-scale methanol production facilities have doubled in size compared to just a few years ago, and when combined with remote natural gas prices, these facilities offer substantially lower costs of production than those in existing plants. However, although methanol can be economically shipped from remote gas areas, the expected growth in demand for methanol for conventional uses does not support the addition of many new plants.

The conversion of methanol to fuel components was accomplished commercially in the Mobil MTG (methanol-to-gasoline) process at a plant located near New Plymouth in New Zealand, but that plant has since been shut down on account of the relatively poor economics of gasoline production. However, Mobil did demonstrate that over a ZSM-5 (MFI) type zeolitic catalyst, methanol could be converted to a largely aromatic product, up to durene, but also with a significant proportion of olefins, principally propylene. Lurgi has recently developed a modified version of this process that minimizes the production of the gasoline fraction and maximizes production of propylene at about 70 per cent; this process is known as MTP for methanol-to-propylene.

Methanol can also be converted to ethylene and propylene via the UOP/Hydro MTO process, thus opening new opportunities for methanol utilization. Ethylene and propylene can then be used to satisfy the growing market demand for polyolefins or can be used in the production of other olefin derivatives. Remote gas strategies for MTO generally consider either the shipping of methanol from remote locations to countries with strong olefin demand or shipping of polymer pellets from fully integrated gas-to-polyolefin (GTP) facilities in remote locations.

Introduction

Figure illustrates various alternatives for the utilization of natural gas in the production of either liquid fuels or petrochemical derivatives. There is at present renewed interest and considerable activity in the planning and construction of large-scale GTL facilities. GTL technology is attractive because it offers great potential for the valorization of stranded gas by taking advantage of the large markets for fuel products. The implementation of GTL and other large-scale gas conversion projects is challenging because investments are high, technologies are often not well proven at the actual scale, there is competition with crude-oil-based products, and plant location is often in remote areas. One way to enhance the economics of GTL projects is to produce products with higher added value. This can include the recovery of normal paraffins for linear alkyl benzene (LAB) production, specialty lube oils, methanol, olefins, and polyolefins. The markets for some of these products can limit the opportunities for production in GTL facilities. The olefin and polyolefin markets, however, are exceptionally large, and these products offer very high added value.

Both GTL and GTP facilities incorporate sizable front-end syngas units for the processing of natural gas, as illustrated in Figure. These units are the major contributors to the relatively high investments required for these complexes. It follows that the integration of these facilities could offer substantial synergistic savings. Potential savings for integrated GTL/GTP complexes may derive from:

- Shared syngas plant,
- Shared utility systems with by-products utilization,
- Shared wastewater treatment facilities,
- Shared administration, laboratory, and maintenance facilities,
- Minimal intermediate-product storage facilities needed.

There are other advantages for integrated facilities, such as advantages of back-integration for polyolefin production:

- Lower cash costs of production,

- ➢ Elimination of costs for intermediate-products shipping and handling,
- ➢ Consistency and better control over feedstock quality.

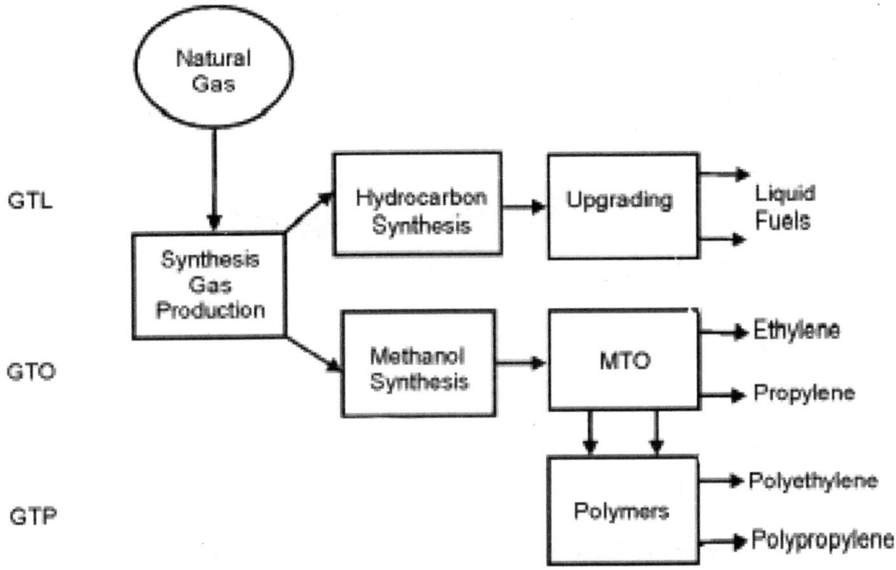

Figure 4.72: Conversion Steps for Natural Gas to Liquids and Polyolefins

When one is considering an integrated facility to convert natural gas to polyolefins (GTP), there are three main process technologies involved. These technologies must fit together at world-scale capacities for an ideal integration. Each of these technologies is discussed further in the next paragraphs.

Syngas/methanol process technology is available from several well-known licensors. Until recently, world-scale capacity for methanol production was considered 2500 to 3000 metric tons per day (MT/D). Now there are a number of projects underway with capacities of 5000 MT/D, and licensors of syngas/methanol technologies are discussing capacities as high as 10,000 MT/D. Most of the new capacity that has recently come, or will soon be coming, on-stream reflects a growing trend in which methanol production has been shifting from heavily industrialized countries to locations with access to lower-priced natural gas.

The combination of large-scale production facilities with low-priced natural gas feedstock substantially reduces the fundamental costs of methanol production, as shown in Figure.

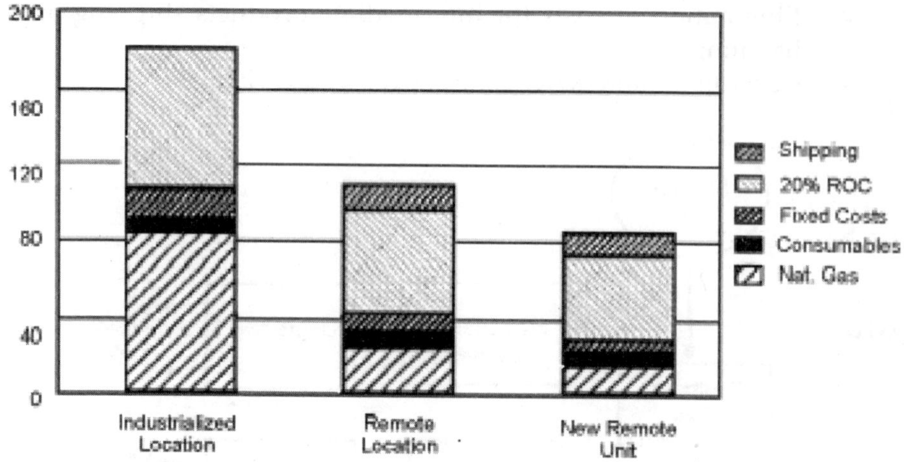

- "Industrialized Location" is based on 1500 MT/D Capacity with $2.50/million Btu gas
- "Remote Location" is based on 3000 MT/D capacity with $0.75/million Btu gas
- "New remote Unit" is based on 6000 MT/D capacity with $0.50/million Btu gas

Figure 4.73: Examples of Methanol Production Costs

Many smaller plants exist today in industrialized locations. The cash cost of production for these plants is typically more than $100/MT of methanol, due primarily to the cost of natural gas. If capital charges are added to provide, say, a 20 per cent return on capital (ROC), then the delivered price of methanol is almost $180/MT. Large-scale plants in remote locations experience a considerable advantage due to low cash costs and economies of scale, even after accounting for the costs of shipping to distant markets. For these remote units the cash costs of production can be less than $50/MT. These units can deliver methanol at about $110/MT, even after adding capital charges and costs for shipping. New mega-scale projects enjoy an even greater advantage and can achieve attractive project economics with methanol delivered at prices less than $90/MT. This enables new applications for methanol such as fuel cells or conversion to olefins and offers large market growth potential.

Polyolefins are widely produced using technologies available from several licensors and may include flexibility to produce several grades of homopolymer and copolymer products. World-scale capacity for polyethylene processes is generally considered in the range of 300 kilometric tons per annum to 350 (kMTA). World-scale capacity for polypropylene processes is generally considered in the range of 250 to 300 kMTA.

THE SYNTROLEUM PROCESS OF CONVERTING NATURAL GAS INTO ULTRACLEAN HYDROCARBONS

Introduction

The term gas-to-liquids (or GTL) refers to a process that converts natural gas into synthetic, ultraclean liquid fuels that contain no sulfur, aromatics, or heavy metals. GTL technology is increasingly being considered as a tool for monetizing stranded gas reserves and for mitigating some of the environmentally driven challenges that are currently confronting the petroleum industry.

From the perspective of energy supply, there are increasing pressures to monetize ever-growing reserves of natural gas. Currently, proven world reserves of natural gas are estimated to be in excess of 5200 Tft^3. More than one-half of these resources are considered stranded—uneconomic to transport due to the proximity of potential markets and/or the size of gas deposits, or subquality because of contamination by nitrogen, carbon dioxide, hydrogen sulfide, aromatics, etc. These stranded reserves represent potentially large additions to oil company balance sheets if they can be monetized.

From the perspective of energy demand, the petroleum industry faces a number of challenges with fuels derived from conventional crude oil—environmental mandates for cleaner fuels, raw materials getting heavier and dirtier, process upgrading investments becoming more difficult to justify, and tightening supplies of middle distillates and naphtha. Coupled with improved economics, the possibility of meeting these challenges with GTL technology is spurring increased interest in its commercialization.

This chapter summarizes Fischer-Tropsch (FT) chemistry and provides an overview of Syntroleum's GTL technology, including descriptions for each major section of the process. Quality of fuels produced from a GTL plant is also discussed, including descriptions of emissions, fuel properties, and applications. The report concludes with discussion on the advantages and economics of the Syntroleum Process.

History of Fischer-Tropsch Reaction

In 1923, Franz Fischer and Hans Tropsch of Germany observed that alkalized iron turnings in an atmosphere of carbon monoxide and hydrogen at elevated temperatures and pressures catalyzed the production of oxygenated compounds and hydrocarbons. After 2 years of experimentation, the first Fischer-Tropsch catalyst was reported and consisted of an ironzinc preparation. After the pioneering work of Fischer and Tropsch in the 1920s, it was several years before Fischer and Kuster first published a report of liquid (slurry) phase carbon monoxide hydrogenation. The earliest known liquid-phase FT patent, awarded to Dr. Mathias Pier of I. G. Farbenindustrie in 1936, was applied for in late 1928, predating Fischer's publications by about 3 years. Subsequent patents issued in the United States and a number of European countries in the late 1930s through the early 1950s indicate a fairly high degree of R&D was occurring immediately before and during World War II. The major players were I. G. Farbenindustrie and Ruhrchemie in Germany, while liquid-phase/slurry FT patents were also issued to a number of U.S. companies (M. W. Kellogg Company, Celanese Corporation, The Texas Company, and Standard Oil Development Company). Today, the leading GTL technology holders are Syntroleum, Sasol, Shell, BP Amoco, and ExxonMobil.

History of the Syntroleum Process

Syntroleum's history began in 1984 when the founder, Ken Agee, began experimenting with the use of air as a source of oxygen for the GTL process. After several years of experimenting, the first series of patents were filed in 1988 that were later issued in 1989. In 1990, Syntroleum began the operation of its first pilot plant, which still operates today. The pilot plant has a production rate of 2 barrels per day (BPD) and is used to test a variety of operating conditions for various catalysts. In 1998, Syntroleum's research and development facilities quadrupled with the purchase of a new building and the surrounding 100 acres. During the following year, Syntroleum, along with Arco (now BP), began the operation of a 70-BPD facility using slurry reactor designs. After approximately one year of operation, the demonstration plant was relocated to Tulsa. The demonstration plant will begin operation in 2003 as part of the U.S. Department of Energy's ultraclean fuels program. Fuel demonstrations are planned in the bus fleets of Washington, D.C., Metropolitan Area Transit Authority and Denali National Park. As of 2001, Syntroleum had licensed its technology to several companies, including Texaco, Marathon, Arco, YPF, Enron, Kerr McGee, Ivanhoe, and the Commonwealth of Australia. Syntroleum has also formed many alliances related to fuel production, testing, and certification; catalyst testing and production; and the design of both onshore and floating GTL plants.

Process Description

The Syntroleum Process consists of the following steps for converting natural gas to finished fuel products: (1) gas pretreating, (2) air compression, (3) synthesis gas generation, (4) Fischer-Tropsch synthesis, and (5) product upgrading. The process converts approximately 10,000 standard cubic feet (SCE) of gas into 1 barrel (bbl) of synthetic hydrocarbons. The maximum capacity of each GTL train is about 11,500 BPD. The process produces several by-products, tailgas, water, and heat, which can be used to generate other products from the plant, such as exportable electricity and steam for desalination. The Syntroleum Process can be adapted for either offshore or onshore applications and can be designed for small gas fields. Figure is a process flow diagram of the inside-battery-limits (ISBL) portion of the Syntroleum Process.

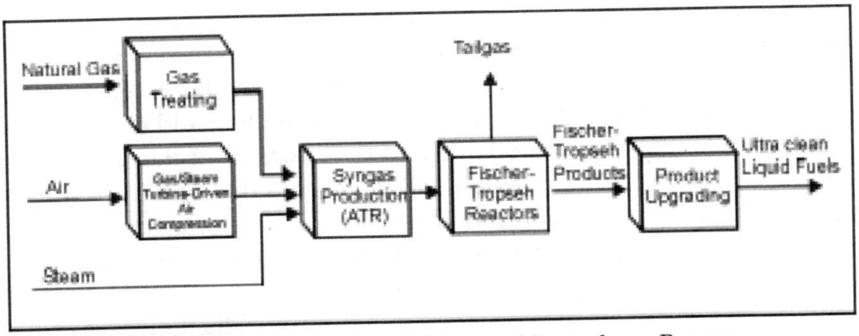

Figure 4.74: Process Flow Diagram of Syntroleum Process

Table 4.25: Syntroleum Gas Feed Specifications

Components	*Limit*
Ethane and heavier	Less than 5 mol %
CO_2	Less than 20 mol %
N_2	Less than 25 mol %
Sulfur compounds	Less than 10 ppm

Feed Gas Pretreating

The feed gas pretreating section is determined by the gas composition and Syntroleum's feed gas specifications, which are based upon those components that are detrimental to the catalytic reactor systems. The gas feed specifications are summarized in Table.

For example, for a gas composition that does not exceed the limitations in Table, minimal gas pretreating is required. In this case, sulfur removal is the only gas treatment needed. Cobalt/molybdenum (Co/Mo) and zinc oxide (ZnO) guard beds are used to reduce the sulfur concentration to 100 ppb for the catalytic processes. When the limitations of CO_2 and/or sulfur are exceeded, the cost of treatment determines whether the gas is feasible for use. Finally, gas that exceeds the specification for ethane and heavier molecules either can be processed to recover and sell natural gas liquids (NGLs) for additional revenue or can be processed to reform the C2+ components into methane for use as feedstock. The choice will depend on the amount of recoverable NGLs and market conditions. Overall, the robust gas specifications of the Syntroleum Process enable many gas fields that are considered marginal to be monetized.

Air Compression

Unique to the Syntroleum Process is its use of air as an oxygen source. A process air compressor driven by a steam or gas turbine supplies the synthesis gas generation section with the stochiometric amount of oxygen (as contained in air) required to generate a synthesis gas ratio of approximately 2:1 (H_2:CO). The inlet air is filtered to avoid corrosion and erosion of the rotating components and deactivation of the downstream catalyst. Gas turbines are commercially available that will burn low-Btu tailgas, a by-product generated from the FT reactors. In addition to providing fuel value, this tailgas stream represents a large mass flow of high-pressure gas that aids in the generation of shaft power from the expander end of the gas turbine. For a steam turbine, the tailgas is utilized for the production of steam.

Shell Middle Distillate Synthesis (SMDS) Process

Gas-to-Liquids Conversion

Most of the world's natural gas (NG) resources are remote from the market. Their exploitation is constrained by high transportation costs and by market outlets. To improve economic utilization of NG, techniques other than transport by pipeline or physical liquefaction (LNG) have been developed. Chemical conversion [gas to liquids (GTL)] to easily transportable and marketable liquid hydrocarbons is now a proven technology.

One driving force for GTL technology is the market pull for clean transportation fuels, in particular gas oil (diesel) fuel. NG conversion produces an exceptional quality of gas oil, free from sulfur and aromatics, with excellent (blending) properties to meet future anticipated gas oil requirements anywhere in the world. Another driving force is the LNG market; limitations to absorb all potential LNG supplies are becoming apparent. Conversion of natural gas to middle distillates provides development of remote gas without dependence on gas markets; the products are marketed into the huge global market for distillates. A large GTL plant with 600 million SCF/day NG intake produces some 3 million TPA middle distillates, some 0.2 per cent of today's 2000 million TPA middle distillate global demand. Current world LNG market is approximately 100 million TPA and methanol 30 million TPA (Figure)

Introduction

Gas-to-liquids technology is a method to convert natural gas into liquid hydrocarbons. The Fischer-Tropsch process for synthesis of hydrocarbon fuels using metal catalyst at low pressure was patented in 1926. It was used in Germany during World War II and in South Africa for coal conversion to replace oil imports. For a long time the process was not economically viable. For natural gas conversion this is now changing with the development of new and efficient technology.

The potential of GTL technology has resulted in a number of companies searching for opportunities to apply the technology for valorization of remote gas reserves. GTL technology offers a new way to produce middle distillates without coproduction of refinery residuals.

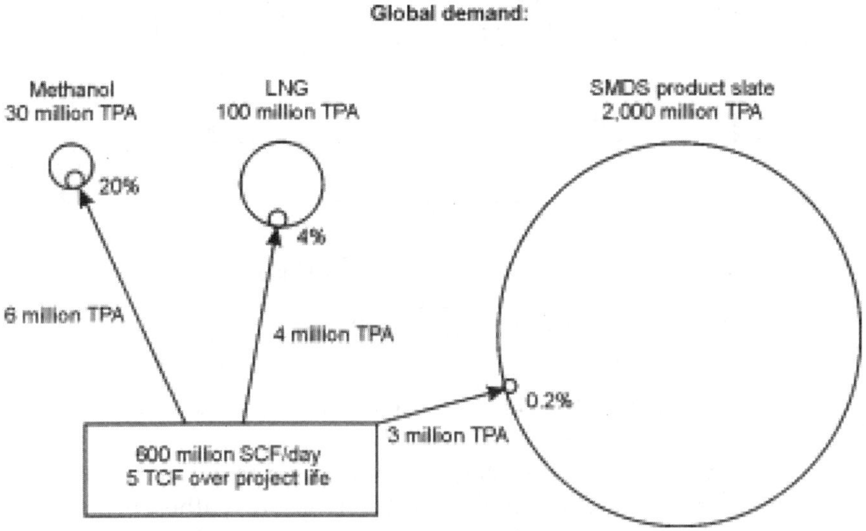

Figure 4.75: Market Impact of NG-derived Products

The first commercial SMDS project was approved in 1989; the plant was constructed in Bintulu, Sarawak, Malaysia (Figure). Production started in 1993, some 20 years after first research efforts were initiated by Shell.

$CH_4 + \frac{1}{2}O_2$ ® $2H_2 + CO$
803 MJ/kmol 767 MJ/kmol
100% 96%

$2H_2 + CO_2$ ® $-(CH_2)- + H_2O$
767 MJ/kmol 621 MJ/kmol
100% 96%
96% 78%

Process Description

The basic conversions of SMDS involve partial oxidation of methane into synthesis gas and subsequent Fischer-Tropsch conversion to paraffins. The theoretical thermal efficiency of this route is 78 per cent on the basis of NG LHV:

The three main process stages are shown schematically in Figure.

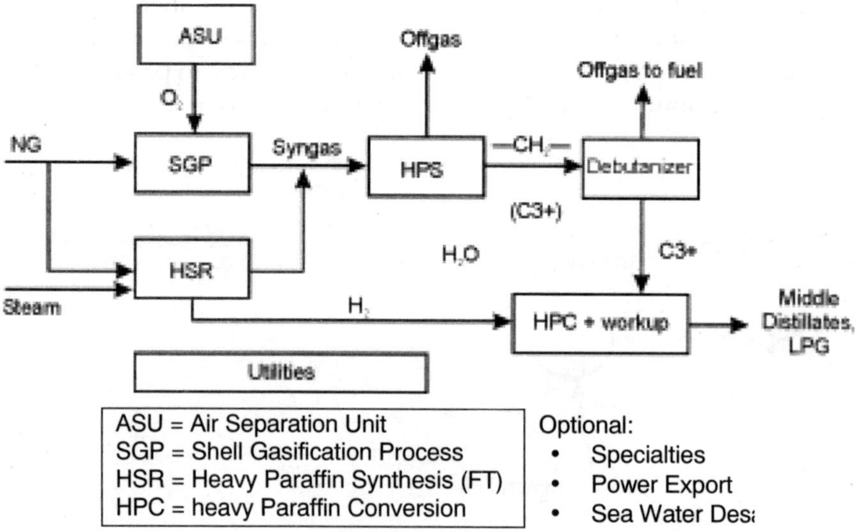

Figure 4.76: Shell Middle Distillate Synthesis Process Scheme

In the SMDS process, these stages are identified as syngas manufacture, heavy paraffin synthesis (HPS, the Fischer Tropsch synthesis), and heavy paraffin conversion (HPC). These stages will be described here. Several support and utility blocks are added.

Syngas Manufacture (SGP, Shell Gasification Process)

Synthesis gas, a mixture of hydrogen and carbon monoxide, is one of the most versatile feedstocks for a wide range of (chemical) processes. In GTL technology, the conversion trajectory from methane to liquid hydrocarbons uses syngas as an intermediate. Direct conversion of methane to hydrocarbon chains with economic selectivity and conversion is not (yet) possible.

Syngas manufacture in SMDS is relatively expensive; between 50 and 60 per cent of total process capital costs are related to syngas production.

Chapter 5

Production Equipment

WELLHEADS AND FLOW CONTROL EQUIPMENT

Wellheads provide the control mechanisms between downhole and surface equipment. As the well is drilled, casing is placed at intervals specified by the well design. The casing setting depth can be determined by abnormally pressured zones, lost circulation zones, sticky formations, or various other reasons as dictated by specific situations.

Wellhead designs must be capable of withstanding wide ranges of temperature and pressure, as well as various corrosive agents. Standard API pressure ratings for wellhead equipment are 2000-, 3000-, 5000-, 10,000-, 15,000-, and 20,000-psi working pressures, While API temperature ratings range from 75° to +250° F (higher pressure and temperature ratings are available for special service). Corrosive environments of carbon dioxide (CO_2), hydrogen sulfide (H_2S) and chlorides (Cl^-) must also be considered in the design process.

API Spec 6A, Specification for Wellhead and Christmas Tree Equipment, establishes the basic requirements for end connections, materials, test procedures, and pressure ratings so that equipment made by various manufacturers will work together.

API Spec 6A also outline the necessary dimensional data that will allow wellheads manufactured by an API manufacturer to be interchanged during field use. Manufacturers do not have to use these standards unless they want to use the API monogram on their products.

Casing and Casing Programs

The wellhead equipment is designed to suit the casing size and number of strings planned for any given well. It is therefore important to gain a basic understanding of casing and tubing use in the drilling and producing operations.

Conductor casing, or drive pipe, is a short string of casing of large diameter (16 to 48 in.) required for offshore operations, swampy locations, and other conditions in which extra wellhead support is necessary. Its principal function is to keep the top of the wellbore open and to provide a means of conveying the drilling fluid returns from the wellbore to the mud pit. The depth to which it is set varies but is usually 100 to 400 ft. The part of the wellhead on the top of the conductor string is the base plate, which is either an integral part of the casing head or a fabricated (welded-on) plate of steel connected to the casing head by gussets.

The surface casing is the first string of casing run on a land well and is generally thought of as the foundation of the wellhead. The size generally ranges from 8 5/8 to 20 in. outside diameter (OD), and the length of this string varies greatly in different areas, from 200 to several thousand feet. It is set at a depth sufficient to protect all fresh water-bearing sands and to prevent an underground blowout. The cement is circulated to the surface on this string. The part of the wellhead at the top end of this string is the casing head, which is either screwed or welded onto the last joint of casing. The casing head top connection flange is then used to install the blowout prevention stack, which allows for deeper drilling.

The intermediate casing may be the second casing string that is run in a well. Its purpose is to allow for deeper drilling by isolating weak formations that could either cause the hole to cave in or lose circulation. The size range for intermediate casing is typically 7 5/8 to 13 3/8 in. OD. Setting depths vary to meet regulations and geological conditions in a given well, but commonly range from 3000 to 8000 ft. A well may have one or two strings of intermediate casing, or it may have no intermediate casing at all. The part of the wellhead at the top end of an intermediate casing string is a casing spool. Some form of isolation seal is used to seal the top end of this casing to the bottom bowl of the casing spool. This seal makes possible the increase in pressure ratings necessary for the higher pressures encountered in deeper drilling. The intermediate casing may be cemented all the way back to the surface, but generally it is cemented back to the end of the previous string or far back enough to isolate a particular formation.

The final string of casing is called the production casing. It is usually set to or beyond the formation that is to be produced. In either case, this will be the effective total depth (TD) of the well. The production casing isolates all the other formations from the producing zone or zones and is generally cemented, like the intermediate casing, back to the end of the previous string or far back enough to provide the necessary isolation of the producing zone or zones. Since the production casing cannot be easily replaced, and since a smaller string can produce the oil in a more efficient manner, tubing is usually installed inside the production casing. The tubing can be sealed off to the inside surface of the production casing by a packer or some other sealing device. Production casing typically ranges in size from 4 1/2 to 9 5/8 in. OD. The part of the wellhead that works over the top end of the production casing is the tubing head. Some form of isolation seal is used between the top end of the production casing and the bottom bowl of the tubing head. This seal allows for pressure rating increase and isolates the production casing from the rest of the wellhead. The strength of the production casing string must be sufficient to contain the full working pressure of the well.

The final string of tubular goods that goes into the well is the production tubing. Unlike casing, it is not cemented in the well. It is supported and sealed by hanging it inside the top bowl of the tubing head, and it can be packed off down below with a packer. The packer seals the OD of the tubing string to the inside diameter (ID) of the production casing. Tubing can be replaced when damaged; also, the well can be deepened or plugged back and a new tubing string used to accommodate the new depth. The annular space between the OD of the tubing and ID of the production casing can also be used to lift the fluids artificially from the well or to inject chemicals when the well is being produced. The parts of the wellhead at the top end of the tubing are the tubing head adapter and the production tree. Tubing size may range from 1 to 7 in. OD, and the bore size of the tree is sized accordingly.

Table (below) summarizes the type of wellhead for each tubular classification and the OD, setting depth and pressure ratings that would generally correspond.

Table 5.1: Tubular Goods and Wellhead Components

String	*OD Range (inches)*	*Depth Set (feet)*	*Wellhead*	*Pressure Range (psi)*
Conductor	16-48	100-400	Base Plate Fitted or Attached to Casing Head	500-1500
Surface Casing	8 5/8-20	200-2000	Casing Head	2000-3000
Intermediate Casing	7 5/8-13 3/8	5000-8000	Casing Spool	2000, 3000, 5000
Production Casing	4 1/2-9 5/8	Total Depth 10,000	Tubing Head	3000, 5000, 10,000, 15,000
Production Tubing	1-7	Near Total Depth	Tubing Head Bonnet and Production Tree	3000, 5000, 10,000, 15,000

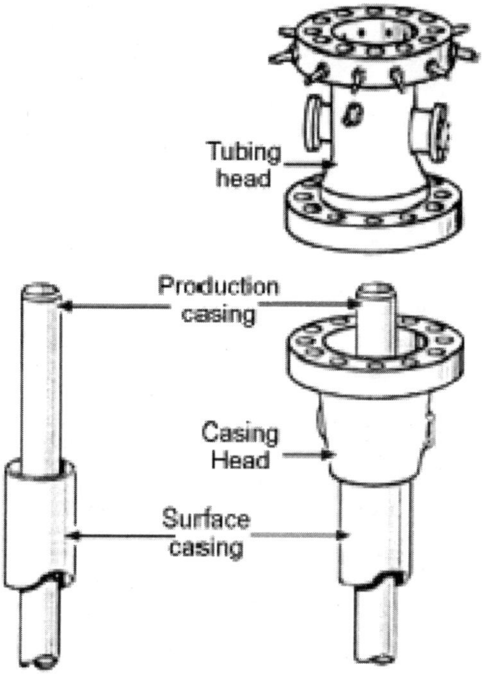

Figure 5.1: Wellhead Configuration for Well with two Strings of Casing

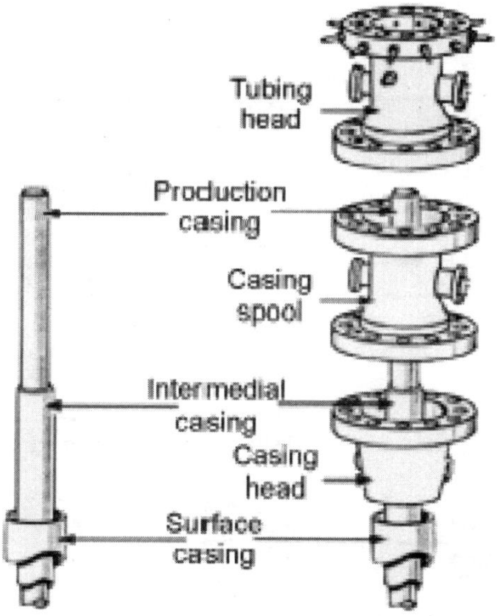

Figure 5.2: With three Strings of Casing

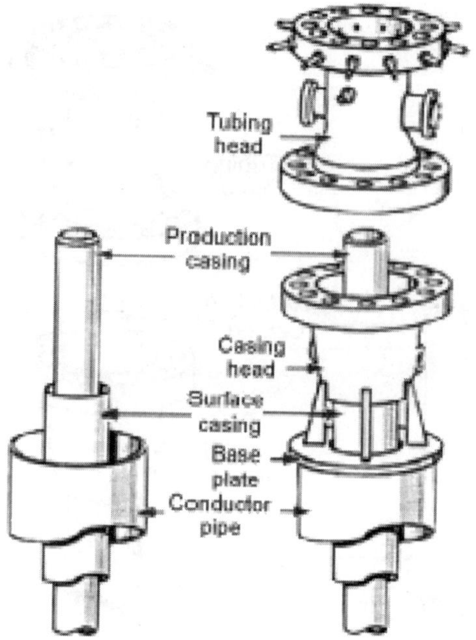

Figure 5.3: With two Strings of Casing and Base Plate

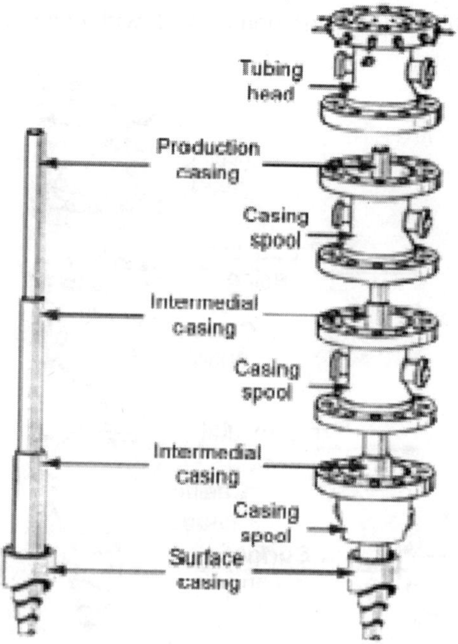

Figure 5.4: (with four strings of casing) Depict four Wellhead Configurations

Casing Heads

The casing head, which is also called the starting head or bradenhead, serves as an intermediate connection between the casing, well control equipment (e.g., blowout preventers) and subsequent casing and tubing spools. The casing program and anticipated pressure ranges will determine the basic casing head design.

Since the casing head is the lowest section of the wellhead assembly, it is subject to the weight of all future casing and tubing strings, plus the weight of any additional surface equipment. The casing head also provides a means by which the next casing string can be centered, supported, and sealed. This is achieved by a load shoulder and controlled bore on which the casing hanger is supported and the annular seal effected. In addition to this, the casing head must provide a means to adapt and connect well control equipment and seal the bore from the atmosphere. And, lastly, the casing head must provide a means of controlled access to the wellbore for pressure control and fluid returns during drilling operations. A base plate may be used to help effectively distribute the weight when extreme loading, due to casing size and hole depth, is incurred. The base plate may be forged as an integral part of the casing head or attached separately with welded gussets. In the case of offshore wells, the base plate is supported by the conductor pipe. On land wells, the base plate may rest on the ground or on a concrete slab prepared for this purpose.

Casing heads are available with either a threaded or slip-on weld bottom. In general, the weld connection is preferred when there is a chance that the casing will stick high. With the welded connection, the casing can be cut at any desired point.

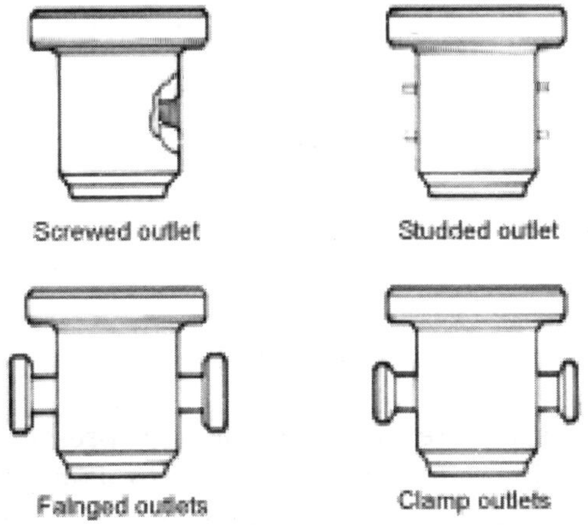

Figure 5.5:

The top connection can be either a flange or a clamp hub. The use of clamped connections is generally attributed to their much faster makeup, lighter weight, and smaller OD.

Casing heads usually have two side outlets that are the same size; in rare instances, one outlet might be larger than the other. Four types of outlets are available: threaded, flanged, studded flange, and clamp hub

Bowl designs vary by manufacturer, but two common designs exist. One includes a load shoulder designed to bear the load of the casing transferred via the casing hanger; and a vertical or near-vertical profile, providing a metal-to-metal seal area to isolate the casing annulus created by the next casing string. The other profile offers a tapered seal area both to provide the seal and carry the load.

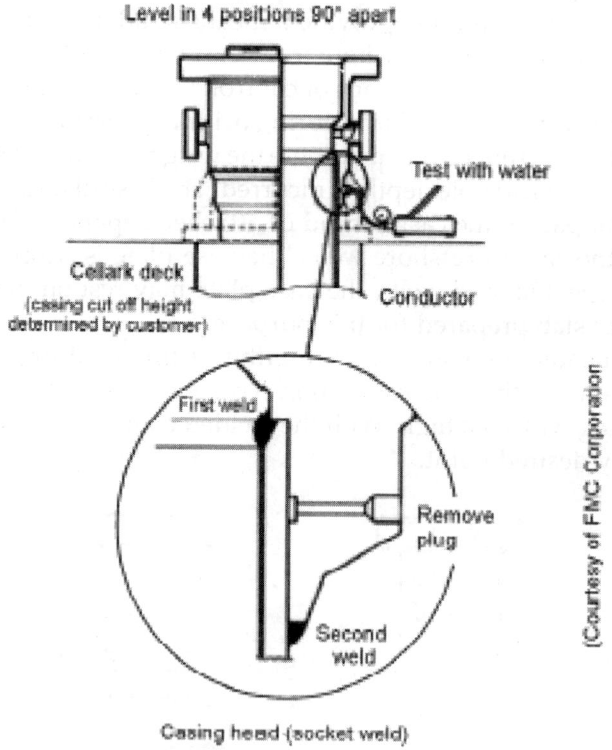

Figure 5.6:

Metal-to-metal seals are preferred for high pressures and corrosive surfaces. However, they are very susceptible to damage from the rough treatment that oil field equipment is often given. While the exact type may vary by manufacturer, metal-to-metal seals are typically interference-energized; therefore, if they start leaking after the well is producing, there is virtually nothing that can be done about it. However, those that are properly installed with undefective components are very reliable and give excellent service. They are particularly applicable where movement of the wellhead components is likely due to temperature variations.

Weld-on heads are usually welded at two points: at the bottom of the casing head and at the joining of the casing top and casing load shoulder of the casing head. After welding and allowing the weld area to cool, the welds are tested by means of a test port connected to the internal area between the two welds. The pressure applied in the test should not exceed 80 per cent of the yield strength of the casing in use.

Casing head options include an inner seal (e.g., an 0-ring) to seal between the casing and casing head, and lockdown screws to hold down bowl protectors.

Casing Spools

Casing spools function in much the same way as casing heads, with two important differences: (1) the spool provides a bottom bowl to seal the previous string of casing, and (2) the bottom connection must be compatible with the top connection on the previous head or spool. Casing spools are manufactured to meet the same requirements as the casing head and are identified by the following: size (bottom bowl normal size as reflected by the casing OD over which the casing spool will normally seal), pressure rating (as determined by the top connection), type (bowl design), bottom connection, top connection, and miscellaneous details (e.g., size and type of outlets, plastic injection ports, and special material considerations).

Like a casing head, the casing spool has a top bowl, which holds the casing hanger that suspends the next string of casing. Again, two side outlets are provided and may be threaded, clamp-hub, flange, or studded. These outlets are most often fitted with gate valves, although a valve removal (VR) plug—which allows installation and removal of valves under pressure by means of a lubricator—and blind flange may be installed.

Unlike a casing head, a spool also has a bottom bowl with a packoff seal and a flange or clamp hub for mounting it on top of a casing head or previous spool. These bottom bowls are designed to accommodate a packoff assembly that seals around the casing stub and forms what is often called a "secondary seal" between the current casing string and the casing annulus.

The lower flange of the casing spool also serves as a test port for pressure testing the casing seals and flange connections.

The top bowl configuration is generally identical to the casing head design for each particular manufacturer. Since intermediate strings usually have a higher string weight, bowls accommodating casing hangers with lesser hanging capacity may not be offered in casing spools.

Casing Hangers

Casing hangers allow the weight or tension load of a casing string to be transferred to a casing head or spool. Casing hangers also center the casing string in the head or spool and provide a pressure-tight seal against the inside of the casing head or casing spool bowl to contain pressure in the annulus between its casing string and the previous string. In some cases, a separate seal ring or packoff bushing is required to provide the seal.

There are two major types of casing hangers: slip-type hangers, which are installed around the casing after it is run; and mandrel-type hangers, which are made up into the string.

Mandrel-type hangers (boll weevils) are threaded top and bottom and are made up directly into the end of the casing string. If no sticking problem occurs while running casing, the mandreltype casing hanger can be used. The main advantage of the mandrel casing hanger is its simple design for hanging and sealing pipe. If the pipe sticks, a slip-type casing hanger is needed.

Slip-type hangers (wraparounds) are hinged or halved to facilitate wrapping around the casing and may be dropped through the blowout preventer (BOP) stack, assuming sufficient clearance is present, given the ID of the stack. This allows the sealing off of the annulus prior to nippling down the BOPs.

Slip-type hangers have slips in serrated segments. The slips have a tapered back that matches the taper on the. inside of the slip bowl. When the hanger is wrapped around the casing, the slip teeth engage the casing. As the casing is lowered, it pulls the slips down with it. The tapered bowl forces the slips against the casing (with a wedging action) as they move down, so their grip on the casing increases as the casing weight increases. The casing hangers should be designed so that the inward force of the slips will not crush or deflect the casing beyond acceptable limits at loads equal to the strength of API round-thread joint connections. For hanging extreme loads from casing heads or spools, two sets of slips in tandem may be required to distribute the load over a larger area of casing.

If the casing weight is sufficient, a slip-type automatic-seal hanger is used with a compression-type seal mechanism that is automatically actuated by the weight of the casing. When casing weight is insufficient to actuate the seal, as in shallow wells when the casing is cemented back to the surface, the hanger design may incorporate a sealing element located above the slips that is mechanically activated with cap screws (Figure 5.7). Manufacturer specifications may vary, but in general, automatic-seal hangers require at least 3 in. of downward casing movement to engage the slips fully and a minimum 40,000-lb load to actuate the compression-type seal.

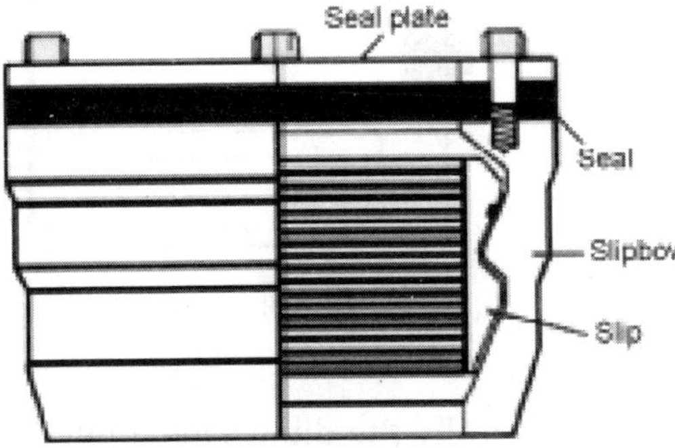

Figure 5.7:

Slip-type hangers can be installed before or after the casing has been cemented. In general, this depends on the length of the casing string. Shallow intermediate strings are usually suspended from the hanger and then cemented all the way to surface. Deeper intermediate or production strings are usually cemented while the casing is suspended in tension from the rig traveling block. Then, after the cement has cured, the traveling block is used to pull a calculated amount of tension on the free pipe above the cement. At this point, the slip-type hanger is installed.

Packoffs and Isolation Seals

The casing packoff serves as an annular seal, which prevents communication between the casing strings and exposure of the flange seal to annulus pressure. Since the casing hanger has already provided a seal between the casing strings, the packoff is called a secondary seal. The packoff seals against the casing OD and the ID of the wellhead. Terminology is sometimes inconsistent, but "packoff" generally refers to a seal acting in either the bottom bowl or the top bowl above the slips.

Each manufacturer offers a variety of packoff assemblies. Typical packoffs include those with interference-type seals (e.g., an 0-ring seal), which would be located in grooves in the lower bowl (Figure 5.8).

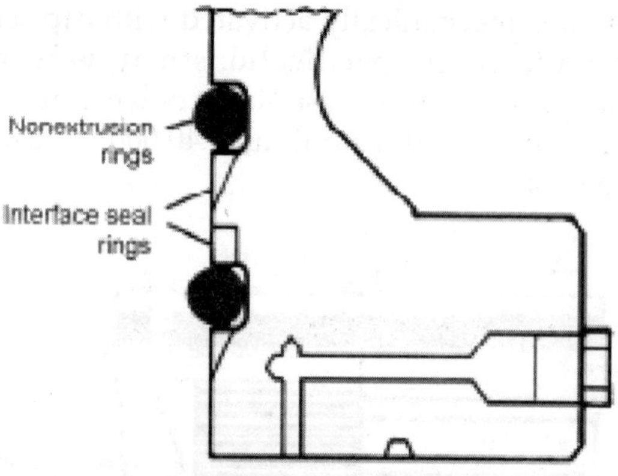

Figure 5.8:

These seals are achieved by the dimensional interference of the seal element ID and the casing OD. Some seal elements are initially activated by dimensional interference and experience limited extrusion of the seal element lip with exposure to pressure. Most manufacturers offer a plastic energized-type seal. These seals are activated by injecting plastic into ports located on the casing spool (Figure 5.9).

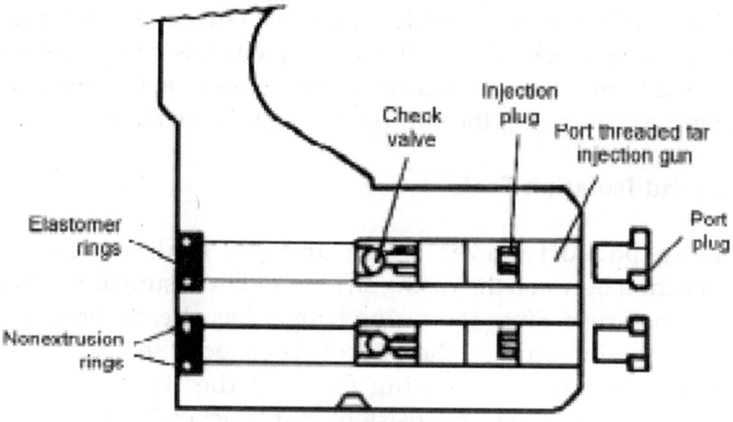

Courtesy of FMC Corporation

Figure 5.9:

In cases where the casing size and the bottom bowl of the next spool are mismatched, so the packoff designed for the casing is too small to seal in the bottom of the spool, a reducing bushing is used in the bottom bowl. This allows a particular casing spool to accommodate numerous casing sizes.

In high-pressure applications, a packoff is often used above the casing hanger in the top bowl to isolate the casing slips. This prevents test pressures applied to the flange area from creating extra forces in the hanger area and collapsing the casing. Since the slips have already applied a force on the casing, it is desirable to avoid any additional stress in this area. Furthermore, the test pressure acts to force the slips further into the bowl, thus increasing the force applied by the slips on the casing. Generally, when testing flanges are not isolated from the slips, the test pressure is reduced to something less than the standard 80 per cent of rated casing collapse pressure

Bowl Protectors

Because of the delicate treatment required for the metal-to-metal seal area of the casing head (and subsequent casing spools), some form of protection is suggested during the drilling process. The bowl protector (sometimes called a wear sleeve, or wear bushing) is designed to protect the entire bowl area of the casing head, casing spool, and tubing from any damage during drilling or workover operations. When positioned in the bowl (Figure 5.10), it shields the sealing surface and the load shoulder from the rotating drillpipe. It may also be advisable to protect the top of the previous casing string. In such cases, a bowl protector with an elongated neck should be used.

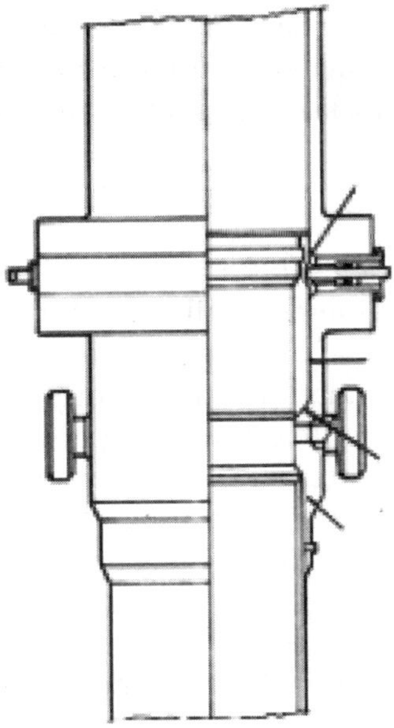

Figure 5.10:

The body of a bowl protector has the same outer configuration as the bowl it will protect. The type of bowl protector required corresponds to the type of casing hanger required for the particular bowl in use.

The inner configuration may be either bit retrievable or full bore. A bowl protector is bit retrievable if its ID is smaller than the OD of the bit and other bottomhole equipment. This type of bowl protector is run and retrieved on the drillstring on top of the bit. A bowl protector is full bore if its ID is larger than the OD of the bit. This type of bowl protector must be run with a bowl protector retriever. Some bit retrievable bowl protectors are also machined to be used with a bowl protector retriever if necessary.

If a bowl protector is designed to be run with a retriever, it has slots or grooves at the top of its body that are used to attach the bowl protector to the retriever. The most common slot is the J-slot: two grooves on the inside of the body, 1800 apart, in the shape of a J.

Running and retrieving procedures vary by manufacturer but generally involve engaging the J-slot with a partial turn to the right. Retrievers may attach to the drillpipe body by means of set screws, or they may connect to the end of the drillpipe.

Hold-down set screws are often provided to hold the bowl protector in place and eliminate rotation. They are located either in the top flange of the spool or in an adapter flange installed directly above the spool during drilling operations. Often only four hold-down screws, 90o apart, are needed. The hold-down set screws are run into a machined groove in the side of the body, pinning the bowl protector in place.

The bit-retrievable type saves a run to retrieve the bowl protector; however, the wellhead bowl is exposed to the drill bit when it is pulled out of the hole, and the lockscrews should always be retracted first.

Test Plugs

The primary function of a test plug is to provide a simple, effective means of sealing the wellbore below the well control equipment (e.g., BOP stack). The test plug seals in the ID of the bowl, not in the casing or tubing. The sealing element varies by manufacturer and plug type but is generally an interference-type elastomer (e.g., an 0-ring or hydraulic packing).

Once the plug is in place, all the connections and sealing areas from the casing head top connection up through the BOP stack can be tested to ensure leakproof integrity during drilling operations. Because the test plug seals in the same area of the casing bowl as the casing hanger, such tests also indicate possible bowl wear.

Test plugs can be ordered with or without weepholes, which are ports that allow communication down the drillpipe and up the outside.

Weepholes are often necessary to allow testing of the lowermost pipe rams.

To select the proper test plug, the bowl size, bowl design, and drillpipe thread dimensions are needed. All test plugs have tool-joint box tops and pin bottoms and are run into the head on the drillpipe. The top and bottom test plug threads must match the drillpipe being used.

Mudline Suspension Systems

Conducting drilling operations with the BOPs at the surface obviously requires some type of bottom-supported platform. The mobile bottom-supported platforms, such as jackup or submersible rigs, can also use conventional wellhead equipment and BOPs at the surface with the use of a mudline suspension system.

When a mudline suspension system is employed, the casing is suspended at or near the mudline, but the casing strings are later tied back to the rig at the surface. Conventional BOPs and wellhead equipment may then be installed and used during the drilling operations.

After the well has been drilled and tested, the BOPs, wellhead equipment, and extension casing from the mudline hangers are removed. If the well is to be completed, a cap is usually installed over the well at the mudline. When the operator is ready to re-enter the well, usually after exploration activities have been completed, the cap is removed and the well completed by either installing a tree on the ocean floor or locating a platform over the well and extending the conductor casing up to the platform. A conventional tree can then be installed at the surface.

A typical mudline suspension system consists of a series of concentric casing hangers, each having an internal profile to provide a support or seat for the subsequent hanger assembly.

Two types of casing hangers are usually incorporated in a mud-line suspension program. Fluted mandrel-type, or boll weevil-type, hangers are generally used for larger size casing suspension, where casing ID and bit OD clearance is sufficient to allow a support shoulder to be provided in the outer hanger. The fluted hanger incorporates a replaceable fluted hanger ring that provides flexibility in the event of a last-minute change in casing program. Expanding-type hangers are used for the smaller casing strings where bit sizes closely approach casing ID, precluding sufficient clearance for a support shoulder inside the outer hanger. Expanding-type hangers use springloaded steel segments that lock the mating downhole hanger.

Both types of hangers provide fluid passage for circulation and cementing returns. Generally, all assemblies may be furnished with circulating ports for washing and displacing cement from around the landing/tieback thread area.

As with conventional mandrel hangers, the hanger body is made up on the casing to suspend it. Most hangers are designed with coarse threads for landing sub and tieback sub connections. Exact landing and tieback procedures vary by manufacturer.

Conventional wellheads may be used with mudline suspension systems. Since casing weight available for the surface casing hanger is limited, some form of packoff in the top bowl of the casing head or spool is common.

If the well operations are suspended for possible future reentry, a plug is placed inside the last casing string. The casing extensions are then removed to the last casing size that it is desired to cap. A cap is then placed, sealing this casing string and all subsequent strings. Any remaining casing extensions are then removed, and the location is marked with a buoy or other locating device.

Tubing Heads

After the final casing string (production string) is in place, a tubing head is installed that will, as did the casing heads preceding it, isolate the respective casing annulus, and provide an internal profile to accommodate a tubing hanger. After the well is completed, tubing is run in the hole and the producing interval isolated at the surface from the tubing casing annulus. This is accomplished at the surface with the tubing hanger.

The tubing head provides for the same design criteria as the casing spool, inasmuch as the previous casing string is packed off in the tubing spool lower bowl and the tubing spool lower connection is compatible with the previous casing spool top connection. In addition, the tubing head facilitates the hanging and/or sealing of the tubing string. All tubing heads have lockdown screws, or lockscrews, that reduce the likelihood of seal movement caused by thermal expansion or annulus pressure. The lockscrew can, when hanger design permits, be used to actuate compression-type seals.

The top bowl of the tubing head provides a load shoulder to support tubing hangers and packoffs, and a controlled bore against which the hanger or packoff can seal. Additionally, the tubing head provides access to the annulus between the tubing string and the production casing.

Generally, the top bowl design is similar to the manufacturer's bowl design in its casing heads. In some cases, the tubing head bowl is identical to the casing head bowl, allowing the well to be drilled deeper. This also allows the placement of hang-off casing in the tubing head, and the setting of another tubing spool on top of the first one to suspend the tubing string. Tubing heads for single and dual completions are identical except that dual tubing hangers require at least two alignment pins.

Like casing spools, tubing heads have two side outlets that may be threaded, flanged, studded, or clamp-hub. In most cases, gate valves are installed on the outlets, but in some cases, a valve-removal (VR) plug with blind flange takes the place of one of the valves. VR threads are standard on all flanged, studded, and clamp outlets.

Tubing Hanger and Tubing Head Adapters

The tubing hanger serves as both a hanging and sealing mechanism. In some cases, it must support the production string, seal the annulus, and at the same time create a high-integrity transition to the tubing head adapter and Christmas tree.

The tubing head adapter is a transition fitting between the Christmas tree and the tubing head. The bottom adapter connection matches the tubing head, and the top adapter connection matches the tree. The top connection may be threaded, flanged, or studded, and bottom connections may be flanged, studded, or clamp type.

The boll weevil tubing hanger is probably the simplest hanging device manufactured for supporting a string of tubing. This hanger is threaded top and bottom and usually has a compressive-type annular seal that may be either weight set (with the hanging tubing weight) or lockscrew actuated. The hanger is screwed onto the tubing string and supported by the tubing head bowl. A basic tubing adapter with a slick bore is used with this type of hanger. This configuration design is primarily for low-pressure completions where downhole control lines or tubing string manipulation are not required.

By adding an extended neck with a sealing element to the standard boll weevil hanger, a seal is formed with the tubing head adapter, thereby isolating the tubing head adapter flange.

This is common in high-pressure wells, gas wells, or sour crude wells. To accommodate the extended neck, sealbore tubing head adapters have an ID machined to provide a controlled bore diameter. They are generally available with a test port and/or a hydraulic supply inlet for downhole control lines. Figure 5.11 illustrates an extended-neck boll weevil hanger with downhole control line capability.

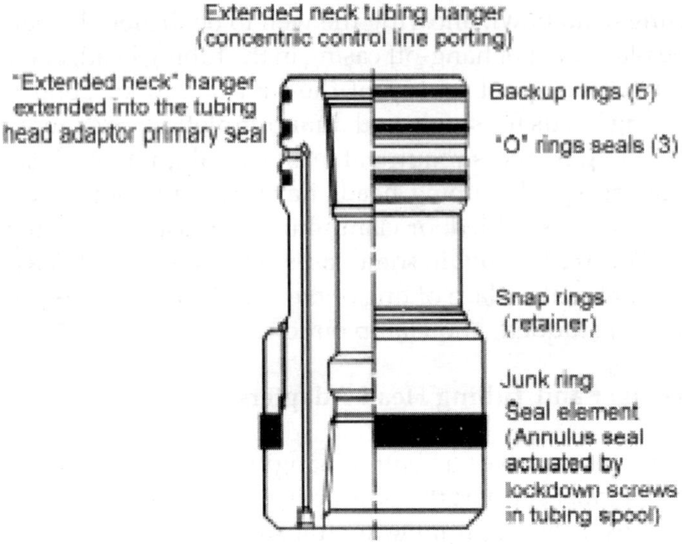

Figure 5.11:

This particular design does not require any specific orientation, as the control line outlet is between two 0-ring seals.

The name "tubing hanger" is a misnomer in some configurations, since the tubing is actually supported by the tubing head adapter and the tubing "hanger" acts only to seal the tubing casing annulus. This is sometimes referred to as a slick joint, or hookwall assembly, and is most useful for moving or rotating the tubing under pressure.

When suspending the tubing from the tubing head adapter, the adapter's ID can be machined to the tubing threads, in which case the tubing is screwed directly into the adapter. This is normally used on low-pressure completions.

When use of back-pressure valves is desired, the tubing head adapter is machined on the ID to accept threads for an adapter bushing placed between the tubing head adapter and the tubing. This adapter bushing contains internal threads to accept a back-pressure valve. Regardless of whether an adapter bushing is used, a split-type wraparound tubing hanger (packoff) or stripper rubber is generally used in this type of configuration.

A wraparound packoff provides for annulus sealing with a compression-type seal actuated by the lockdown screws. A stripper packoff facilitates the running or pulling of tubing under low to medium well pressures. The packoff is installed in the tubing head and retained by the lockdown screws. Annular sealing is achieved by an interference seal on the OD and a molded seal element that is pressure energized on the ID.

In a dual completion, where tubing strings are run to two separate zones, a dual tubing hanger is required for the independent suspension and sealing of the tubing strings. In addition, the hanger can be manufactured to provide access for downhole equipment, back-pressure valve grooves, and interface sealing between the hangers and the tubing head adapters.

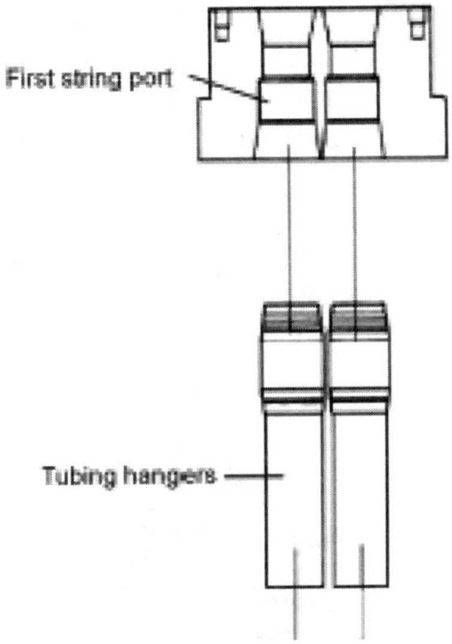

Figure 5.12:

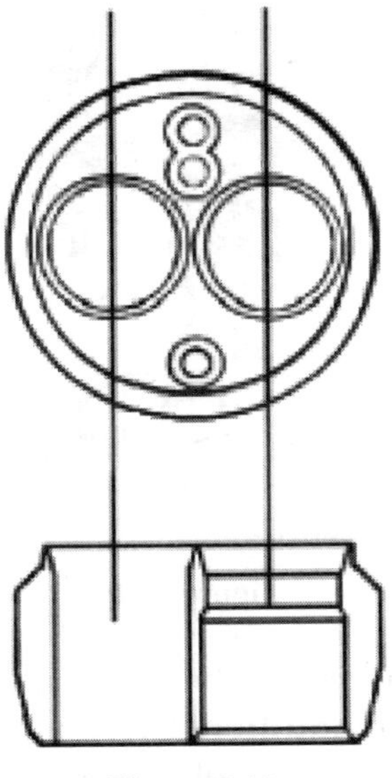

Figure 5.13:

The two basic types of dual hangers are the mandrel and the split type. With the mandrel design, a parent hanger suspends two independent tubing hangers. The parent hanger provides sealing on the ID of the tubing head bowl. The independent hangers then provide sealing between the hanger OD and the respective parent hanger ID. Figure 5.12 and Figure 5.13 is an example of a mandrel-type dual hanger.

This type of dual completion is efficient and reliable, but it does not have the versatility of split-type duals, which offer greater clearances.

A variation of the mandrel dual hanger involves hanging the long string from the parent hanger, similar to the boll weevil single hanger, and then landing the short string with a mandrel in the parent hanger.

In the split-type hanger, the parent hanger is generally a nonsealing support ring that supports the individual mandrels but relies on compression-type packoffs to seal the tubing annulus. Figure 5.14 illustrates a split hanger example.

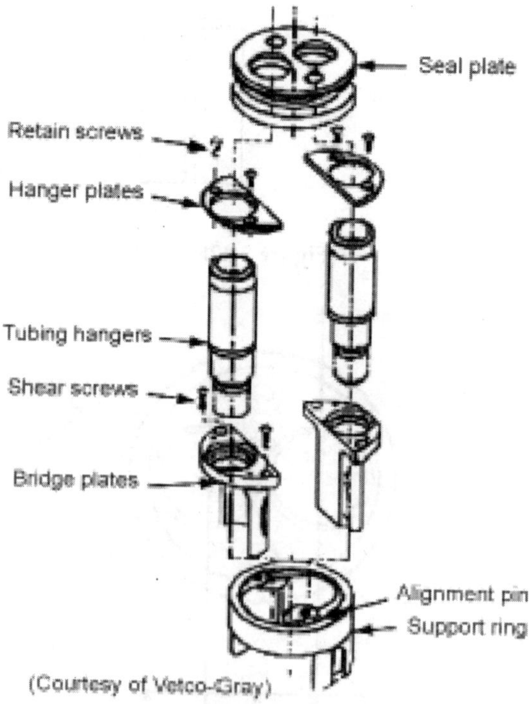

Figure 5.14:

An excellent application for the split hanger is in running gas-lift mandrels because of the additional clearance it provides. It is used exclusively where downhole control lines are required.

Both the mandrel and split hanger types are often machined to accept back-pressure valves and use extended-neck hangers to seal in the tubing head adapter.

A primary means of preventing blowouts in producing wells is by using surface-controlled subsurface safety valves. These valves are normally set several hundred feet below the surface in the tubing string and can be either wireline or tubing-retrievable types. These downhole safety valves require the installation of hydraulic control lines for opening and closing operations.

The various wellhead manufacturers offer several methods by which this control line can exit at the surface and be connected to the surface control manifold, but they generally provide a cavity within the tubing hanger with an extended neck at the top to seal in the tubing head adapter. The tubing adapter has a port that exits to the surface. Some downhole safety valves require two control lines; therefore, designs are available to accommodate two control lines for each string in a dual hanger.

Figure 5.15 depicts a single completion with a downhole control line.

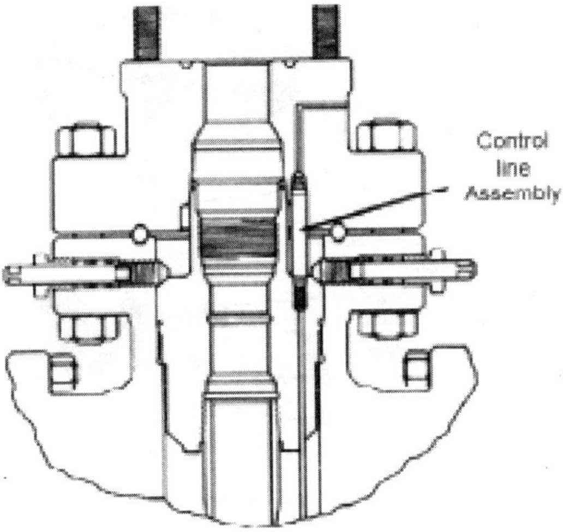

Figure 5.15:

In this case, the control line seal nipple is installed in a threaded port provided in the hanger. Metal-to-metal seals are combined with hydraulic packing to seal between the nipple and tubing head adapter. The seal between the 1/4-in. control line and seal nipple uses a tapered ferrule cone that, when engaged by the threaded gland, forms a metal-to-metal seal between the control line and the top of the seal nipple.

The backpressure valve (BPV) is a device for plugging the tubing string in the tubing hanger. It is commonly used in removing the BOPs and installing the Christmas tree, moving the completion or workover rig off location, repairing the Christmas tree, or indefinitely shutting in a well.

While effectively plugging the well, the BPV allows for pumping down the tubing for such functions as displacing drilling fluids before connecting the tree to the tubing head or displacing the tubing through the tree after installation. The fluid used for flushing must not contain abrasive material, so as not to erode the BPV.

The common types of BPVs either screw into the hanger or latch into a mating profile in the hanger. Running and retrieving procedures vary by manufacturer, but all use some type of lubricator for running and retrieving under pressure.

Gate Valves

A gate valve (Figure 5.16) is a full-opening valve in which the closing device is a slab of steel, called a gate, with a hole drilled in it.

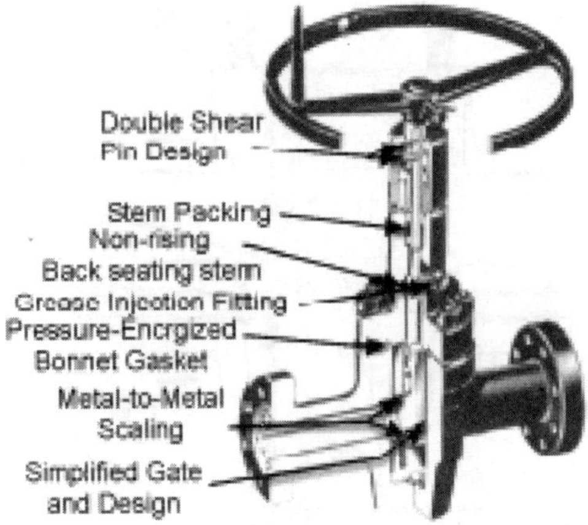

Figure 5.16:

The hole in the gate is pushed or pulled across the valve body seat or hole opening to allow fluid to pass. To stop the flow of fluid, the solid gate section is placed across the valve seat.

Gate valves provide the primary on/off control on the wellhead. They are used for annulus control on the outlets of casing heads, spools, and tubing heads. They are the prominent fixture in the run of the Christmas tree used as master valves, crown valves, and wing valves. Gate valves should always be operated fully opened or fully closed.

The gate valve is an assembly of several parts, each designed to perform a specific function and to operate in conjunction with the others to provide a controlled flow passage for line fluid or to shut off that flow entirely. (Figure 5.17) is a parts diagram of Cameron Iron Works' J gate valve, which will provide a reference for the upcoming text.)

Parts List

1. Handwheel Assembly
2. Grease Fitting
3. Bearing Cap
4. Stem Adapter
5. Double Shear Pin
6. O-ring
7. Bearing Race
8. Thrust Bearing
9. Bearing Mount
11. Packing Gland
12. Stem Packing
13. Nut
14. Stud
10. Bonnet
15. Grease Injection Fitting
16. Pressure-Energized Bonnet
17. Stem Gasket
18. Gate Neck Attachment
19. Retainer Plate
20. Reverse Bore Gate
21. Seat
22. Seal Ring
23. Body
24. Nameplate

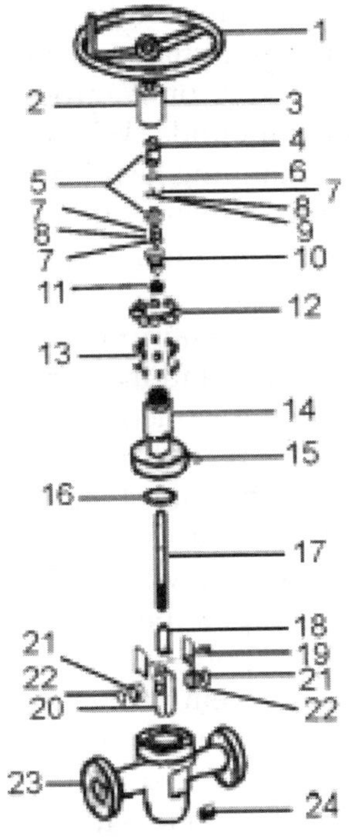

Figure 5.17:

The main component in the valve assembly is the valve body. The body is a pressure vessel, which means that it is designed to contain the fluid under maximum service pressure. The body is provided with a bore or conduit through which the fluid passes when the valve is in the open position. At each end of the bore are the end connections, which provide for attachment to other valves, fittings, or pressure vessels. The connections generally offered are flanges, threads, clamped hubs, and weld ends.

Midway between the end connections is the valve cavity, which runs at 900 to the conduit and houses the gates and seats. The cavity is closed off by the valve bonnet, which bolts to a studded connection at the open end of the cavity. The seal mechanism between the cavity and bonnet is commonly a metal-to-metal, pressure-energized type. The bonnet also provides a housing for the stem, stem packing, and bearing assembly.

In order to shut off the flow of fluid through the valve, the valve is equipped with a gate and a pair of seats. The gate is solid at one end and has a bore through the other end. When the gate bore is aligned with the body bore, the valve is open and fluid can pass through the valve. When the gate is moved to block the body bore, the passage of fluid is stopped.

The stem packing works to isolate cavity pressure from the atmosphere between the valve stem and the packing bore in the bonnet. Packing is often the chevron, or W, cross-sectional type, composed of fabric and elastomer compounds. When the stem packing leaks, the seal may be renewed by pumping heavy grease or sealant onto the packing area. This temporary measure buys time for maintenance or replacement.

There are substantial variations in gate valve designs offered by different manufacturers. Gate valves may have rising or nonrising stems, balanced or nonbalanced stems, slab or split gates, directional or bidirectional flow, and may or may not allow body pressure.

As the gate is pulled across the sealing seats in the valve bore, the valve stem of a rising stem-type gate valve extends externally through packing inside the valve bonnet when the valve is operated from open to closed position. Rising stems normally do not rotate. The stem is driven by a threaded driver, which acts as a screw jack to make the stem rise and fall to push and pull the gate inside the valve, creating the sealing action across the valve port.

A nonrising stem pushes and pulls the gate across the seats to open and close the valve just as the rising stem does, but it does not allow external travel of the stem. The stem does not rise or fall outside the valve body. A nonrising, rotating stem with no external travel is accomplished by having the thread driving assembly built into the gate, rather than into the bonnet of the valve.

To reiterate, the valve stem acts to push and pull the gate across the sealing surface of the valve bore. By attaching another stem of equal diameter to the opposite end of the gate, a balanced stem effect is accomplished. With a single-valve stem, the bottom of the gate has a larger area exposed to the working pressure, thus increasing the friction pressure between the stem threads and the threaded driving assembly. With the balanced stem, the valve stem is only working against the friction created on the packing by the pressures inside the valve.

A slab gate is a one-piece gate normally used with a "floating seat" design. The slab gate must be allowed to float to seal against the valve bore seat. If we apply pressure from one side of the valve bore against the gate, it will force the gate against the sealing seat and will not allow the pressure to pass from the valve body to the downstream side of the valve bore. The seat pockets machined in a gate valve body must allow enough tolerance between the bore of the seat pocket and the seat OD to allow seat travel as pressure is applied to a closed gate. Some floating seats with a slab gate are designed to seal in the seat pocket bore of the valve on both sides of the gate, creating a piston action that seals both seats to the gate when the valve has pressure in the valve bore. This type of seat design also allows the bleeding of body pressure to zero.

Split gates have two sections to the gate, and most employ a fixed, or stationary, seat. The manufacturing tolerances between the valve seat pockets and the OD of the seat itself are so close that the seal is accomplished by pressing the metal surfaces together. The split gate design varies significantly by manufacturer, but generally some type of expanding device to force the gate against the seat is included. Most split gates also require the use of a sealing compound.

A bidirectional valve can be placed in a flowline, on a Christmas tree, on a manifold, or wherever it may be used with either flange attached to the pressure side. A valve design that requires the sealing of the gate on a downstream seat only is called a directional valve. When it is installed on a manifold or on a Christmas tree, the upstream side of the valve bore, in relation to the gate and seat inside the valve, must be designated so that the upstream seat side can be installed toward the pressure source.

Composite valves are manufactured as single units that contain two or more valves built into an assembly. The primary reason for using a composite valve is to save height and space. On dual trees, each gate valve assembly generally has a valve for each string.

Most manufacturers have designed valves to maintain the same pressure inside the valve body and in the valve bore at all times. As mentioned previously, some designs allow bleeding of pressure within the valve body to zero. This is advantageous should a leak develop in one of the valve body seals (grease port, body fitting, bonnet seal, stem packing, etc.), since the body pressure would bleed to zero without loss of source pressure. However, on large bore valves at higher pressures, total loss of body pressure would result in excessive friction of the gate pulling across the piston load of two seats, increasing tremendously the torque required to open the valve, until the pressure was returned into the body.

Some fixed seat valves may maintain body pressure even after the adjacent line pressure is bled off. This can result in "pressure locking" the valve. Once the gates are moved far enough to break the seal, the pressure bleeds off and the torque drops.

Gate valves are designed to reach their fully open or closed position with a fixed number of turns of the handwheel. This number is sometimes specified on the valve by the manufacturer. There is no need to add more torque to the stem after reaching the fully open or closed position. It is generally considered good practice to back off one-quarter turn to relieve excessive torque on the stem after fully opening or fully closing these valves. Expanded gate valves are an exception to this rule, since they require additional stem torque to open or close the valve fully.

A gate valve requires lubrication of the valve body and stem operating mechanism on a regular basis. This varies depending on the amount of use but is necessary at least every three months. Some manufacturers require a special sealing compound as well as lubricating compound; therefore, proper grease or sealing compound for each manufacturer's valve should be obtained.

An actuated valve is an externally powered safety valve used to provide automatic valve closure when abnormal conditions occur. Although the API uses the term "actuator," the term "operator" is also quite common in the field. The actuator is fitted atop what is otherwise a standard gate valve to open or close by using compressed air (pneumatic actuator) or pressurized, hydraulic fluid (hydraulic actuator).

The type of valve most commonly used in this context is the rising stem type, and its operating principle is quite simple. During normal operation, pressure provided by the actuator forces a piston, and thereby the valve stem, down into the valve, whereby the holes in the gate are aligned with the bore of the valve. The point to remember is the stem is being forced into the cavity of the valve body and the pressure in the valve exerts a constant piston effect on the valve stem. This piston effect is used to close the valve when the pressure in the cylinder of the actuator is released. A coil spring is placed under the cylinder piston in the actuator to close a valve with no body pressure.

Generally, pneumatic actuators are designed to operate at relatively low control pressures (e.g., 75 to 350 psi), whereas hydraulic actuators are generally designed to operate at control pressures of 1500 to 3000 psi. To close a pneumatic actuator valve, the compressed air is simply vented to the atmosphere by control valve mechanisms. In hydraulic actuators, the hydraulic fluid must be returned to the hydraulic pump reservoir. In both cases this release of pressure is caused by an automatic release valve reacting to a signal from the control panel when the monitored critical pressures go outside acceptable ranges. The pressures to be monitored (e.g., flow-line or separator) are determined by the user, as are the acceptable ranges. Actuated valves can be designed to close or open in the event of abnormal conditions; the fail-closed type (which is designed to close upon loss of actuator pressure) is the most common. Once a valve has moved to its fail-safe position, it generally does not return to normal operation until the controls are reset, permitting repressuring of the control system.

An important consideration for actuated valves is the closure time. A gate valve is most vulnerable in the opening and closing operation because the sealing surfaces are exposed to the abrasive turbulence in the flow of the well. If the valves are to last, they must be opened and closed quickly.

Overrides and lockouts can be attached to most actuators so that wireline and other well operations can be performed in complete safety.

The valve removal/reinstallation tool used in conjunction with the VR plug facilitates the removal, reinstallation, or repair of side outlet valves under pressure. The wellhead connection (to which the valve is attached) must have inside threads machined to accept the VR plug. By installing the VR plug in the wellhead outlet, well pressure is effectively contained, allowing the valve to be bled down and allowing repair or removal to be performed. This is done with the use of a lubricator that attaches to the outer connection of the valve to be removed.

Production Chokes

Production chokes are Christmas tree components that precisely regulate the flow of oil or gas to achieve a carefully calculated rate of recovery. By maintaining the correct backpressure, chokes can increase the ultimate recovery of hydrocarbons from a formation, control the rate of formation pressure decline, reduce sand production and the migration of fines, possibly control water coning and gas fingering, and minimize tree damage from erosion caused by turbulence and cavitation.

Chokes come in two basic types: adjustable and positive. Adjustable chokes are often used during completion operations to allow the operator to clean and flow test the well. Once the optimum flow rate is determined, the adjustable choke is usually replaced with a positive choke for production. Both styles are available in a variety of sizes, connections, and pressure ranges to meet various requirements. They are almost always located just downstream of a wing valve on the Christmas tree to facilitate isolation for service or orifice changes.

Adjustable chokes consist of a choke body, adjustable choke bonnet and stem, and stem seat. Counterclockwise rotation of the choke handle moves the stem tip out of the stem seat, thereby increasing the effective flow area (Figure 5.18).

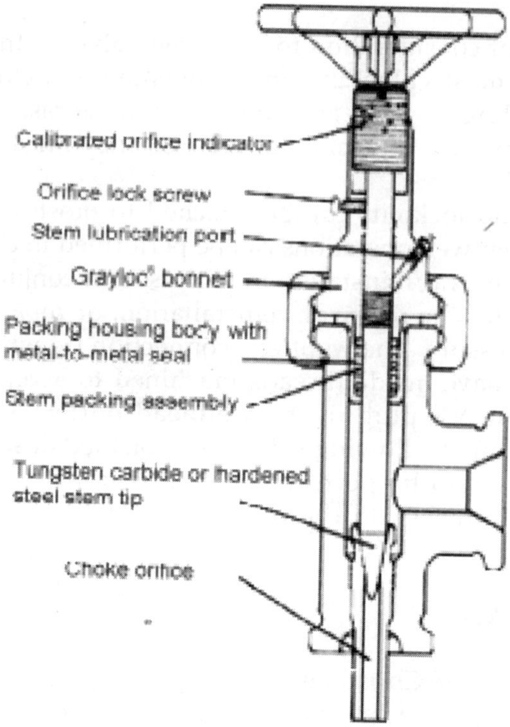

Figure 5.18:

The stem tip consists of some type of hardened steel to resist abrasion. Located on the bonnet assembly is a position indicator sleeve, which is commonly calibrated in 64ths of an inch, from zero to full bore opening. The adjustable choke is not designed to stop the flow completely, although it may do so at low pressures. As with gate valves, the stem packing acts to isolate choke body pressure from the atmosphere. Almost all chokes, and particularly those with higher pressure ratings, have an injection port for the stem packing.

A positive choke consists of a choke body with a constant orifice size. Choke nipples were once used with a carefully calibrated bore. While their initial cost was low, their use has been virtually abandoned because changing them is difficult and time consuming. Modern chokes have internal threads and a seating shoulder in the bore to accommodate choke beans with a specific size orifice. The bean can be inserted and removed quickly with a special wrench.

Many low-pressure adjustable chokes can be converted to positive chokes by replacing the stem seat with a choke bean and the bonnet/stem assembly with a seal cap.

Wellhead and Christmas Tree Assemblies

A Christmas tree is the assembly of gate valves, chokes, and fittings that controls the flow of oil or gas during production. Tree design is based on the number of tubing strings used for completion, tubing bore size, maximum anticipated production pressure, trim requirements, and flow rates. The bottom connection of the tree matches the top connection of the tubing head adapter. The tree and adapter are usually made up and installed as a unit immediately after tubing is suspended.

The tree consists primarily of a series of gate valves and a production choke. Gate valves located between the tubing head adapter and the production tee are called master valves. Christmas trees always have at least one master valve; usually two are used. Dual master valves allow the use of the top master valve for normal use, thus reducing wear on the lower master valve, which is the most difficult to replace. Replacement of the top master valve can be accomplished with relative ease by isolating the upper portion of the tree with the lower master valve.

A gate valve called a crown valve or working valve is often placed above the production tee. This valve facilitates installation or dismantling of a lubricator without shutting in the well.

A gate valve is almost always placed immediately off of the production tee. This valve, called a wing valve, can be used to shut off flow to the production facilities and still allow work down the tree or tubing.

A tree cap is sometimes installed on top of each tree above the crown valve to provide quick access to the tubing bore for bottomhole testing, installing a backpressure valve, swabbing, or paraffin scraping. Most tree caps are tapped for a pressure gauge and have internal lift threads to facilitate the installment of the tree.

All valves, connections, tees, and other equipment in the vertical run of the tree must be greater than or equal to the nominal ID of the tubing; however, valves in the wing section are sometimes a smaller size.

Line Pipe Fabrication and Specifications

The first efforts to standardize pipe sizes date back to the second quarter of the 19th century, when nominal sizes for iron pipe and pitches of thread were established. However, these initial standards were so broad that interchangeability of different manufacturers' products was practically impossible. About 50 years later, Robert Briggs, who was at one time superintendent of the Pascal Iron Works in Philadelphia, prepared a paper giving detailed information about American pipe and pipe thread practice. Briggs' paper established a definite formula for external pipe threads that became known as the Briggs Standard, and it has been used ever since. Today it is incorporated in the American National Standards Institute (ANSI) Standard B2.l for pipe threads and is now officially known as the American Standard Pipe Thread.

As the pipeline industry grew, the need for stronger screwed joints developed, and the Briggs thread was lengthened and larger couplings were developed for pipeline use. These are known as the line pipe thread and the line pipe coupling, as differentiated from the standard thread and standard coupling used on standard pipe.

Numerous piping codes and standards for practically every kind of service have been issued or are being considered by various industry associations. To facilitate their use, the standards are channeled through one central coordinating organization that adopts, classifies, rejects, or modifies the codes and standards developed by the other associations.

Many different codes and standards promulgate the basic requirements peculiar to each industry. They define construction materials, manufacturing or fabricating methods, inspection or testing requirements, dimensional tolerances, etc., and have thus become American National Standards. After it was determined that the American National Standards were in conformance with ANSI procedures, they were adopted by ANSI.

ANSI, formerly the American Standards Association (ASA), is a federation of national associations and government departments that dates back to 1918, when it was organized by five engineering societies as the American Engineering Standards Committee. Today, more than 200 organizations representing industry, government, and consumers and 900 companies participate in its work, which is dedicated to the issuance of standards that serve not only as voluntary guidelines to ensure uniform and reliable products, but as the basis of federal regulations. The Occupational Safety and Health Act (OSHA), which was enacted in law by Congress in 1970, relies extensively on ANSI standards.

The Institute does not develop any standards itself; rather, it provides a vehicle for standards review through the formation of sectional committees that comprise representatives of the various organizations or groups that have a substantial interest in a particular standard. One of the important ANSI requirements for the acceptance of any standard is complete agreement among all interested parties.

The American Society for Testing Materials (ASTM) has formulated a considerable number of standards that deal with specifications for piping, tubing, and bolting.

Working closely with the guidelines of the ASTM and ANSI, the API publishes piping specifications that are the recognized standards in the oil industry.

API Specification 5L, Specification for Line Pipe sets forth standards for seamless and welded steel line pipe, including:

- Standard-weight and extra-strong threaded pipe;
- Standard-weight plain end, regular weight plain-end, special plain-end, extra-strong plain end and double-extra-strong plain-end pipe;
- Bell and spigot and through-flow-line (TFL) pipe;
- Grades covered by API Spec. 5L include A25, A, B, X42, X46, X52, X56, X60, X65, X70, X80, and grades intermediate to X42 and higher. Minimum yield strength ranges from 25,000 psi for A25 to 80,000 psi for X80, while minimum ultimate tensile strength ranges from 45,000 psi to 90,000 psi for these grades.

Manufacturing processes for steel line pipe have proliferated, as have the chemical formulations used in making steel. These developments were needed for more efficient production of pipe with higher strengths and other special properties so that pipe weights and wall thicknesses could be kept within reasonable and economically feasible limits. Table 5.2 gives the approved processes of manufacture and the chemical requirements of pipe conforming to API 5L, Specification for Line Pipe.

Table 5.2: API-Approved Processes of Manufacture and Chemical Requirements (courtesy API) Ladle Analysis, per cent

	Ladle Analysis, percent					
	Carbon	*Manganese*		*Phosphorus*		*Sulfur*
Process of Manufacture	*Max.*	*Min.*	*Max.*	*Min.*	*Max.*	*Max.*
Seamless:						
Grade A25, Class I	0.21	0.30	0.60		0.030	0.030
Grade A25, Class II	0.21	0.30	0.60	0.045	0.080	0.030
Grade A	0.22		0.90		0.030	0.030
Grade B	0.27		1.15		0.030	0.030
Electric-weld or submerged-arc weld:						
Grade A25, Class I (electric weld only)	0.21	0.30	0.60		0.030	0.030
Grade A25, Class II (electric weld only)	0.21	0.30	0.60	0.045	0.080	0.030
Grade A	0.21		0.90		0.030	0.030
Grade B	0.26		1.15		0.030	0.030

Table 2 Notes:

1. For each reduction of 0.01 per cent below the specified maximum carbon

content, an increase of 0.05 per cent manganese above the specified maximum is permissible, up to a maximum of 1.45 per cent.

2. For cold-expanded seamless pipe in sizes 20 inches and larger, the maximum carbon content shall be 0.28 per cent.
3. Other chemical analyses may be furnished by agreement between purchaser and manufacturer.
4. Either columbium, vanadium, titanium, or a combination thereof, shall be used at the discretion of the manufacturer.
5. For grades 65 in sizes 16 inches and larger with wall thickness 0.500 inches and less, the chemical composition shall be as shown or as agreed upon between the purchaser and manufacturer. For other sizes and wall thicknesses, the chemical composition shall be as agreed upon between the purchaser and manufacturer.
6. Either columbium or vanadium or a combination of both shall be used at the discretion of the manufacturer.

High-test line pipe has more exacting requirements; this means that fewer manufacturing processes qualify but more post-manufacturing treatments are needed to meet the varying requirements. Variations in chemical content, sometimes including the addition of elements such as columbium, vanadium, and titanium in very small proportions, are needed to provide the strength and other properties of high-test line pipe. Table 5.3 gives the chemical requirements for ladle analysis for each grade and process of manufacture.

Table 5.3: Chemical Requirements for Ladle Analysis (Courtesy API)

Process of Manufacture	*Grade*	*Carbon*	*Manga-nese*	*Phosph-urus*	*Sulfur*	*Colum-bium*	*Vana-dium*	*Titanium*
Seamless:								
Non-expanded	x42	0.28	1.25	0.030	0.030			
Non-expanded	x46, x52	0..31	1.35	0.030	0.030			
Cold-expanded	x42, x46, x52	0.292	1.25	0.030	0.030			
Non-expanded or Cold-expanded	x563,x603	0.26	1.35	0.030	0.030	0.0054	0.024	0.034
Non-expanded or Cold-expanded	x65		(by agreement)		0.030			
Electric-weld or submerged-arc weld:								
Non-expanded	x42	0.28	1.25	0.04	0.05			
Non-expanded	xX46, x52	0.30	1.35	0.04	0.05			
Cold-expanded	x42, x46, x52	0.28	1.25	0.04	0.05			
Non-expanded or Cold-expanded	x563,x603	0.26	1.35	0.04	0.05	0.0054	0.024	0.034
Non-expanded or Cold-expanded	x605	0.26	1.40	0.04	0.05	0.0056	0.026	

There are basically six different methods of pipe manufacturing that produce most of today's steel piping. Similar manufacturing methods are used to produce other metallic piping.

Butt-welded (furnace-welded) pipe is manufactured from flat strips of steel called "skelp," with square or slightly beveled edges. The skelp is mostly produced from steel with high phosphorous content, which is best suited for furnace welding. It is furnace-heated full length to welding temperature and then passed immediately through forming and welding rolls, simultaneously forming the tube and welding the edges of the strip. Additional rolling then straightens and finishes the pipe.

The butt-welded joint cannot be expected to be as strong as the plate from which the pipe is made, and a "joint factor" of 60 per cent must be used in calculating the pipe strength. This reduces allowable internal pressure to 60 of that for some other processes. This pipe is normally manufactured in sizes from 1/8 in. through 4 in. and is the lowest in cost among the various types of steel piping available for use in pressure systems. The process is included in API pipe standards as being suitable for Grade A 25 pipe, which has a minimum tensile yield strength of 25,000 psi (Figure 5.19).

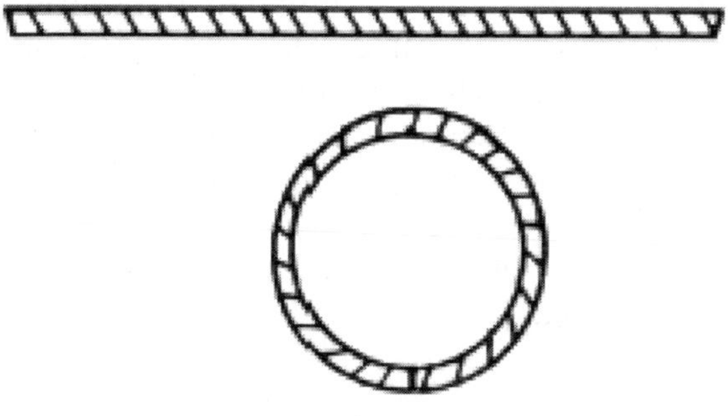

Figure 5.19:

Lap-welded (furnace-welded) pipe is also manufactured from skelp, but the ends, which have been scarfed, overlap instead of being butted together (Figure 5.20).

The skelp is first heated and shaped into a tubular form, then reheated to welding temperature, slid over a mandrel, and welded through the compression of two grooved welding rolls that compress the pipe and achieve a furnace weld. Additional rolling completes the manufacturing process. Pipe sizes normally range from 4 to 16 in., and most manufacturing is done to meet ANSI/ASTM specifications A53 and Al20. The joint factor for this pipe reduces allowable pressure to 80 per cent of that for some other processes.

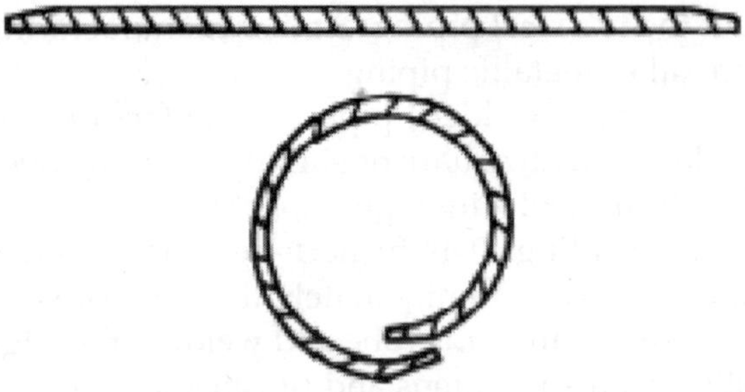

Figure 5.20:

Electric fusion-welded pipe is made by a process in which a plate with suitably prepared edges is first hot- or cold-rolled into a tubular shape. The resultant longitudinal opening is then welded together, with or without additional filler material being deposited at the same time. Electric arc welding can be manually or automatically performed and may be of the single- or double-weld joint type, depending on plate thickness. Minimum size for this type of pipe is normally 4 in., but there is practically no upper size limit.

Electric resistance-welded pipe is similar to the electric fusion-welding process in that a plate is first rolled into a tubular form. The welding operation is then performed while the tube is being compressed by two or more pressure rollers. The whole operation can be performed without preheating the plate or pipe. This type of pipe is usually available in sizes from 1/2 in. to 30 in..

Seamless piping may be produced by two different processes: hot piercing and hot cupping.

The hot-piercing process starts with a round bar, billet, or bloom (all different names for a similarly unfinished steel product), which is heated to more than 2000° F. It is then pierced and forced over a short mandrel by revolving rollers. The initial product is a short, thick-walled pipe that, through a continuing process of either hot rolling or hot drawing, is finally brought to the desired size.

In the hot-cupping method, a steel plate heated to forging temperature is placed against a bottom die, and a round-nosed plunger is pushed through. The emerging cup is repeatedly heated and forced through smaller dies while a closed end remains. This closed end is finally cut off, and the resultant pipe is straightened.

The primary aim of these seamless piping processes is, as the name implies, the production of a joint of pipe with no welding seam.

Availability of seamless pipe was a significant factor in the growth and development of the pipeline industry. Elimination of the longitudinal seam permitted fuller utilization of the pipe strength. The ability to raise pumping pressures and thereby increase pumping capacity with little change in pipeline construction costs resulted in lower costs for pipeline transportation of petroleum products. For many years, seamless pipe was preferred for its strength and comparatively low maintenance costs.

The advent of the electric-weld process for making longitudinal seams as strong as the other parts of the pipe wall has eliminated many of the advantages of seamless pipe. Seamless pipe is generally limited to sizes under 25 in. in diameter.

Spiral-welded pipe is formed with steel strips that are spiralwound to form long cylinders. The edges of the steel, which may butt or overlap, are then butt-welded or fillet-welded together by the electric arc method. This pipe, mostly manufactured as a thin-walled product, is available in diameters up to 48 in. and in lengths up to 60 ft.

Carbon steel is used extensively for large steel tubing in the drilling and maintenance of oil and gas wells. In particular, API has promulgated standards that are applicable to the use of casing and tubing in the drilling industry.

Wrought-iron piping is a derivation of carbon steel pipe but has better corrosion-resisting properties. The word "genuine" is often used to ensure that the correct material is being referenced. The manufacturing process uses either butt or lap welding. Low carbon content and the inclusion of about 3 of iron silicate (slag) tend to give the finished pipe a tough, fibrous structure and lend substance to the claim of better corrosion resistance. ANSI standards for the manufacture of this type of piping (A-72) are now obsolete since alternative products have replaced the use of genuine wrought-iron pipe.

Intermediate alloy steel pipe is often used for low-temperature service, as are stainless steel and aluminum pipe. The composition of these alloys includes nickel and/or various other metals. The manufacturing method may be either seamless or welded, and special testing (known as the notched-bar impact test) is required during and after manufacture. Rigid specifications must be met in order to achieve satisfactory results.

Intermediate ferritic alloy steels have the common nomenclature of "alloy piping" and comprise chrome-moly alloy steel piping, which is the workhorse for high-pressure, high-temperature piping in power plants and oil refineries. The various types of this pipe are usually identified by their P-number, such as P-11 or P-22, or by the per centage of chrome and molybdenum content, such as 1 1/4 CR-l/2 moly or 2 1/4 CR-l/2 moly. The various grades all contain some molybdenum, and most contain chrome.

Austenitic stainless steel pipe is the group of chrome-nickel alloys also known as Series 300 stainless steels. Piping manufactured from this alloy is produced in both welded and seamless forms. Various manufacturers have adopted special welding and finishing methods, which include grinding down weld beads in welded pipe. Most piping requires annealing and pickling after fabrication, and these requirements are incorporated in the piping specifications.

The most common grades of austenitic stainless steel, such as 304, 316, 321, or 347, contain various amounts of chrome, nickel, and other metals. There are also different material compositions within certain types. For instance, by reducing the carbon content in the metal, the material is qualified to be graded extra low carbon (ELC). ELC material allows better weldable quality.

The size range for stainless steel pipe today is no more limited than that of carbon steel pipe. Stainless steel is used in chemical plants, paper mills, and numerous other applications because of its resistance to corrosive attacks.

Pipe Size Nomenclature

Nominal pipe sizes (NPS) range from 1/8 in. to 48 in. Sizes 1.8 to 12 in. are still referenced by the old iron pipe sizes, as originally cataloged by Briggs, but sizes 14 in. and larger use the pipe OD as the base. The three classifications that refer to wall thickness are standard, extra strong, and double extra strong. Standard wall thickness for sizes 1/8 in. to 10 in. denote the various applicable dimensions, but for sizes 12 in. and larger, standard wall thickness means a uniform 3/8 in. Extra strong wall thickness from 1/8 in. through 6 in. varies with each size, but for sizes 8 in. and larger, it denotes a uniform 1/2-in. wall thickness. Double extra strong standards are recognized for sizes from 1/2 in. to 1 in., with the wall thickness for 10- and 12-in. pipe being the same 1-in. thickness.

The piping tables in ANSI B36.10 list schedule numbers for piping from 1/8 in. to 36 in., as well as wall thicknesses for NPS-related sizes. Most piping sizes designated as NPS standard weight or extra strong are also identifiable by an appropriate overlapping schedule number.

Most noteworthy of this overlap is the designation standard weight pipe for all piping sizes 1/8 in. to 10 in., where this equals Schedule 40 pipe. Similarly, piping described as extra strong is the equivalent of Schedule 80 pipe in sizes 1/8 in. to 8 in. Because of this similarity in certain size ranges, it is still common practice to refer to Schedule 40 piping as standard and Schedule 80 piping as extra strong. Schedule numbers are not available for the entire size range indicated above but are restricted to specific sizes. The schedule numbers, which start with 10 for the lighter (thin) walls and go up to 160 for heavy (thick) walls, are limited to the following:

Table: 5.4:

Pipe Schedule	*Size Range, inches*
10	14 - 36
20	8 -36
30	8 -36 (but not 26)
40	0.125 - 36 (but not 22 or 26-30)
60	8 - 24
80	0.125 - 24
100	8 - 24
120	4 - 24
140	8 - 24
160	0.5 - 24

API Standard 5L provides dimensional requirements for line pipe. This standard-weight pipe is comparable to ANSI NPS standard weight pipe, with the exception that API lists several wall thicknesses for 8-, 10-, and 12-in. pipe.

Stainless steel piping schedules sometimes overlap with piping schedules for carbon steel piping but are always identified with the suffix -S. Schedules 40S and 80S both range from 1/8 in. to 12 in., and wall thicknesses equal those of their respective Schedules 40 and 80 in all sizes except 12-in. Shedule 40.

Installation

The first step in pipeline installation is preparation of a right-of-way. This includes installation of temporary gates or passageways in fences; clearing by cutting and disposing of timber, brush, and crops; grading to permit passage of pipe-stringing trucks and construction equipment; bridging of small streams; and making all other arrangements required for safe and expeditious conduct of the work.

Widths of required clearing vary with pipe size, but widths of 50 to 100 ft are most common. Disposal of materials such as timber, tree stumps, brush, and waste must be in accordance with agreements between right-of-way purchasers and owners, or in accordance with other effective, safe, and legal procedures. Burying any sort of waste material in the pipe ditch is not an acceptable means of disposal.

The path of the center line of the pipeline ditch should be described with respect to the location of survey lines as shown on alignment maps and survey stakes. "Depth of ditch" is usually defined as the depth required to provide the specified amount of cover when the ground surface is returned to its normal level. In some cases the depth of burial is governed by laws and regulations. The nature and use of the land has a bearing on the amount of cover needed. In soil so rocky that it is not likely to be cultivated and where the pipe is not likely to be exposed by soil erosion, the amount of cover can be reduced. In many such areas, 12-in. cover is considered to be adequate, but applicable regulatory requirements should be checked. In soil, the cover should be enough to prevent plowing and other farm operations from causing damage to the pipe and its protective coating. Since excavating equipment is a principal source of damage to pipelines, the trend is toward deeper burial. Specified ditch widths are commonly 1 1/2 to 2 times the pipe diameter, above a minimum width of 12 in.

Crossings for railroads, highways, roads, streets, and other public passageways usually must be constructed in accordance with industry standards or governmental regulations. Openings under roadbeds for installation of pipe casing are most often made by boring horizontal holes from one side to the other.

Pipe laying consists of the operations required for joining individual pipe lengths, valves, and fittings into a continuous conduit for fluid flow. Individual pipe lengths are bent as needed to conform to vertical and horizontal changes in direction of the ditch. Bends should be made to avoid metallurgical and dimensional damage to the pipe and minimize damage to its protective coating. Pipe ends are prepared for welding by inspecting them for damage, repairing any damage, and cleaning by wire brushing or, if necessary, by abrasive grinding. A swab should be run through each pipe length just before it is placed in alignment for welding to ensure removal of large objects, much of the loose dirt, and mill scale.

Pipe with longitudinal welding seams should be placed so that the seams are kept above the center line of the pipeline. The seams in adjacent lengths should not be aligned.

Welding is done in several passes. A first pass is carried out to assemble the pipe joints that are held, during this operation, by clamps. Companies differ in their welding specifications. Some are explicit in naming welding rod grade and size for each bead and the number of beads per weld for each pipe grade and wall thickness. Others specify only that weld strength, ductility, and other properties should equal or exceed those of the pipe when tested in accordance with standard methods, such as those set forth by the API.

For many years, nearly all field joints were made by electric arc welding using direct current and coated-stick electrodes.

This method is still the most widely used; however, other currently available methods produce good quality welds and may have advantages over electric arc welding, such as reducing the required number of beads or passes, speeding up the process, reducing cost, or producing better welds. When welding procedures are chosen by the contractor, inspection and test procedures should be used to ensure weld quality.

Line pipe may receive a protective coating in coating plants prior to being hauled to the right-of-way. In this case, the contractor's work consists of removing the portions of the coating damaged during hauling and construction, recoating these areas, and coating the pipe ends left bare for welding. In another commonly used procedure, the contractor applies the entire coating on the right-of-way after the pipe has been welded into long sections.

In either case, the operation consists of two stages. In the first stage, the pipe is brushed clean and receives a primer coat of paint applied cold. In the second stage, melted enamel is poured around the pipe. The thickness of the coat depends mainly on the temperatures of the product and the pipe. Glass fiber and kraft paper, whenever necessary, are rolled on at the same time by a spool-carrying arm rotating around the pipes (with mobile machines) or by a fixed arm while the pipes themselves rotate in a coating yard.

Coating material may be described in generic terms and limiting values of characteristics such as hardness, melting point, application temperature, and so forth. In some cases, the materials may be identified by brand name and reference may be made to the manufacturer's information and instructions.

Lowering consists of placing the completed pipeline in the ditch. Even the most carefully conducted lowering operation sometimes results in damage to the pipe or its coating. Since such damages are not likely to be detected after the pipe is lowered, it is important that someone with authority be present to authorize repair of damage and modification of the procedure if necessary.

It is usually specified that pipe-handling equipment be designed and constructed to keep unit pressure and bending stress on the pipe and coating within safe limits.

Backfilling consists of covering the pipe with the earth removed from the ditch or with other specially prepared material, and completely filling the ditch so that normal ground level is restored. The usual method is that of pushing or pulling the excavated earth back to the edge of the ditch and allowing it to fall in. The specifications may require that this procedure be modified if there are rocks or large, hard clods of earth that might damage the pipe or coating. Ways of limiting the force of such impacts may be specified. Continued uneven pressure against the coating by rocks in the backfill may cause considerable damage; a backfill of rock-free earth or crushed rock to a height of about 4 in. above the pipe may be specified.

Natural settling and compaction is sufficient in most cases, and the surplus earth may be left in a rounded mound over the ditch. However, compacting the backfill to a density approximately equal to its density before excavation is sometimes required. This is usually needed on or near roadways to prevent the formation of depressions or ridges that would be hazardous to traffic.

In some cases, special procedures are required to backfill the spaces between the ditch bottom and the lower quarters of the pipe. When the ditch bottom is flat and hard and the backfill does not readily flow into these spaces, the weight of the pipe plus that of part of the backfill may be supported by a narrow strip along the bottom of the pipe. Later settling of the backfill could also cause shear stresses in the coating, causing it to flow downward. If the pipe size and soil conditions are likely to result in such potential damage, backfilling methods that fill these spaces more completely are generally specified.

Offshore pipelines obviously require the use of highly specialized equipment, generally lay barges designed for the purpose. These barges, while laying, are anchored by several anchor lines on the sea bottom. Their movements are carefully controlled by pulling and releasing the anchor lines as required to make them progress along the route of the pipeline. The principle is to lay the pipes on the seabed by letting them sink, in a flexed curve, from the laying barge out to the seabed.

The specific pipeline equipment of these barges is mainly composed of a pipe-laying platform along the entire length of the vessel with the working positions required for the welding and preparation of the line prior to its launching into the sea. The pipe moves past these different working positions as the barge moves forward, adjusting its anchor lines in step with the progress of the welding of the pipe joints. These working positions include, in particular:

- one for positioning the pipe joints and preparing the bevels;
- multiple (three, four, or five) welding stations, usually equipped with automatic welding machines, which are particularly well suited for this purpose;
- a weld inspection station;
- a station for the application of anticorrosion and concrete coating on the weld zone to ensure the continuity of the coating all along the line.

Testing and Inspection

Hydrostatic tests are most often specified as the means of proving that the complete pipeline is capable of withstanding, without leakage or failure, an internal pressure higher than the proposed operating pressures. The operation of filling the pipeline with water provides an opportunity for determining whether obstructions, reductions in pipe diameter, or other unknown abnormalities exist. This can be accomplished by having the water push ahead of it a scraper, or pig, fitted with a metal plate slightly smaller in diameter than the pipe ID.

The test pressure is usually in the range of 100 per cent to 150* of the minimum yield strength of the pipe and must be maintained without change for a given period of time (e.g., 24 hours) to indicate pipeline integrity.

The pipe line should be tested after it is backfilled and all construction work that might affect the pipe is complete. Certain sections of pipe should be tested while still readily accessible for repair, particularly those assembled separately for installation under water or in other locations where test failure repairs would be unusually difficult or expensive once the pipe is in place and backfilled. Such sections must be retested along with the remainder of the complete pipeline.

Pipelines often must be tested in segments because of differences in ground elevation and weight of the test water. The pressure needed to test pipe at higher elevations in a test segment might overstress the pipe at lower elevations because of the hydrostatic head. Segment lengths and locations must be chosen with the purpose of keeping differences in hydrostatic head within tolerable limits.

Pipelines are unique among industrial plants in that they may be tested after construction is complete in a manner that permits discovery and removal of defective parts. This effective test procedure is possible because of the simple tubular shape of the pipe. Each part of the pipe is equidistant from its longitudinal axis and thus is equally stressed by internal liquid pressure. The uniformity in dimensional and other properties of steel plate used in making pipe and the quality of longitudinal and circumferential welds ensure that the strengths of all parts of the pipeline are equal, within narrow tolerance limits.

This susceptibility of pipelines to proof testing has been largely responsible for the industry's ability to use such a large portion of the pipe's strength.

Two weld inspection methods are available: nondestructive and destructive testing. Nondestructive tests, as the name implies, may be performed on welds without affecting their properties or future usefulness.

This method is applicable to production welds, including field welds. Its availability and general use have resulted in great improvements in weld procedures and weld quality. Destructive tests include determining tensile strength of welds or adjacent pipe metal by exerting tension on samples and measuring the force required to break them, then examining the samples for defects. Bend tests are performed to determine ductility and to determine the presence or absence of certain defects. Destructive tests are required for welds made as a part of welding procedure qualifications. Tests for welder qualification may, at a company's option, be by nondestructive or destructive means, or both.

Radiographic examination is the principal mode of nondestructive testing. In essence, a radiograph is a photographic record of a test specimen exposed to the passage of x-rays or gamma rays. X-ray inspection is more often used in manufacturing facilities where an unwieldy machine with an x-ray tube may be permanently installed. For inspection of field welds, however, portable radioactive isotopes, such as cobalt 60 or iridium 192, provide an ideal source of radiation.

In either case, the radiation proceeds in straight lines to the test object. Some rays are absorbed and others pass through the material being examined. Less dense areas of the test object, which may result from slag inclusion, gas bubbles, porosity, or other internal weld defects, absorb less radiation and result in darker spots on the film, indicating their size and exact location.

Radiography operators must undergo extensive training for equipment operation and film interpretation. Although API standards for acceptability apply, the operator's judgment is an important element in the interpretation of radiographic film.

One type of radiography requires attaching or placing the film on the outside of the pipe and having the radiation source at the opposite side of the pipe. The rays pass through both pipe walls to provide film exposure that only encompasses a segment of the weld. This method requires an average of three exposures, each encompassing approximately 120° of a pipe circumference.

Another radiography method places the radiation source inside the pipe, permitting exposure of the total weld at one time. This requires an access hole through which the radiation source may enter the pipe.

A different method of inspection, magnetic-particle examination, uses the magnetic properties of the material to be inspected; thus it is limited to the ferritic metals and cannot be applied to austenitic stainless steels and most nonferrous metals.

Wherever magnetic lines of force are interrupted, poles are formed. Such interruptions are created by flaws in welded areas, and miniature poles are thus established at these flaws.

The weld arc to be examined must be smooth and clean, and grinding is usually required for such a preparation where nonautomatic welding processes are involved.

Iron particles, mostly dry magnetic powders, are dusted or sprayed onto the area to be examined. The flaws—or rather the resultant magnetic poles—even if below the surface, exert sufficient pull on the surface-applied powder to align it in such a way that cracks are clearly indicated. Wet powders contained in fluorescent liquids or colored magnetic particles are also used in this type of examination and are often preferred because of the easier identification of the flaw lines.

To ensure the most effective detection of weld flaws, two examinations at right angles should be performed. Operators for magnetic-particle inspection can be trained relatively quickly, and the interpretation of results is relatively simple; however, irrelevant indications of magnetic permeability are not uncommon.

Ultrasonic examination of piping or tubing is standard procedure in many manufacturing processes and is often a requirement of the various applicable standards. This process involves directing a beam of ultrasonic energy into the specimen, and the energy transmitted through it is indicated.

Ultrasonic inspection uses high-frequency mechanical vibration and sound frequencies between 200 kHz and 25 MHz. Today there are a number of examination techniques available, and various special instruments have been developed to furnish examination data and assist in their interpretation. Ultrasonic examinations employ only low-amplitude stresses and do not affect the material undergoing such tests.

Advantages and disadvantages of this method are as follows:

Advantages

- Readout and interpretation of test results can be accomplished almost instantly;
- The accuracy of fault detection is not hampered by the direction of a particular crack;
- Even minute cracks and flaws that might defy other examination methods can be detected.

Disadvantages

- This method does not normally provide a permanent record of the inspection, although tape recordings of a test can be made;
- Only trained and experienced operators who are well versed in

the use of the special equipment can be expected to use it effectively and provide a true analysis of test results;

- Good contact between test material and the search unit is essential, and air is normally excluded from such contact areas through the application of a suitable coupling.

Ultrasonic examination is usually restricted to small areas and should not be confused with low-frequency resonance methods, in which a large test specimen is vibrated at sonic frequency. Since ultrasonic testing equipment is portable, it is well suited to field examination of circumferential pipe joints (butt welds) and is being used extensively on large-diameter overland transmission lines. Ultrasonic is generally used to test welds on pipe too small to radiograph.

Pipe Insulation

Insulation is normally applied to piping systems to prevent any heat exchange between the fluid carried in the pipe and the exterior surroundings. It may be used to prevent heat loss from a pipe carrying warm or hot fluid, or to protect low-temperature fluid in the pipe from higher temperature surroundings.

Another use of piping insulation is personnel protection. Piping systems that carry hot fluids, even in places where heat loss is irrelevant (or even desirable), are often insulated to prevent accidental flesh burns where such piping is within easy reach of personnel.

In the past, asbestos-derived materials, which are known for their fire- and heat-resistant properties, played a major role in general insulation practice. Major thermal insulating materials today are calcium silicate, diatomaceous earth, polyurethane foam, perlite, and an assortment of felts, glass fibers and plastics, as well as cellular glass for cold insulation.

Insulation is often available as a preshaped product where two halves are slipped over the straight lengths of pipe. However, fittings, flanged joints, and valves require a craftman's skill to cover their divergent contours, unless custom-shaped materials are available. Pipe insulation in industrial applications is often covered with metallic aluminum or stainless steel, which provides additional protection and better appearance, while another type of insulation might use a canvas jacket. However, polyurethane foam application differs markedly; it is either poured, blown, or sprayed into a partially enclosed area before being covered with a protective covering.

Temperature Differentials

Regardless of the use of insulation, temperature differentials in piping systems should be taken into account during the design phase of any system. Pipe movement resulting from temperature differentials—whether created by the transmitted fluid or by the surroundings—must be controlled.

In the Arabian desert, where outside temperatures vary widely between day and night, the solution to this problem is very simple. Surface pipelines are not laid in a straight line but are installed in a continuous zigzag. Any expansion during the day or contraction during the night is compensated by the increasing or decreasing of the angles at these bends.

A more sophisticated approach, the installation of expansion loops, has been a standard practice in piping systems for a long time. These loops may be welded using 900 elbows at the four corners, or may consist of bent pipe.

Pipe bends are usually fabricated to the expansion U-bend pattern or as a double-offset U-bend (Figure 5.21 and Figure 5.22).

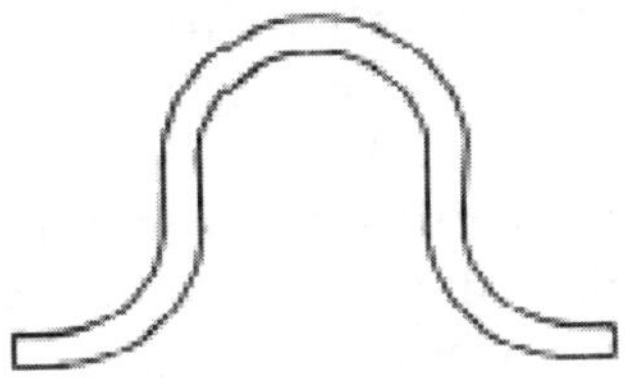

Expansion U-bend

Figure 5.21:

Since these expansion loops require a considerable area for installation, manufactured expansion joints are necessary in piping systems with space restrictions.

There are two basic types of expansion joint: the sleeve, or slip, and the bellows. Piping systems that incorporate manufactured expansion joints must be restricted in their movement, so that any expansion or contraction is not deflected but is confined to the expansion joint. The best way to achieve this is with properly located pipe anchors and alignment guides.

The sleeve-, or slip-type expansion joint consists of three major parts: an external sleeve connected to the piping on one side, an internal slip connected to the piping on the other side, and a stuffing box or packing-gland arrangement to hold the pressure. This type of expansion joint is manufactured to allow for expansion or contraction from an anchor point in either one or two directions along its axis.

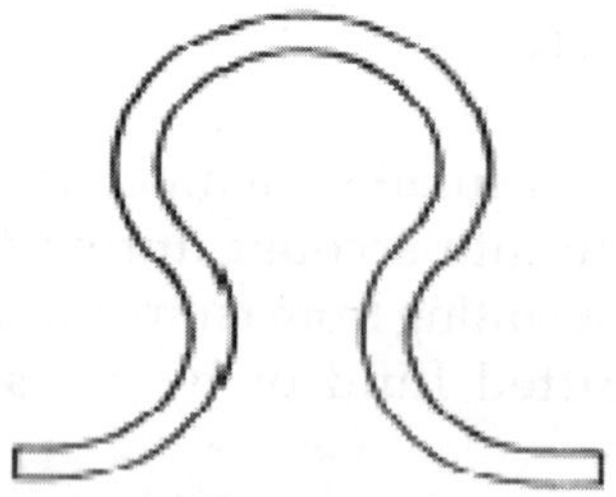

Double Offset Expansion U-bend

Figure 5.22:

The bellows-type expansion joint is manufactured as either equalizing or nonequalizing. The number of bellows incorporated in any expansion joint can be in accordance with the requirements of any particular system and may range from a single bellow to more than 20.

This nonequalizing type consists of bellows only, with suitable ends for installation into a piping system. After expansion or contraction, there is no control available to return the bellows to their original configuration.

The self-equalizing type employs control rings to distribute the compression equally among all bellows. It limits the amount of compression to which any one bellow is subjected and thus ensures equalization after expansion is relaxed.

Internal and external guides are offered for equalizing and nonequalizing types to limit movement to the axial direction. Liners that protect the bellows from early failure caused by erosion, corrosion, or exposure to high fluid velocities may thus sometimes serve as both guide and protection. Liners also protect the bellows from high-temperature fluid flow.

The nonequalizing bellows-type expansion joint is preferable in situations where no axial expansion is present but a lateral offset or angular displacement of the piping may occur. There is a limit to this type of application for expansion joints. Where more severe displacement is expected, another form of pipe equalization must be used—for instance, flexible piping.

Another way to provide for pipe expansion is to incorporate flexible connectors, or swivel joints, into a piping system. Swivel joints are manufactured primarily from steel, stainless steel, bronze, or similar materials, and may have threaded or welded connections.

Swivel joints are manufactured from two or more components, depending on the degree of flexibility required. They permit, in a simple configuration, a rotation of 360°. By adding components to the swivel joint, its flexibility may be increased from the single plane to a variety of configurations.

Problems in piping systems or equipment connections are always created through vibration, thermal expansion and contraction, shock, or swing connections. These can best be solved by using flexible piping that is designed to withstand the rigors of continuous or frequent movement.

Rubber hose is available in many variations, from plain to multilayered and heavily reinforced (with fabric or steel). Inner liners made from various plastics, depending on the fluid being handled,- are also an integral part of many rubber hoses.

Corrosion Control (External)

Unlike other materials, such as cast iron or plastics, steel is liable to corrode when buried in the ground. Thus, steel pipes used to transport products such as hydrocarbons need protection from attack by various corrosive factors in the soil—e.g., chemical ions, bacteria, conditions favoring current circulation through the pipe, and stray currents. The protection must eliminate, for all practical purposes, the danger of leaks for the life of the pipeline.

In the final analysis, corrosion of steel in the ground is invariably caused by an electrochemical effect. This corrosion is the result of an electrical current formed with the connection of an anode and a cathode, both located on the pipe, using the pipe itself as the conductor. To complete this circuit, a continuous electrolytic path must exist between the same anode and cathode, which is usually supplied by soil moisture or water in which the pipeline is immersed. Simply stated, current flows from cathode to anode in the pipe and from anode to cathode in the electrolyte.

Various conditions determine the creation of these anodes and cathodes, as well as the creation of the electrical potential necessary for this flow of current.

With the electrical current flow, corrosion occurs at the anode only and is directly proportional to the amount of current. Corrosion is reduced by increases in resistance to electric current flow in any part of the circuit and is stopped completely when the current is broken at any point.

Soils with lower electrical resistivity are more likely to develop anodic areas. Soil resistivity measurements have been used to forecast corrosiveness of soils before pipeline construction and to locate corrosive areas on existing pipelines. However, it has been proven by experience that even high- resistivity soils create some corrosion in the lower resistivity sections of the line. Moreover, on right-of-ways where the soil resistivity is in the low range (100 to 1000 ohm-cm), corrosion is often limited again to the lower resistivity range. In other words, there appears to be no critical value of resistivity separating corrosive from noncorrosive soils, but differences in resistivity are more important.

The nature of corrosion immediately suggests two means of its control. One is to break the electrical circuits of existing or potential corrosion cells. The other is to make all parts of the structure cathodic, since metal is not damaged at cathodes. Both means are used and, as would be expected, are most effective when used in combination.

The covering of metal with a waterproof, electrically nonconductive coating is an age-old method of corrosion control. Pipeline coatings need special properties. An industry of considerable size has grown up to develop, produce, and apply such coatings. It should be apparent to anyone familiar with the realities of pipeline construction that a completely waterproof, electrically nonconductive coating for an entire pipeline is a goal that may be approached but never fully achieved. Coating materials and application procedures are available that provide an acceptable degree of protection.

One type of coating is formed into sheets about 0.010 in. to 0.025 in. thick and attached to the pipe by means of adhesives. For application, the sheets are formed into rolls of tape of appropriate width with adhesive on one side. Application is by conventional wrapping machines, generally the best method when thc coating is to be applied in the field.

Another approach has been to extrude molten thermoplastic materials as seamless sleeves or sheaths over cleaned, adhesive-coated pipe. Upon cooling, the sleeve shrinks to a tight fit. Sheath thicknesses vary from about 0.010 in. to 0.030 in., depending upon pipe size or purchaser's specifications.

Coatings most closely approaching the ideal, insofar as coating structure is concerned, are the thermosetting and the thermoplastic resins applied to new pipe under carefully controlled conditions at coating plants. These processes are accomplished in different ways, but in both cases a film is formed directly on the pipe surface. Film thicknesses can be precisely controlled and are usually in the 0.010- to 0.030-in. range.

Two principal points should be emphasized for nonspecialists in the coating field who are selecting coating materials or application procedures:

1. Consider the conditions to which coatings are to be exposed while in service, plus special conditions applicable to the individual project.
2. Check proposed coating materials and application procedures to determine whether they will produce coatings with the needed properties.

As previously stated, another method of combating electrochemical corrosion is to make the entire pipeline cathodic. Cathodic protection does exactly that. It may be visualized as collecting all the anodes of all existing and potential corrosion cells and transferring them to a place of the pipeline operator's choosing. This is accomplished by controlling the action of large corrosion cells. These cells are formed by connecting the negative terminals of DC power sources to the pipeline and connecting the positive terminals to expendable solid conductors buried in the ground. These solid conductors are called anodes, and interconnected groups of them are often called "ground beds." The electrical circuits are completed by current flowing through the soil from the anodes to the pipeline and along the pipeline to the current source. A diagram of such an installation is shown in Figure 5.23.

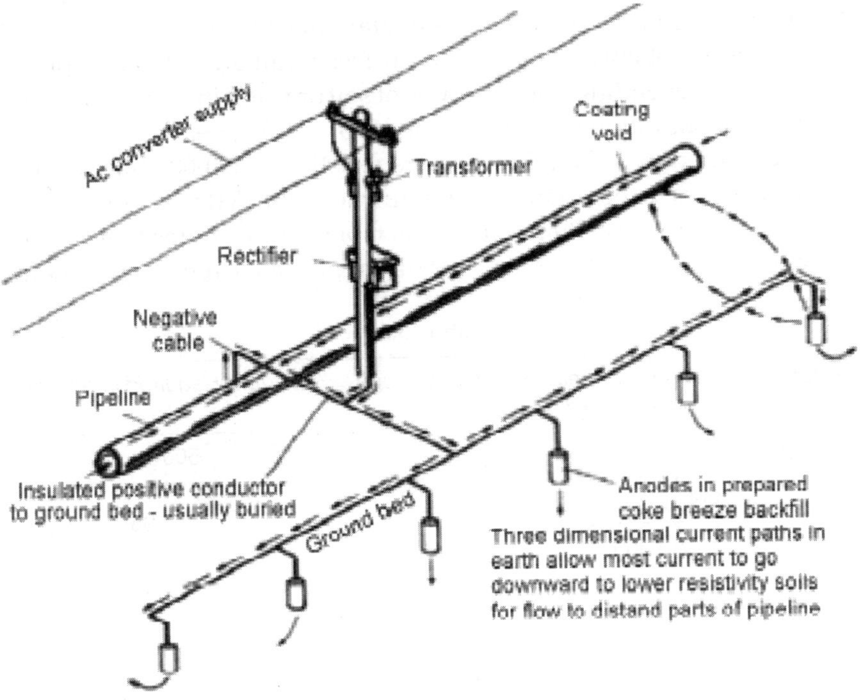

Figure 5.23:

Unlike protective coatings, this means of corrosion control requires continuing expenditures for operation, maintenance, and monitoring of its effects. An extensive field of technology has been developed for measurement and evaluation of cathodic protection requirements, the design and installation of facilities, measurement and interpretation of results obtained, and other operational and maintenance problems. Many engineers and scientists work in this field, and cathodic protection work is usually assigned to professional corrosion engineers. This discussion is intended to be a very brief introduction.

Cathodic protection is based on such a simple, fundamental, and obviously applicable principle that the question naturally arises as to whether it should be the primary means of corrosion control. The answer is that there are limiting factors that would make its use impractical on the large scale if it were the only method. It is ideal as a supplement to protective coatings, and the combination can provide effective and economical corrosion control for buried pipelines.

Factors that need to be determined in a cathodic protection installation are current and power requirements and power sources. A look at the factors affecting current requirements supports the fact that cathodic protection should not be the sole combatant of corrosion.

The amount of current needed to protect a unit area of bare pipe steel varies widely, depending upon the type of soil or water in which it is buried. A common assumption is the requirement of 1 mA for each square foot of bare steel buried in soil having average resistivity of 1000 ohm-cm. Table 5.5 indicates the current required, as a function of the effective pipe coating resistance, for 10 mi of 36-in. pipe. This table demonstrates that coatings, even poor ones, make a tremendous difference in current requirements.

Table: 5.5:

Effective Coating Resistance For One Average Square Foot in Ohms	*Current Required in Amperes*
Bare Pipe*	500
10,000	14.91
25,000	5.964
50,000	2.982
100,000	1.491
5,000,000	0.0298
Perfect Coating	0.000058

The extent to which coating and cathodic protection is used depends upon the intended life of the pipeline. The use of low-quality coatings with the logic of requiring low amperage of adequate protection should be avoided, since the long-lasting properties of corrosion protection are dependent on the quality of the installed coating. Coating deterioration could increase current requirements beyond economic and even physical limits early in the life of the pipeline.

A perfect coating is assumed to be perfectly holiday-free, to consist of

3/32 inch layer of material having resistivity of 1013 ohm-centimeters and current required is that needed to cause a 0.3 volt drop across the effective resistance between pipeline and remote earth. Polarization effects are neglected. Lower values of coating resistance result from quantities and degrees of imperfections in coatings.

Bare pipe is assumed to require a minimum of 1 milliampere of current per square foot.

External corrosion prevention, is generally effected during pipeline installation. Corrosion is also a potential problem for the ID of the pipe. Internal corrosion may be caused by the corrosive nature of the product (sulfur-containing crude, for instance) or most often by traces of water, which are always present. The water is deposited on the pipe wall or condenses on the wall out of the gas. For both liquid and gas, the first step is always the elimination (or drastic reduction) of water in the pipeline.

A method of choice in gas pipelines is to apply a fine layer of plastic paint to the internal pipe walls. This method is relatively inexpensive, and its cost can be offset by the saving in compression energy.

When oil is in turbulent flow in pipelines, brine droplets are prevented from accumulating into large drops that would separate from the oil. Any loose material is swept along with the oil stream. The result is that the corrosive agents in the oil are kept out of contact with the steel.

Maintaining continuous turbulent flow as an internal control measure has been used to some extent in sour crude oil gathering systems. The use of smaller pipe, pumps instead of gravity to induce flow, the scheduling of operations to make flow more nearly continuous, and preventing air from entering the piping system can all reduce internal corrosion.

While turbulent flow aids in the prevention of internal corrosion, the most popular method of internal corrosion control for transporting potentially corrosive liquids is the use of corrosion inhibitors. The inhibitors now used in pipelines are of two main types:

1. Water-soluble, hydrocarbon-insoluble inhibitors tend to form a separate phase from the product transported in the pipeline, and to wet the entire surface of the internal wall of the pipe by capillary effect;
2. Hydrocarbon-soluble inhibitors are injected directly into the product transported in the pipeline. They are usually organic products with a polar group in the molecule, which tends to fix them on the internal pipe wall.

The inhibitor generally acts as a sort of coating. This means that in the initial stage of the operation, the inhibitor contained in the product generates the protective layer; later, the concentration is merely that which is necessary to keep this layer in place on the metal surface.

The inhibitor used most often in aqueous solution is sodium nitrite in alkaline solution. It is introduced into the pipe simply by gravity, using an instrument designed to set the rate of injection of the solution by the use of a regulating device such as a needle valve. This injection rate varies with the conditions in the line, the type of service, and the length of pipe to be protected. The injection rate, a few parts per million, is monitored by analyzing the water collected at the exit of the protected section. This water should still contain traces of nitrite, and its pH must be between 8 and 8.5. One advantage of this type of inhibitor is that it separates from the product through normal decantation in the storage tanks at the outlet of the pipeline and thus cannot affect its properties. One disadvantage is that it is somewhat less effective than some of the oil-soluble inhibitors.

Many types of oil-soluble inhibitors have been developed. They are usually organic compounds with a hydrocarbon chain, containing polar groups, amino groups or fatty acid groups. Besides their inhibitory effect, they must not affect the properties of the products into which they are injected. Inhibitors added to crude must not contain even trace amounts of substances that might poison the catalysts used during refining. The injection rate again varies with the condition in the pipeline, the type of the inhibitor, and the corrosive properties of the products, and is of the order of a few parts per million.

The effectiveness of corrosion protection may be monitored in three ways: the value of the line friction factor, the amount of sediment entrained by the products, and, more directly, by inspecting the corrosion of specimen samples placed in the pipeline.

In a new pipeline, inhibitors do not result in significant improvement in the friction factor during the initial cleaning phase. At the end of this phase, when the amount of dirt begins to diminish, the friction factor generally shows a marked decrease and, after a certain time of operation, tends to stabilize at a relatively low value, which should be maintained afterward.

Large amounts of sediment in the scraper traps or filters indicate an ineffective inhibitor, assuming it is not accompanying the product at the inlet.

These control tests are supplemented by inspecting the corrosion of samples. These samples are steel pieces left in the pipeline for one to three months and then taken out so that their corrosion status can be observed. They should be placed out of reach of the scrapers on an easily dismantled flange, or on a support mounted across a lateral branch of a conduit, so that they can be taken out and replaced without having to dismantle anything else. The steel should be the grade most susceptible to corrosion, which roughly corresponds to reinforcing steel. The piece should be electrically insulated from its support and from the pipe steel to prevent any galvanic effect that could distort the results.

The degree of corrosion in the sample is determined through either a comparison of weights before and after exposure to corrosion, or an examination of the surface of the specimen. In the first case, the test piece should be carefully washed and all traces of oxide removed from its surface before the second weighing. In the second case, the surface aspect is classified following a special procedure and comparing it with a standard scale of appearance; the preparation of the specimen is also highly important, with all specimens polished in the same way to ensure comparable results.

Cleaning and Inspection

A pig is a cylindrical device that is inserted into a pipeline at a "trap" or launcher, an then pumped downstream to a recovery point. Its purpose is to help maintain the integrity of the line. Depending on the type of pig used, this may involve cleaning of the pipeline interior, inspection for cracks or damage, or other functions.

Pigs can be grouped into two main categories. Utility pigs, or scrapers (Figure 5.24) and (Figure 5.28), are used primarily to clean paraffin, scale and other materials from the pipeline interior, and to separate fluids.

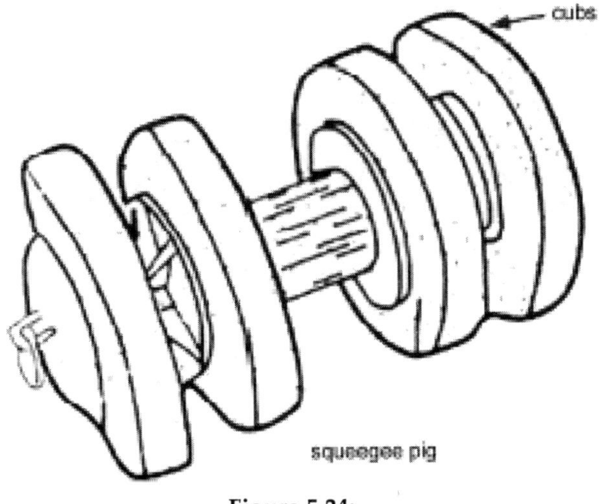

Figure 5.24:

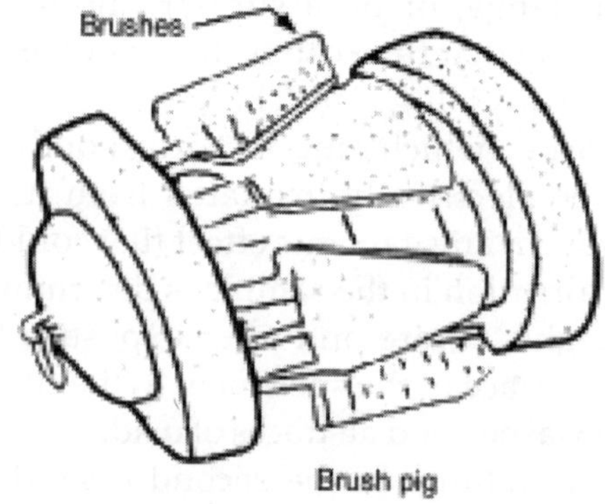

Figure 5.25:

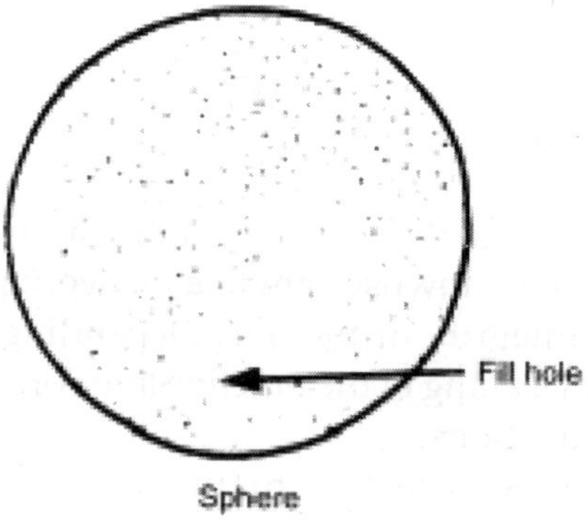

Figure 5.26:

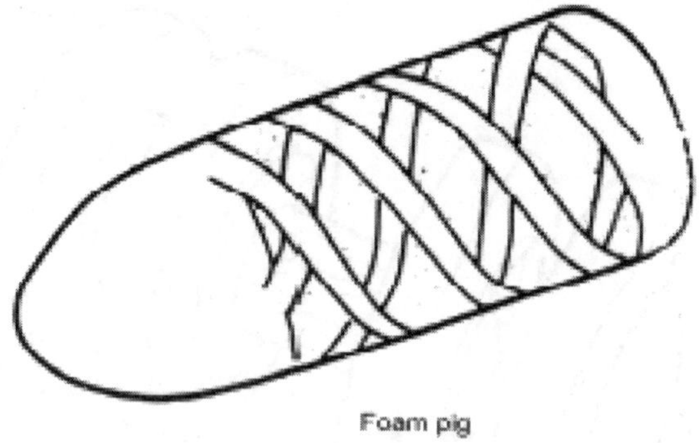

Figure 5.27:

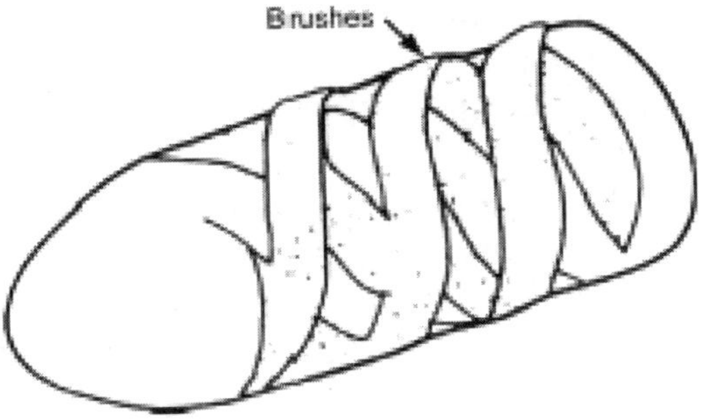

Figure 5.28: Foam Brush Pig

They generally have few moving parts. Intelligent pigs are much more sophisticated, in that they may include sensors and recording packages for monitoring the condition of the pipe, along with batteries or some other means of self-propulsion.

Many types of scrapers are similar in design to those used in the earliest days of pipelining. Generally, they consist of a centralizing element and either blades or brushes to scrape deposits from the pipe surface. The shaft on which the scraper parts are mounted is frequently fitted with a universal joint near its midpoint so it can go through pipe bends. Development of tough, oil-resistant, synthetic rubber like materials has made it possible to make longer-lasting and more effective disks or cup-shaped propulsion members so that there is little bypassing of the oil stream.

Scrapers for large pipe are quite heavy and tend to wear on the bottom side, permitting the oil stream to bypass. Efforts to overcome this tendency include use of strong centering springs that press wire brushes with wear-resistant bristles against the pipe. Sometimes this centering action depends on the stiffness of the bristles. The force and the brushes also improve the cleaning action, which is the primary reason for running the scrapers. Sometimes a portion of the oil stream is bypassed through the scraper and directed through nozzles in directions tangential to the pipe circumference. This is intended to produce slow rotation and even wear of the scrapers. The bypassed oil stream also tends to disperse displaced solids and keep them from being compacted into solid plugs that could not be moved by safe pump pressures. Such stoppages are unusual but have occurred where large quantities of rust scale and/or wax have been displaced. Rates of travel and arrival times at specific locations cannot be predicted accurately when scrapers bypass a part of the oil stream.

For large diameter scrapers, the center member is usually tubular for greater strength and rigidity. This tube or cylinder may be sealed and filled with air to make the scraper buoyant, and pressurized to prevent collapse caused by oil pressure.

Sometimes buoyancy can be adjusted so that scrapers are almost weightless in the oil stream.

Spherical scrapers are made of oil-resistant plastic and are inflated with liquid. The size of the spheres can be adjusted within narrow ranges by changing the liquid content and pressure. Their shapes can be distorted to considerable degree, and they can pass through some fittings that other scrapers cannot. They are often used as batch separators to reduce interfacial mixing of different oils, but they also serve to sweep water and solids along. Spherical scrapers are preferred by some operators and maintenance people for nearly all purposes. Water is ordinarily used to inflate and size these spheres, but lighter liquid may be used to reduce weight.

Scrapers made of plastic foam are also used. The foam is an open-cell type that becomes saturated with oil and so retains its shape under pressure. The front and back portions of these scrapers are usually sealed to prevent bypass of oil and to provide propulsion. These scrapers are compressible and can pass through smoothly contoured fittings of reduced cross-sectional area. They can be used in pipelines of varying diameters. Other scrapers of special design are also dual-sized for pipe of different diameters.

The launching of pigs has been automated in some instances. Some traps are arranged so that one or more pigs may be loaded into launching traps. They may then be launched at any desired time by remote-controlled valve changes. At unattended locations, an arriving pig may signal the appropriate time for launching or may initiate a programmed automatic-launching operation.

Intelligent pigs extend the traditional role of pigging to include pipeline inspection. Pipe wall thickness, for example, can be measured using pigs that generate magnetic fields and record variations in flux density as they travel along the line. Other pigs record wall thickness using ultrasonic measurements. Dents or other impositions on a pipe's inside diameter can be detected using a pig with multiple-arm calipers and a recording device. Other areas of development in pigging technology include leak detection, monitoring protective coatings, improved cleaning and debris removal and inspection of small-diameter lines.

Maintaining the interior of pipelines consists mainly of preventing deposits of materials on the pipe surfaces that would interfere with oil flow. Internal corrosion of the pipe can complicate this maintenance. Internal corrosion may be accelerated by deposits that tend to hold corrosive brines in contact with the pipe. Determining the location and extent of internal corrosion and repairing the damage are extremely difficult. These difficulties make prevention and control of internal corrosion all the more important. Linings are sometimes applied to pipe in place, thus extending its effective service life. Inspection devices, which can be propelled through pipelines in the same way as scrapers, have been developed. These devices measure and record pipe metal thicknesses and indicate effects of both internal and external corrosion. It is expected that efforts will be made to expand applicability of such inspection to more pipelines. In general, control of internal corrosion can best be effected by construction features and operational methods

Gate Valves

Gate valves used in pipeline applications have many features in common with wellhead gate valves. Gate valves are generally multiturn valves that, in their basic construction, consist of a valve body, seat and gate, stem, packing, and rotating wheel (Figure 5.29).

The seat is located at the bottom of the valve and combines with the disc to provide the actual valve components that regulate the flow of any fluid through the valve body. The contacting surfaces of the gate and body seats must be smooth and flat for effective sealing.

The fluid pressure on one side of the gate presses it forcefully against the opposite seat, thus improving the seal. The same force tends to reduce the seal on the pressure side, with the result that the bonnet cavity and the valve stem packing are nearly always subjected to maximum pipeline pressure. Gate seat rings and body seat rings are often removable for refacing when they become damaged. Seat ring materials are selected for resistance to wear, corrosion, and sliding friction.

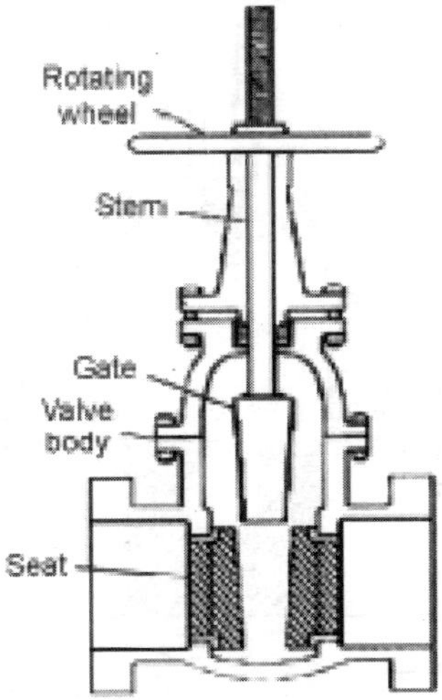

Figure 5.29:

In order to actuate a gate valve, the disc is either lowered or lifted by means of a stem that projects outside of the valve body and is actuated by a handwheel. The protrusion of the stem to the outside atmosphere requires some method of retaining the fluid in the pipeline, which is accomplished by installing a gland packed with a fluid-resisting barrier to prevent leakage.

Gate valves are used primarily as stop valves, to fully shut off or provide full flow. They are ideally suited to wide-open service, such as at an outlet of a storage tank for liquids in oil and gas pipelines. The flow can move with almost no resistance in a straight line when the disc is fully raised. Seating in a gate valve is at a right angle to the line of flow, which makes the valve impractical for throttling operations and makes close regulation of flow nearly impossible.

Though many valve accessories have been improved throughout the years and a great many variations of gate-valve designs are commercially available today, the basic components have not changed. The valve body is still surmounted by a bonnet. In small valves intended for low-pressure applications, it is common practice to use a U-bolt, a screwed joint, or union-type joint to connect the bonnet to the valve body. In larger valves for use in higher pressure and temperature situations, other means must be used to retain the fluid. The obvious solution is the use of a flanged joint and gasket.

Some gate valves are designed with a rising stem, where an indicator riding on the valve spindle can show whether and to what degree the valve is open. Valves with nonrising stems are also available (Figure 5.30).

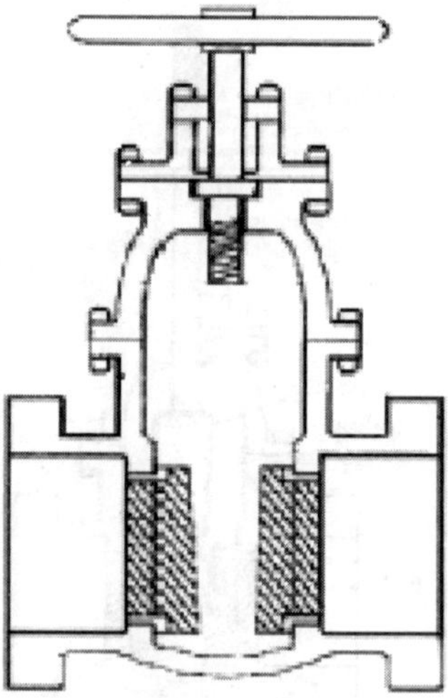

Figure 5.30:

Repacking gland boxes can be accomplished on many valves, even when they are under pressure and in the open position, thanks to a feature called backpressure seating. This effectively seals the packing retaining chamber from the pressure of the fluid flow.

One major variable in gate valve design concerns the different types of wedges available.

The single-wedge disc gate valve is usually solid and fits into tapered valve seats, which may be replaceable, or into an internal part of the valve body. This single-wedge design is particularly well suited to overcome misalignment and dimensional changes within the valve body caused by temperature variations. The tapered construction of the disc and seats, with the resulting large seating area normally results in a wide and true contact between the disc and the corresponding faces of the valve seat (Figure 5.1 and Figure 5.2).

A variation of the solid wedge is the flexible disc, which is solid only through the center so that some movement of the faces in relation to one another is possible. This flexibility, which is attained without additional parts, can assist greatly in ease of operation and ensuring valve tightness, not only on the inlet seat but also on the outlet seat. Another design used in conjunction with tapered valve seats is the split wedge.

In double-disc gate valves, the two seating surfaces can move relative to each other. This gives good shutoff even with misaligned seats or different seat angles. This design also provides a tight seal without relying on fluid pressure, making it particularly suitable for steam, gases, and light oils.

Two wedges between the discs multiply the closing force. They act to force the discs apart and against the valve seating surface only after the bottom wedge strikes the valve body upon closing. Upon opening, the top wedge moves first and relieves the wedging action against the gates. Seat wear is minimized because the discs contact the seat without sliding.

Valves with port openings somewhat smaller than their pipe size are available for use where pipeline scrapers are not used.

Large valves, when installed in a horizontal position, are often required to have an internal track and rollers that assist in opening and closing operations.

Flanged cleanouts can also be provided to assist in the cleaning of valve seats without dismantling the entire valve during maintenance operations.

Conduit gate valves are a form of gate valve with elongated gates. There is a full-bore port in the lower part of the gate assembly that provides a smooth conduit for the fluid. The smooth, full-bore nature of this port permits passage of pigs, or scrapers. The upper parts of the gates are solid and close the valve when they are lowered. The double wedges built into the gates provide tight seals against the seats when the valve is fully open or closed. As a result, the valve body is sealed off from pipeline pressure, and it may be drained for replacement of valve stem packing or other repairs. The drain valve may be kept open, and any fluid flow may be observed. This feature, called "block and bleed," can show whether the valves are providing liquid-tight shutoff of fluid flow.

A variation of the conduit valve has a single gate made of a single steel plate containing a port. The sides of the gate are smooth and parallel. Body seats remain in contact with the gate at all times and seal against fluid flow into the body cavity when the body cavity drain is open

Plug Valves

Like all valves, plug valves come in a variety of types and materials. The two main groupings are lubricated and nonlubricated valves. There are also differences in the shape of the plug and its opening for fluid flow, as well as the general valve housing.

Plug valves exhibit positive closure with practically no possibility of seat leakage. Another outstanding feature of the plug valve is its quick opening and closing operation. A quarter turn fully opens or closes the valve. With the help of a stop collar on the valve, the operator does not have to rely on the feel of a wheel's resistance to tell if the valve is closed. Valves may be wrench-, wheel-, or gear-operated, and almost all wrench-operated plug valves have a flat or square wrench head on the plug stem.

The plug valve in its basic form consists of only three parts: body, plug, and cover. Of these three parts, only the plug is nonstationary and is used during valve operations to open or close the flow of fluid through a 90° rotation. This movement either aligns the porthole that is part of the plug with the direction of flow or shuts off any flow. The porthole, or flow opening, in the plug may be round, oblong, or diamond-shaped, and it may permit either full or restricted flow through the valve.

Nonlubricated plug valves are made for a wide range of applications, which may vary from the small drip cock at a gauge glass or the shutoff cock at a gas range to the large, so-called cone valves that are used in water-supply installations.

Depending on the manufacturer, the plug may be inserted from the top or bottom into the valve body. The use of cylindrical plugs is often preferred in nonlubricated plug valves since they are less likely to gall or freeze than conical plugs. Plastic seals are often molded into grooves of the plug to provide better seals, and bottom springs assist in their operation.

A variation of the standard plug is the eccentric plug, which is only about one-third of a full plug in area. It permits a full flow in the open position and closes with less contact of the seat on body walls. One manufacturer encapsulates in plastic all parts of the valve that come in contact with the fluid so that a corrosion-resistant valve is created. Most manufacturers use their own unique longitudinal sizes, so that interchangeability of these valves is practically impossible. Similarly, design details for most valve parts vary so that design and purchase requisitions have to be directed toward one particular valve model.

Lubricated plug valves are even more varied than nonlubricated plug valves, but the main difference is indicated in the nomenclature. A lubricant is forced into various grooves in the plug body to minimize friction and thereby prevent sticking, and also to assist in sealing surfaces and valve stem. The lubricant pressure is also used to unseat the plug from its position and through this action nullifies any adhesion that may have taken place. While nonlubricated valves do not require frequent maintenance, the lubricated design offers this advantage as a means of freeing valves stuck by a prolonged setting in one position.

The plug itself may be conical (tapered) or cylindrical in shape, and each type has its advocates. Proponents of the tapered form refer to its quicker and simpler valve operation due to the tapered rotary principle (Figure 5.31).

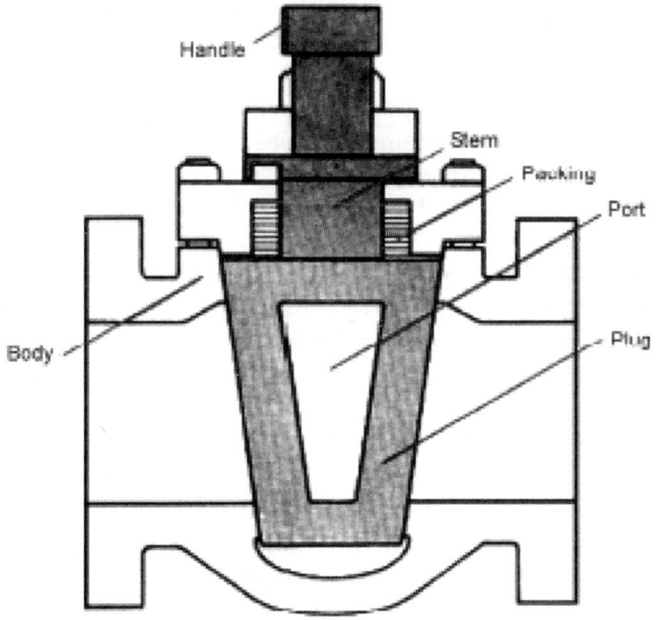

Figure 5.31:

Another advantage of the tapered form is the combination of the tightly fitting, wedge like seating and the sliding movement of plug rotation. Proponents of the cylindrical shape cite its special seating construction—such as spring-loading or teflon bearings—which also assists in ease of operation. Lubricants can be forced into its various distribution channels by a special gun that fits a buttonhead fitting on top of the plug, or they can be inserted in stick form into the lubrication opening of the plug shank and then spread by pressure applied through the turning down of an elongated lubricant screw.

A lubricant check valve normally prevents any lubricant from leaking to the outside. Most other accessories vary, depending on the manufacturer's design.

A valve maintenance program is necessary to provide adequate lubricant levels in each valve whenever the valve is in operation. Frequency of lubrication depends on the type of service.

Multiport valves are yet another variation of both the lubricated and nonlubricated plug valve. The basic structure of the plug valve permits a very simple adaptation of multiport construction. In the three-way form, a single L-shaped passage connects any two ports; with a T-shaped opening, three ports can be simultaneously connected, or two ports connected while blanking off the third. The four-way valve type may have two cutouts on opposite sides that connect any two adjacent ports of four, or it may connect two out of four available piping systems.

Multiport valves provide for a wide variety of interconnection of multiple piping systems. In lieu of having the plug available with many port openings, an eccentric-type plug can be used that either seals off only one of the various piping systems or leaves all valve ports open. Multiport valves are adaptable to numerous valve port possibilities, but the purchase requisition of the required valve has to give exact details of open and closed port requirements.

As previously stated, most plug valves are lever operated, since a 900 turn of a lever is sufficient to open or close a valve; in larger valves, however, even that little movement is often difficult to achieve. Valves of 6 in. and larger can usually be furnished with a gear-type operator that facilitates operation.

Plug valves are often furnished with flanged or threaded ends; however, some valves are also furnished with weld ends. Welding connections to such valves must be made under controlled conditions. If too much heat is dissipated from the welding area, the body or plug of the valve might be affected and the valve seating could be damaged.

Repairs and maintenance are usually easy to achieve with disassembly of the valve, and in most valves, the plug can be removed from the valve body without difficulty. However, cleaning before reassembly is absolutely necessary, and a cleaning solvent is the best method.

Ball Valves

Ball valves have features similar to those of the plug valve, such as quarter-turn operation and complete flow shutoff in both directions. In fact, it might be said that the ball valve evolved from the plug valve. A ball is kept in place inside the valve by two resilient seal rings in simple body form with a full diameter bore through one axis. The ball-shaped internals provide good seating, especially when used in conjunction with plastic surfaces, and thereby reduce the possibility of leakage through the valve when in the closed position (Figure 5.32). The ball may be single or double port.

Ball valves are also manufactured with easy-disconnect features. That means that the actual valve body can be removed from the piping system just by removing a few bolts. Such valves may have end fittings that remain part of the piping system, while the valve body is either flanged or clamped in-between these end fittings and thus is easily removed for repair or replacement. Some of the other classic designs permit removal of the ball-type disc through the top and are top-entry valves, as opposed to the end- or side-entry valve, which must be disconnected to remove the ball disc.

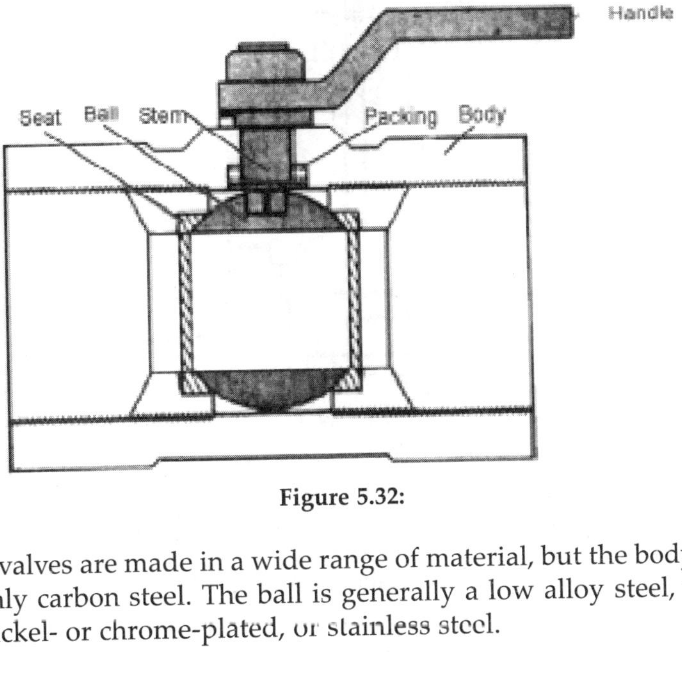

Figure 5.32:

Ball valves are made in a wide range of material, but the body is most commonly carbon steel. The ball is generally a low alloy steel, which is either nickel- or chrome-plated, or stainless steel.

The sealing elements are usually simple rings of elastomeric or plastomeric material (e.g., rubber, neoprene, or nylon), with the choice dictated by fluid compatibility and maximum service temperature requirements. A more complex seal design may be used on larger valves or for special operation service.

Some of the advantages of ball valves, such as unimpeded flow when fully open and fast 900 closure, can be of prime importance in installations where closure might be affected and speeded up through remote control in case of an emergency.

This valve is not recommended for throttling, since partial flow conditions expose localized areas of the ball and seals to high-velocity flow. Because of their popularity, they are manufactured in a wide range of sizes.

Globe Valves

Globe and angle valves have similar seat and disc arrangements. The only difference is that the globe valve provides for continuous straight flow (Figure 5.33), while the angle valve provides for a 900 change of flow direction through the design of its valve body.

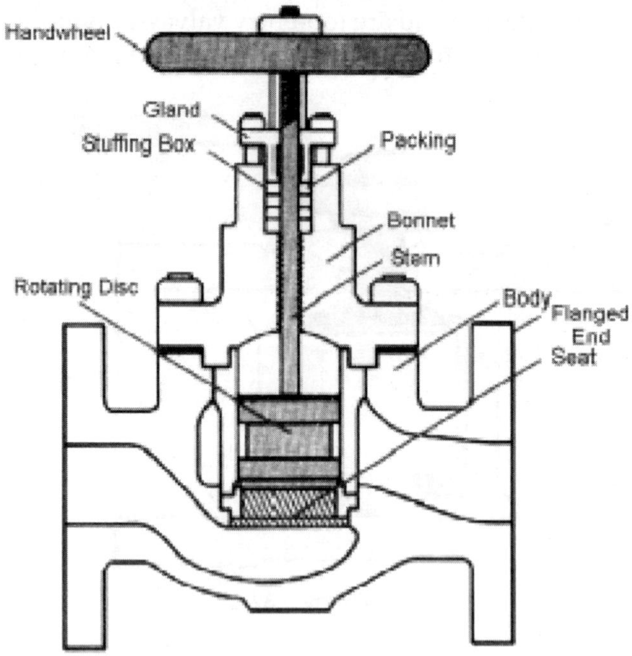

Figure 5.33:

The Y-shaped globe valve is widely used, particularly in high-pressure systems. Because of its particular body formation, it has less flow resistance than the regular T-shaped globe valve.

The bonnet, gland, and stem design of globe or angle valves is in many respects similar to gate valves, but the valve internals are markedly different. While gate valves are primarily designed to shut off fluid flow completely or provide an unimpeded port opening when fully open, the globe or angle valve is better suited to perform a throttling operation or to permit flow through only a partially open port.

Globe-valve seating is parallel to the line of flow, and there is no contact between seat and disc as soon as the valve is opened. Due to the obvious advantages of implementation in the partially opened or throttled condition, globe valves are often used in industrial applications such as the feeding of liquid or steam to equipment with different flow requirements at different times—e.g., supplying water to a boiler or steam to a heater.

As with gate valves, there are also various globe or angle valve designs, and there is practically no limit to the materials of construction. Bronze, iron, and steel are preferred valve body materials, but valves in stainless steel, aluminum, polyvinyl chloride (PVC), and other materials are also abundantly available. The various valve designs are generally identified through variations in internal—or, more specifically, disc—design.

The original ball-type disc is the oldest kind of globe valve. Lately it has been largely replaced by the conventional type, which has retained most features of the ball-type with the exception of the convex shape of the disc. The basic design feature is a flat-surfaced (though internally slightly tapered) valve seat fitted with a disc of convex configuration that uses the taper in the seat for closing. This type of seating has only a narrow line of contact, which normally assists an easy, pressure-tight closure; however, the deposit of foreign particles on the narrow seat ledge, which is not an uncommon condition, often prevents such tight closure. The valve should therefore be used in less demanding applications and in low-pressure service. The seat and disc are mostly metallic, and the disc is usually replaceable. The seat can be reground without removing the valve from service.

The plug-type disc is best suited for its throttling application and is also best able to withstand the rigors of high-pressure and high-temperature service. A long, tapered metallic plug fitted into a corresponding seat provides a wide area of seating contact combined with a proper selection of metals. This is most effective in resisting erosive effects of close throttling. Because of the side seat-bearing area, foreign matter in flow can seldom damage a seat or plug area large enough to cause leakage.

Both seat and disc can be replaced in most plug-type valves.

Needle-point disc valves are designed to give fine control of flow in small-diameter piping. The name is derived from the sharp-pointed, elongated plug provided in lieu of a disc that fits into a matching orifice-like seat area. Even when fully open, the needle-point valve does not permit a full flow, since the open seat area is only a fraction of the piping flow area. The stem threads are finer than in any other comparable valve, so that considerably more rotations of the stem are required to permit full flow-through or full closing of the valve.

This specialized construction permits application of the valve in many flow-control situations that require close regulation, as is often necessary when calibrating instruments; for this purpose, needle-point valves are available with an indicator showing the number of stem turns made, so that fine adjustments can be achieved and can be repeated after the valve is opened or closed.

While globe valves normally impede the flow of fluids because of their varied seating construction, angle valves generally reduce the number of pipe fittings required. By their very construction, they perform the functions of both a globe valve and an elbow. Small globe valves have a major advantage over gate valves in that it is very easy to replace a disc or seat, or even regrind an integral seat, without removing the valve from the system. This makes the globe valve ideally suited in applications where ease of repair and maintenance are essential and minor flow restrictions are unimportant.

Globe valves should always be installed so that the line of flow exerts pressure from below the disc and assists in lifting the disc when the valve is opened. The direction of flow is usually indicated by an arrow embossed on the valve body.

Butterfly Valves

Butterfly valves go back to the shutterlike damper that was initially used in applications where no tight shutoff was required but flow restriction might be necessary. Today's valves, which are mostly outfitted with rubber or elastomeric seats, provide a shutoff service like any other valve. The valve design is particularly suitable for installations where space considerations are important; this makes this type of valve a favorite also for very large piping systems, since there is practically no size limitation.

The valve basically consists of the valve body, stem and butterfly disc, sealing gland, and valve actuator. The body and disc are usually constructed of cast iron. The disc profiles and seal design vary, but the discs are usually convex and streamlined to minimize head loss. The valve body, which is also the valve seat when the butterfly disc reaches a perpendicular position, is often lined with rubber or plastic materials to provide a pressure-tight shutoff. Where the stem protrudes from the valve body, a sealing gland is provided to eliminate fluid loss. The valve actuator for the simple quarter-turn operation may be manual, like a simple lever; gear-operated (for a large valve); electrical; hydraulic; or pneumatic.

These valves are difficult to render completely tight when closed because of the absence of any wedge effect. An additional drawback is some resistance to flow in the full open position. This can result in the collection of fibrous or stringy material in contaminated fluid. For this reason, these valves are most suited for clean fluid. The valve's low actuating power requirements are its primary attraction.

The simple valve design has been diversified by introducing three different valve bodies without variations in the interaction between seat and disc.

The flanged butterfly valve has a short valve body and is flanged at both ends. If necessary, welding ends, in lieu of flanges, can be provided. However, the welding butterfly valve is neither a standard nor a desirable connecting method because of possible damage to the seating surfaces.

The lug-wafer butterfly valve has a shortened valve body with protruding lugs whose bolt circle matches adjoining flanges. Lugs can be connected to the mating flanges through full-length bolts that squeeze the wafer between two pipe flanges. Tapped holes can be provided and cap screws can be used to fasten the lugs individually to each flange, thus permitting the valve to be used as a dead-end valve.

The wafer butterfly valve consists of a short body like the lug wafer, but without the lugs. This valve can be inserted between two adjoining flanges but must be centered exactly, since there is no guidance provided for its location. Gaskets may be molded onto the body or may have to be inserted for a satisfactory flanged joint.

Torque characteristics play an important role in the operation of butterfly valves, and the various friction effects, as applicable to bearing, seal, and seating, should be determined when selecting a butterfly valve.

Diaphragm Valves

The diaphragm valve takes its name from the importance that the diaphragm plays in the valve design. Like the plug valve, the valve consists of only three basic elements, namely the valve body, valve diaphragm, and the operating mechanism, which might be referred to as the valve bonnet.

The valve body is designed so that fluid flow is directed over an indented weir, which stops the flow when the diaphragm is brought down and compressed against it by the operating mechanism. The operating mechanism is a convex compressor disc that is raised or lowered by a handwheel-operated stem, or spindle. In lieu of this handwheel-operated stem, an air actuator can also be used to open or close valves by applying compressed air with or without the assistance of helical springs.

The main valve feature, the diaphragm itself, is available in a variety of elastomeric materials or rubber, depending on the valve service requirements. The resilient diaphragm provides a cushioned, leak-tight closure and is designed so the fluid cannot penetrate it. It thereby isolates the bonnet and operating mechanism from the fluid being handled by the valve. This simple construction eliminates the need for glands or valve-stem packing. Thus the flanged connection between bonnet and valve body and, of course, the actual valve connections into the piping system are the only weak spots where a leak might occur. Because of its elastomeric composition, the diaphragm and the valve itself can handle fluids within only a limited pressure-and temperature-rating range.

Through its ease of assembly from the three basic components, the diaphragm valve has also been developed as a basic valve that can be lined with nearly every pipe-lining material available today. It has been established as an integral part of many lined piping systems where valving is required.

Valve endings are normally either threaded or flanged, but other specialized endings, such as the socket-bonding end for PVC and butt-weld for stainless steel, are available. The diaphragm valve must be welded carefully and with the bonnet and diaphragm removed so as not to warp the valve body.

In order to prevent diaphragm rupture caused by an inadvertently applied overbearing force during closing, a special indicator may be used.

The diaphragm valve has a multitude of applications, and the various lining materials and their corrosion or abrasion resistance have greatly attributed to its widespread acceptance in the process industries. It is equally useful as a shutoff valve or as a flow-control or throttling valve. Sizes of diaphragm valves normally range from 1/2 in. to 16 in.

Maintenance work or diaphragm replacement is easily accomplished without removing the valve from the piping system by simply removing the valve bonnet.

Check Valves

Check valves are designed to permit flow in only one direction. In the absence of normal flow, the valve is usually closed, and it opens automatically when pressure is applied to the valve by fluid flow. The valve also closes automatically if pressure builds up against the direction of flow, thus preventing any backflow or pressure buildup beyond the design requirement of a given component of a piping system. Since fluid flow is permitted in only one direction, it is imperative that the valve be installed correctly. An arrow on the outside of the valve body indicates flow direction; counterflow installation can easily impair or damage delicate parts of a piping system. In the open position, check valves effect a relatively high resistance to flow as well as significant turbulence in the valve area.

Although there are only two basic categories of check valves, swing and lift checks, each has many variants. As with all other valves, body materials and end preparation of check valves can be made to suit any piping system, and there are two basic types of valve bonnets: flanged and threaded.

Swing check valves are so named because of their particular mode of operation. They are by far the most widely used in general industry since they offer little flow resistance and are virtually foolproof. Like gate and globe valves, check valves have a valve seat and disc, or a variation thereof, which in this case is the only moving part. The disc, which is hinged at the top, seats against a machined seat in the tilted bridge wall opening (Figure 5.34).

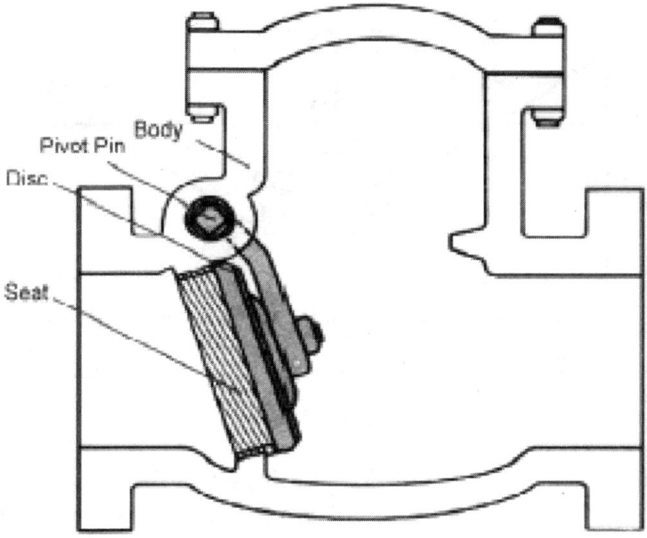

Figure 5.34:

The disc swings freely in an arc from a fully closed position to one providing unobstructed flow. Discs can be furnished with metallic or nonmetallic facings, depending on operational and maintenance requirements. To increase sensitivity to flow, an outside lever and weight can be attached to assist in the valve's operation.

Lift check valves have an internal construction similar to globe valves, and the same body casting is often used for both. The disc, which is seated on a horizontal seat, is equipped with guides above and/or below the seat and is guided in its vertical movement by integral guides in the seat bridge or valve bonnet. The disc is free-floating and rests on the seat when it is not operating. Both seat and disc are metallic and can be easily replaced after removal of the valve bonnet (Figure 5.35).

Ball check valves are of similar construction to lift-check valves. Instead of a guided disc, a ball serves as the flowcontrol medium. When operating, the ball is constantly in motion, reducing the effects of wear on any one area of its sphere.

Spring-loaded ball check valves are well suited for vertical downward flow. Swing checks cannot be used in these circumstances, since the flapper tends to hang open due to gravity and may not close in the event of flow reversal.

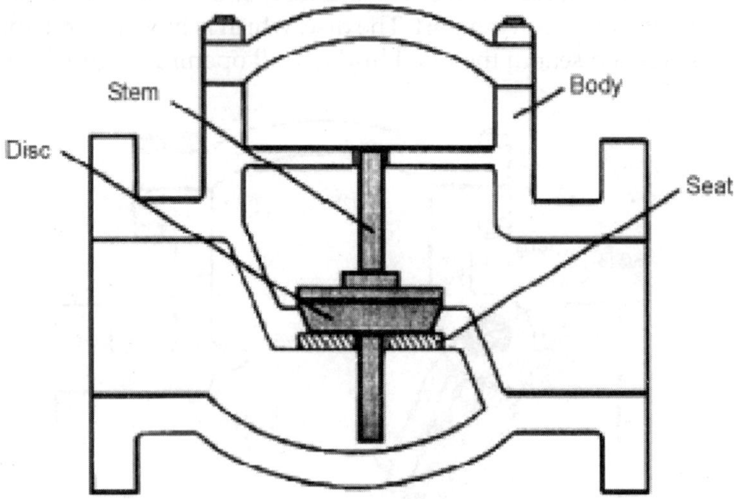

Figure 5.35:

A butterfly check valve employs a hinged, rather than pivoted, disc, with a sealing ring around its edge. With forward flow, the two halves are swung together to trail downstream. With reverse flow, the two halves open to approximately 450, sealing the bore. Its advantages are rapid action and a resilient seal.

Like the conventional butterfly valve, the butterfly check valve requires a short length, approximately equal to a standard pipe flange. The check valve model is also produced in wafer form for clamping between two pipe flanges. Body length need only be sufficient to accommodate the hinge post and two valve plates in closed position.

In general, check valves, like gate and globe valves, rely on iron, steel, and bronze as materials of body construction, while seats and discs are furnished in a variety of metallic alloys and nonmetallic materials. However, the availability of check valves in other materials for incorporation in piping systems that are not steel-based is practically unlimited. Most check valves have removable seat rings, and discs are also easy to replace after the valve is opened.

Valves in horizontal service have either a screwed cap or bonnet, or a flanged top that can be opened without removing the valve from the line. The resultant opening is big enough to allow the valve internals to be lifted out. Valve seat removal usually requires special tools and is not a job for the uninitiated, but the disc can be changed quite easily. The disc on swing check valves is supported by a hinge that, in most valves, can be removed by loosening the screws on both sides of the valve body. Check valves should be inspected or tested periodically, since inoperative or leaky check valves may otherwise not be detected until a serious malfunction takes place.

Pipe Fittings

Flanges and fittings are a vital part of any piping system. While flanges are used only for pipe or equipment connections, fittings may change the pipe direction, make a branch connection, reduce or increase the continuation in size, or terminate a pipe run.

Flanges and fittings are available to suit many pipe materials and various pressure ratings or schedule numbers. Flanges are initially identified by pressure ratings, and the particular connecting method generally does not dictate their material composition. Fittings are mostly identified or specified in accordance with the method of connection, such as threading, butt welding, or socket welding, since different materials or manufacturing methods apply to each. The various fittings that are generally available for all connecting methods are the following (Figure 5.36 and Figure 5.37)

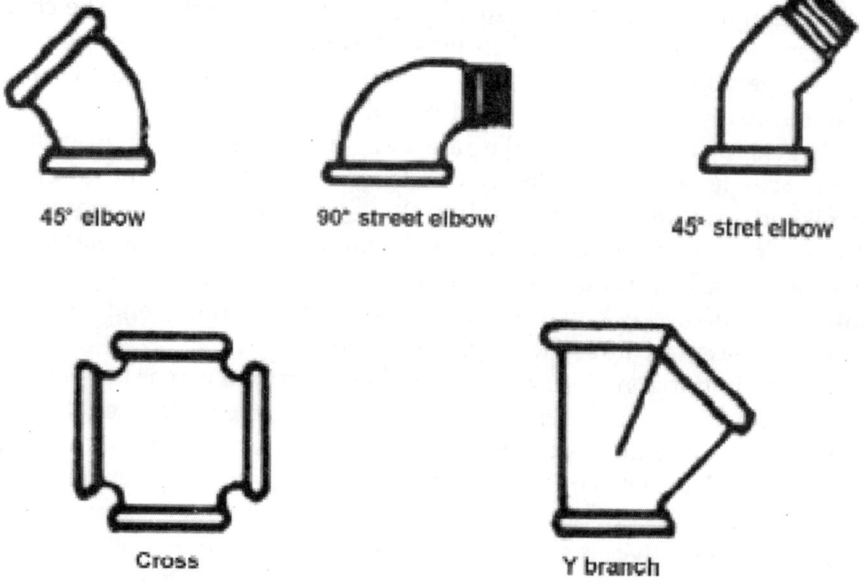

Figure 5.36:

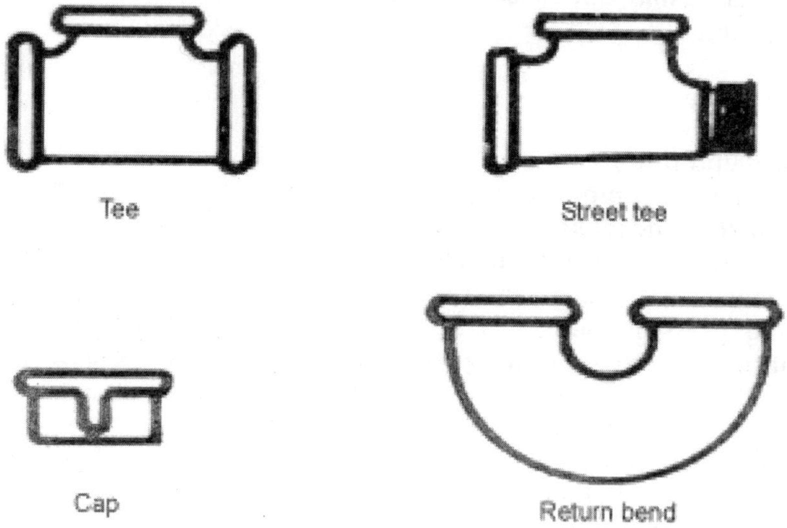

Figure 5.37:

- below 90° or 45°
- return bend (180°)

- reducing elbow (90° or 45°)
- tee or reducing tee
- lateral or reducing lateral
- concentric or eccentric reducer
- cap
- cross or reducing cross

Fittings are available in cast iron, malleable iron, brass and copper, cast steel, and wrought steel.

Cast-iron and malleable-iron fittings are limited to low-pressure service (usually no more than 250 psi) and a maximum temperature of 300° F. Brass and copper fittings are also suited to low-pressure applications and are not appropriate for standard flowline service.

The most common fittings are those of steel compositions, and these are available in carbon and high alloy, depending on the application. Numerous ANSI/ASTM/API standards apply to the various fitting dimensions and material composition.

Threaded fittings are available for most materials and pressure classes.

Welding fittings are manufactured to provide fittings that are, by means of material composition and end preparation, suitable for welding. The material composition of these fittings is usually similar to that of the pipe to which they are being connected. Since the manufacturing process for fittings is different from that for pipe, different specifications apply.

Metallic welding fittings are normally furnished with 37 1/20 beveled ends so that a V-shaped groove is provided for depositing weld metal wherever a welded connection is being used. When pipe or fittings with a wall thickness of 7/8 in. or more are being used in a piping system, a double-V or U-shaped bevel is usually provided.

Socket-welded fittings are mostly restricted to sizes 4 in. and smaller. These fittings are provided with a bell-shaped end that is internally machined to encompass the external diameter of the pipe that fits into it. Actual ID of the fitting matches the external diameter of the pipe to which it connects.

During construction, a minuscule space should be left between the pipe and the internal shoulder of the fitting to allow for expansion during the welding process.

Most socket-welding fittings are similar to threaded fittings, except that the nomenclature "reducing insert" is used in lieu of "bushing" (Figure 5.38).

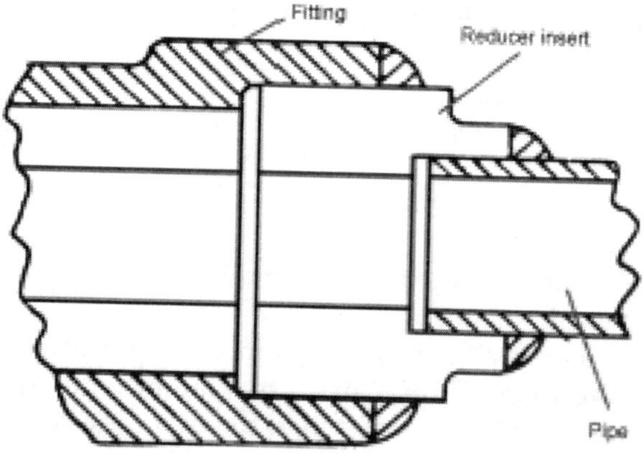

Figure 5.38:

While fittings function on a more or less permanent basis, flanged connections permit easy opening of piping systems for repairs or maintenance and are most often used as connectors from piping to equipment.

Like piping, flanges are manufactured from a multitude of materials and in a variety of types for different pressure ratings. The basic carbon steel flange types are as follows (Figure 5.39 and Figure 5.40) .

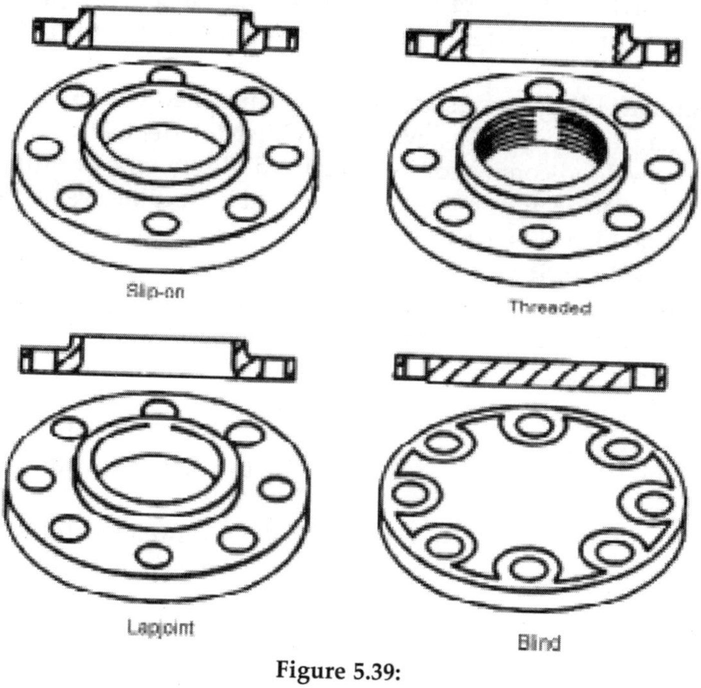

Figure 5.39:

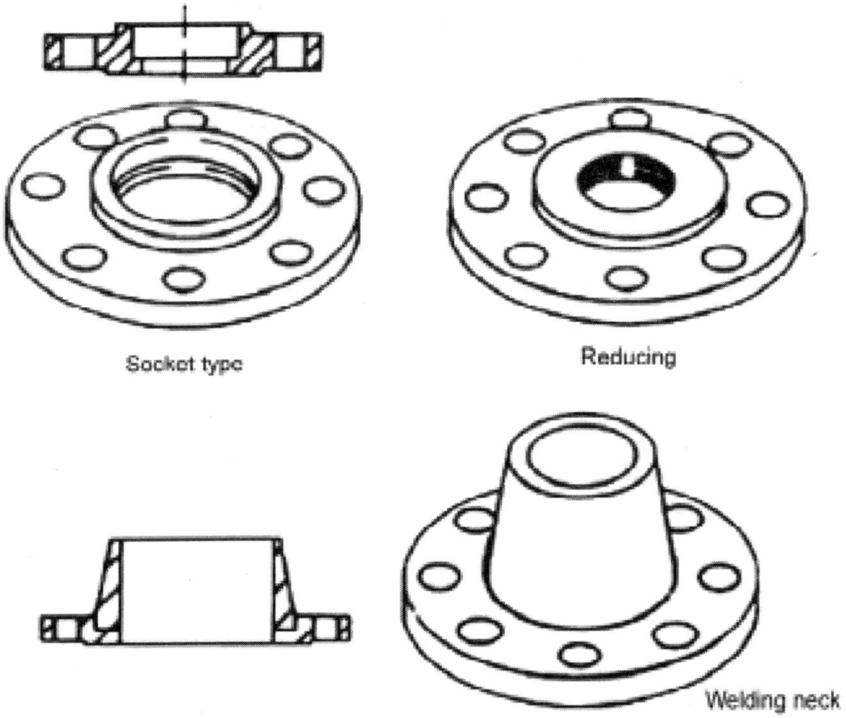

Figure 5.40:

- slip-on
- lap-joint
- threaded
- welding-neck (regular or long)
- socket-welding neck
- blind
- reducing (threaded, socket weld, or slip-on)
- orifice

The various types of carbon steel flanges in the 150- and 300-psi pressure classes are normally furnished with a 1/16-in. raised face. For higher pressure ratings, the raised face is usually 1/4 in. Several different flange facings, such as large or small male and female facings, oval or octagonal ring joints, or large or small tongue and groove, are standardized and commercially available.

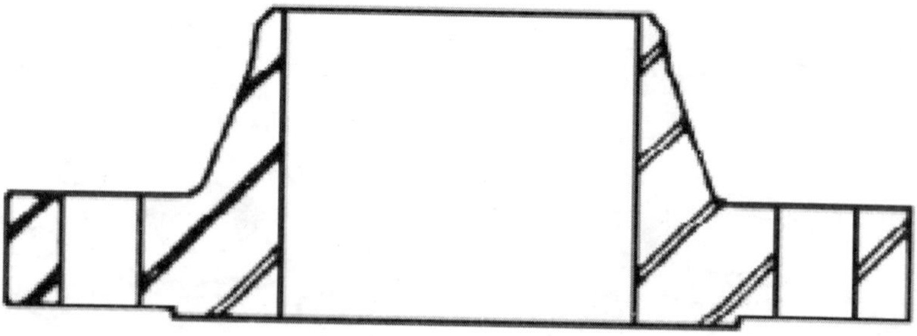

Figure 5.41: Raised Face

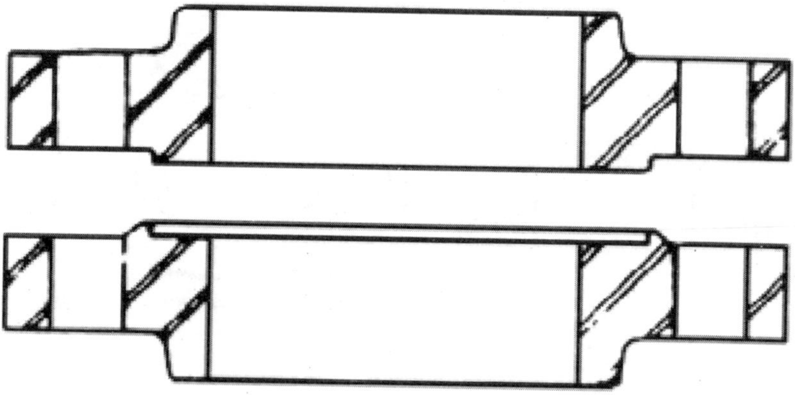

Figure 5.42: Large Male and Female

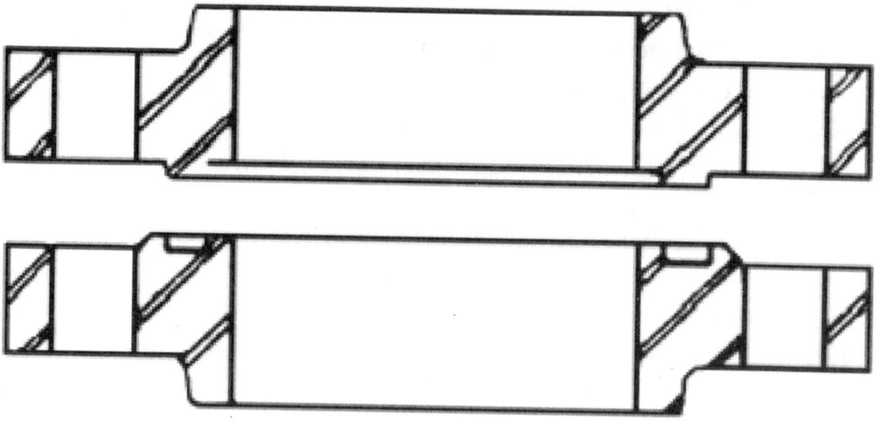

Figure 5.43: Large Tounge and Groove

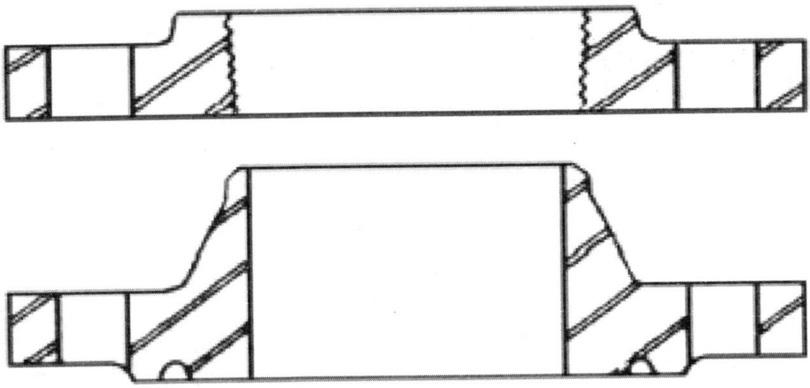

Figure 5.44: Flat Face and Ring Joint

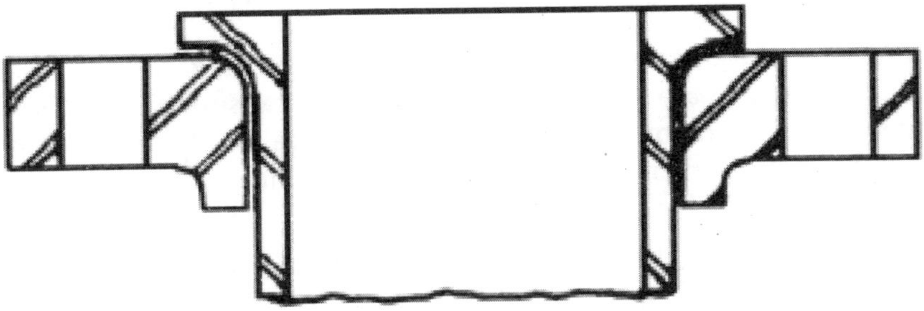

Figure 5.45: Lap Joint

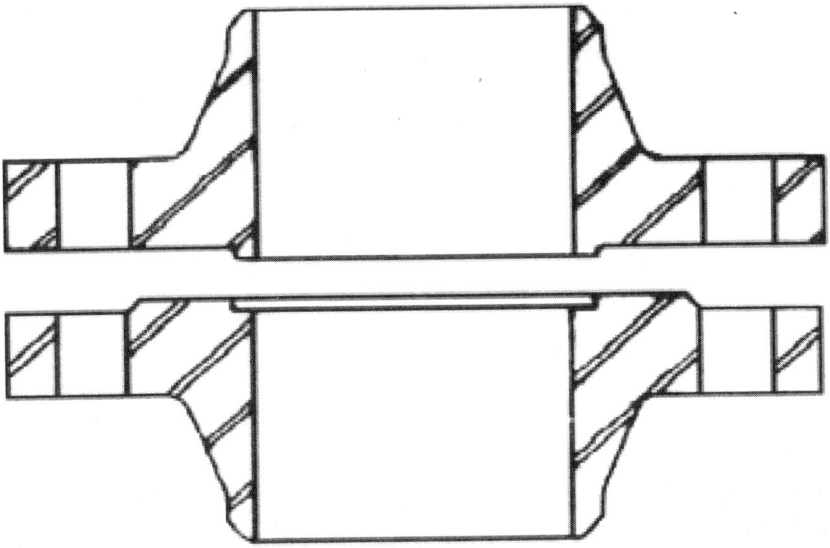

Figure 5.46: Small Male and Female

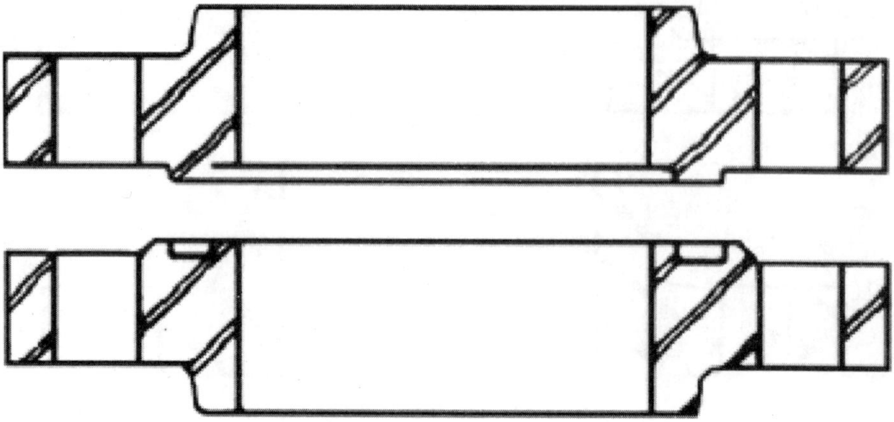

Figure 5.47: Small Tongue and Groove

Where metal gaskets are used with these facings, the gasket area should be reduced to increase the gasket compression. Because of the small gasket contact area, a tight joint may be secured with the ring-type facing using low bolting loads, thereby resulting in lowered flange stresses.

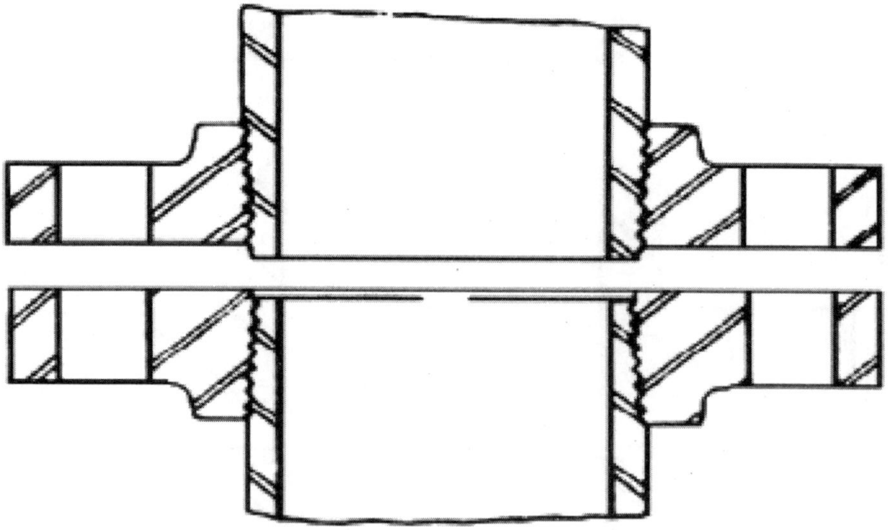

Figure 5.48: Small Male and Female

ANSI provides dimensional data and operating pressure ratings for several flange classes, 150 through 2500, for various steel and alloy flanges, along with specifications governing material compositions.

Basic Manifold Elements

A manifold enables the production from several wells to be combined before being routed to production equipment such as an separators, treaters, or storage tanks. A manifold can also isolate one well's stream from those of other wells and send it to a test separator. A header is a pipe conduit passing through the manifold that, through appropriate valving and connections, can accept fluid from any one of several inlets.

Manifolds vary in complexity, but a standard assembly might include a production header, a test header, and the necessary valves for operation. If wells of appreciably different flowing pressures are routed into a manifold, it becomes necessary to segregate the wells into two or more streams, and to include both high-pressure and low-pressure production headers. Other installations might require segregation on a different basis, (e.g. water content of produced oil). Should it be necessary to blow down flowlines from individual wells, a blowdown header would also be included. Other capabilities might be required depending on operating conditions.

Headers

The high-pressure header takes production from individual high-pressure wells and routes the combined flow to a production separator. The kind of production coming into the header determines the valving necessary on the individual wells upstream of the header. In most cases, a check valve is required on each flowline, as well as some type of isolating valve. This header must be sized so that the flow velocity stays within the prescribed limits. The pipe must be of adequate grade and wall thickness to safely withstand the maximum operating pressure

The low-pressure header takes production from low-pressure wells and routes the combined stream to the production equipment. Depending on the relative amount of production, this header may be larger or smaller than the high-pressure header. For the same amount of production and the same GOR, this line would have to be larger than that of the high-pressure header to maintain velocities within prescribed limits. Valving and piping are usually the same as for the high-pressure header except for size and working pressure.

A blowdown header would be necessary for blowing down flow-lines from individual wells. This may be required if paraffin, sand, salt, or other material is likely to precipitate in the flow-lines from the wells. Heavy amounts of accumulation might require the inclusion of equipment for "pigging," and wirelines, as well as other equipment. About the only auxiliary equipment necessary for this header would be a valve from the individual well flowline to the header. Since high velocities will most likely be encountered when wells are "blown" through this header, sturdy hold-downs must be supplied to prevent movement.

It is usually necessary to periodically test individual wells. This requires a test header that enables one well at a time to be turned into the header and routed to the test equipment.

Other Manifold Components

Chokes may be installed on the inlet to the individual well flowline stringers of the manifold or at the wellhead. There are advantages and disadvantages to each location, but on most low- to medium-pressure installations, chokes are frequently installed on the inlet to the manifold.

Occasionally, a diaphragm operator is installed on a choke to automatically control the pressure downstream of the choke. The downstream pressure is sensed by a pressure controller, and an output pressure proportional to this pressure is transmitted to the diaphragm operator on the choke. The diaphragm operator positions the choke stem to maintain the set downstream pressure.

Diaphragm operators are most commonly used on low-to-medium pressure applications, except where it is necessary to open and shut-in the wells by remote control. The more common application is on gas wells, where to increase liquid recovery, the producer tries to take as much pressure drop as possible across the choke without forming hydrates. If hydrates begin to form in the choke, indicated by a drop in downstream pressure, the diaphragm operator opens the choke further to blow out the hydrates. When the condition has been corrected, the choke is repositioned automatically by the operator.

Valves to prevent backflow are installed primarily to prevent the stronger wells from backflowing into the weaker ones. Check valves in this application are normally installed upstream of the choke, so their working pressure must be at least as great as that for the choke. The opening should be as near to full bore as possible to prevent excessive pressure drop and the accumulation of paraffin, salt, or other types of solid material.

An inlet block valve is customarily installed on the inlet to each well flowline stringer to permit shutting-in each well separately. This function could alternatively be performed by shutting the individual block valves to the production and test headers; however, it is usually desirable to have one valve perform this function.

The block valve may be either manual or automatic. An automatic valve has a pneumatic, hydraulic, or electrical operator. The signal that activates the operator may come from either the individual well flowline or from a remote location. In either case, the valve body must be capable of withstanding full well pressure unless pressure-actuated devices that prevent full well pressures from being transmitted to the manifold have been installed at the wellhead. Even if such devices are used, it is a good idea, for safety purposes, to use valves that can withstand full well pressure. Automatic valves preferably have trim, commonly described as "quick opening," in which maximum capacity is reached at a relatively small proportion of the total available travel.

The types of manual valves most commonly used for inlet block valves are ball, lubricated plug, and gate.

Header block valves must be provided in the manifold to route production from each well either to the production header or the test header. The valves may be either manual or automatic, but in either case should have as large a port opening as possible.

Other Apparatus

A pressure gauge is usually provided on each well flowline stringer for local indication of the individual well pressures at the manifold. If these readings are to be transmitted to a central location, pressure transducers are used either instead of or in addition to the pressure gauges.

For widely fluctuating pressures, it may be necessary to install damping devices on the pressure gauges.

Relief valves may be necessary between the valves of the manifold to ensure that excessive pressures are not built up in the lines when fluid is trapped between the valves. These relief valves are usually not required except in cases where only a very small amount of gas is produced with the oil. A nominal amount of gas ensures a sufficient cushion to prevent excessive stresses on the lines.

Sufficient drain connections should be provided to allow draining of all portions of the manifold. This is particularly important if the manifold is shop-fabricated. Test water trapped in the manifold can be very troublesome, especially if the manifold is shipped to a colder climate.

Sufficient and properly spaced sample connections should be provided to allow sampling of both the individual well streams and the combined production.

Manifold Fabrication

Although simple manifolds may, under ideal conditions, be fabricated satisfactorily in the field, prefabrication is preferable in most cases. Some of the advantages of prefabrication are as follows:

- Working conditions for the fitters and welders are usually more satisfactory.
- Better welding and handling equipment is more likely to be available.
- Supplies such as welding rods, fittings, pipe, etc. are probably more readily available.
- The job usually goes faster.
- Facilities for testing the various components and the completed manifold are likely to be better.
- Inspection facilities are probably much more satisfactory.
- Drawings, which make future changes easier and faster to accomplish, are much more complete. Often in the field, drawings are not even made.

About the only real advantage of field fabrication is the ability to maintain a close liaison between the operator and the fabricator or contractor.

Manifold Equipment for Special Applications

Certain types of equipment, although not found on all manifolds, are used quite often for special-purpose applications, such as in automated and remote-control operations.

The use of butterfly valves is increasing, particularly in low-pressure (less than 200 psi) applications. Like the ball valve, the introduction and use of new synthetic rubbers for seating is the primary reason for this increased interest.

As shown in Figure 5.49, the butterfly valve consists of a body, disc, shaft, seat, and 0-ring. With comparable materials, the butterfly valve costs less than the ball, gate, or plug valve.

It gives tight closure and has very low pressure drop characteristics. These valves cannot be used where unobstructed full opening is needed, and, as mentioned before, working pressure is also a limitation.

Rotary selector valves consist of a body or production bowl, a rotor element with seals, production or flow outlet and test outlet ports, seven usable flowline inlet ports, and one home port. A well is placed on test by turning the rotor body to a position opposite a well inlet port. The seal to isolate the well to be tested is accomplished by force exerted by a wavy adjustable spring.

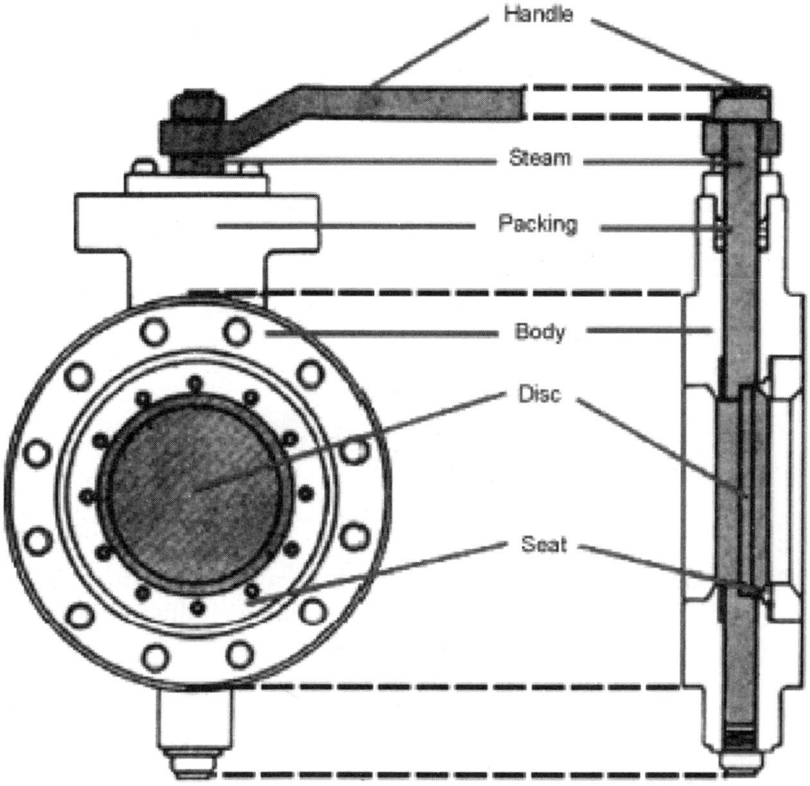

Figure 5.49:

The use of the rotary selector valve has been limited as compared with the prefabricated three-way, two-position, and three-way ball valve on automatic well test (AWT) headers. A careful comparison should be made of the rotary selector valve with other types if it is being considered for a field automation project.

One type of three-way ball valve is shown in Figure 5.50.

This is a floating ball, 180° rotation-type valve. Other types available are floating ball, 90° rotation, and trunnion-mounted 90° rotation. The three-way ball valve has been used in automation projects more extensively than the rotary selector valve. However, it still is not used as much as the three-way, two-position plunger and stem divert valve.

Again, it is recommended that a careful comparison be made between the three-way ball valve and three-way, two-position plunger and stem-type AWT headers before a final decision is made.

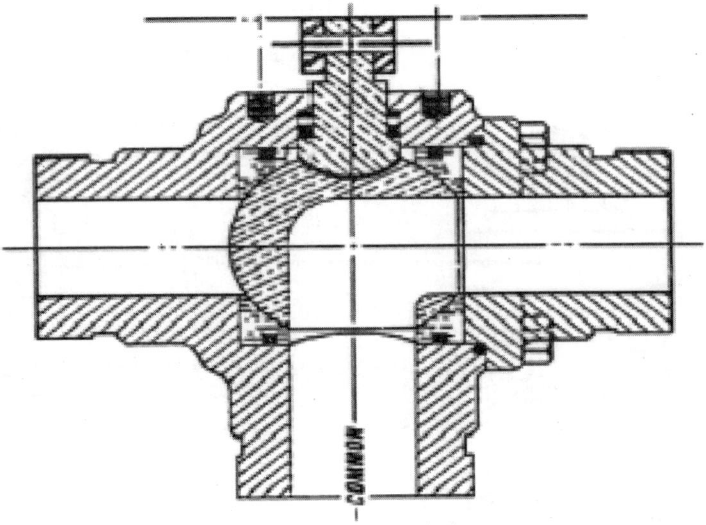

Figure 5.50:

Three-Way, Two-Position Plunger and Stem Valves A springclosed to lower port, diaphragm-actuated, three-way valve is illustrated in Figure 5.51.

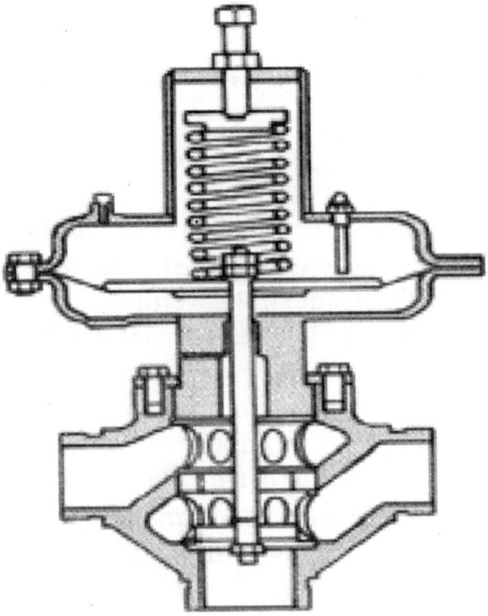

Figure 5.51:

With no pressure on the diaphragm, the flow through this valve would be from the common port out the upper port. By applying pressure to the diaphragm, the upper port is closed and flow would be from the common port out the lower port. There are many of these valves on the market, and marked performance variations can result from subtle design variations among valves.

Two-way Plunger and Stem Valves The control valve shown in Figure 5.52 shows a single-port, diaphragm-actuated, spring closed, throttling-type valve. This valve is normally used to control liquid dumps where pressure drop through the valve is not critical or may even be desired.

If minimum pressure drop is a desired feature, a double-ported plunger and stem valve should be considered.

The terms safety valves and relief valves are often used interchangeably to designate valves that protect against excessive pressure. There is, however, a difference between the two. A safety valve (sometimes called a pop valve) is an automatic pressure-relieving device actuated by pressure under the seat and characterized by full-opening or pop action upon opening. The construction of a safety valve includes a disc that overhangs the seat, and the seat is usually surrounded by an adjustable ring to form a huddling chamber, so that after the valve begins to open, the pressure is applied to the additional exposed surface. and results in faster rise of the disc to the full open position. Safety valves are primarily used with gases (compressible fluids) and, therefore, require the full opening, pop action to give immediate relief.

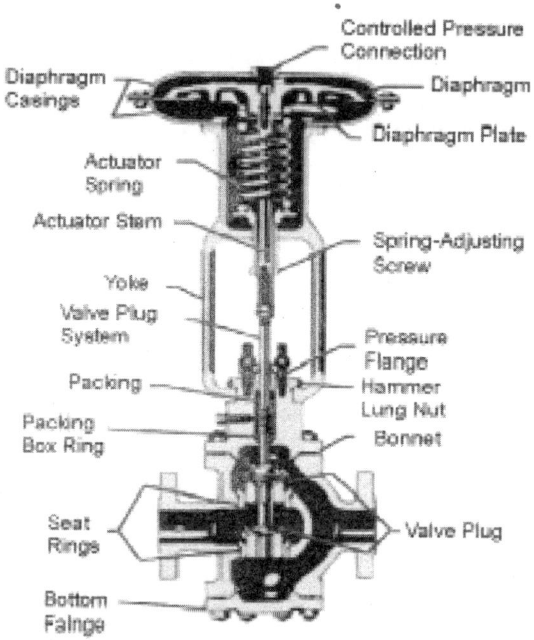

Figure 5.52:

Relief valves are used primarily with liquids. Since a relatively small discharge of liquid (noncompressible fluids) provides relief, it is not necessary that relief valves arrive immediately at the full open position. Hence, they slowly open further with any increase over the initial opening pressure. With a pure relief valve, the area exposed to the overpressure is the same whether the valve is open or closed. However, by constructing the valve with an overhanging umbrella shape or diaphragm for lift and using full or semi-nozzle-type seats, we have what are often called safety-relief valves. This is the type of valve that is normally used in oil production service because it can be used for either compressible or incompressible fluids.

The selection and sizing of safety-relief valves is fully covered in many manufacturers' catalogs and sizing manuals. When used on pressure vessels, the relief valve is usually sized so that the relieving pressure does not exceed 10 per cent of the vessel maximum working pressure when operating at designed capacity.

Gas back-pressure valves are self-contained devices that prevent pressure in a system from exceeding some preset level by relieving it into a low-pressure line. This is done by a spring-loaded disc or piston that lifts to open the valve when the pressure in the system rises above that exerted by the spring. Usually the gas back-pressure regulation is not accomplished by directly applying the pressure to the disc or piston but through assistance from a diaphragm that acts on the spring; the spring then moves the disc to open or close the valve. Backpressure regulators are quite similar to relief valves but contain design refinements that give better control in continuous service. Relief valves, on the other hand, are made for emergency service only and, therefore, do not normally meet the accuracy required in practically all continuous operations.

Chapter 6

Refinery Filtration

The charts below and the adjacent schematic represent typical flow systems within the refinery. Systems are detailed within the drawing and include major filter locations/applications. The numbered items in each system coincide with the recommended filter/separation application tables below:

Table 6.1:

Particulate/Backwash Filters		
Application	*Problem*	*Benefits*
1. Amine Sweetening	Liquid hydrocarbon and particulates cause foaming and fouling problems within an amine plant.	Reduced amine losses, foaming, and flaring; improved energy consumption; reduced maintenance, labor, and disposal costs.
2. Tail Gas Treating Unit	Many tail gas units contain an amine unit.	Same as Amine Sweetening (1)
3. Aromatics	Liquid hydrocarbon and particulates cause foaming and fouling problems.	Process similar to amine; reduced solvent losses and foaming; Improved energy consumption; reduced maintenance, labor, and disposal costs.
4. Catalyst Protection	Solid particulates plug and deactivate catalyst bed (5-10 μm); water can deactivate some catalysts.	Improved conversion efficiency and profitability; fewer catalyst change outs; improved conversion/ yield; lower maintenance, labor, and catalyst costs.
	Particular Filters	
5. Final Product Filtration	Refinery final products contain particulates and water.	Improved product quality; fewer reprocessing and contamination costs; lower maintenance and disposal costs.

(*Contd.*)

(*Contd.*)

	Application	Problem	Benefits
		Backwash Filters	
6.	Fluid Catalytic Cracking Unit slurry oil	Catalyst fines reduce value of slurry oil.	Improved product quality and revenue; less downstream equipment maintenance.
		Liquid/Gas Coalescers	
7.	Amine and Sulfur Recovery Unit	Carried-over amine contaminates catalyst at the sulfur recovery unit. Carried-over hydrocarbon and treating chemicals initiate foaming in amine contactor.	Fewer catalyst change outs; improved sulfur conversion, less equipment fouling. Reduced amine losses, foaming and flaring.
8.	Refinery Fuel Gas	Fuel gas composition changes rapidly, and contains condensable hydrocarbons; liquids and solids will foul, plug burners and combustors; low NOx burners very sensitive to plugging.	Improved reliability; lower maintenance costs; improved burning efficiency.
9.	Hydrogen Compressor Protection	Hydrogen composition changes rapidly and contains condensable hydrocarbons; liquids and solids will foul internals of compressors; must know composition of liquids being recovered by coalesce.	Improved reliability; lower maintenance costs; improved efficiency.
		Blowback filters	
10.	Catalyst Recovery from Gas Streams	Catalyst fines discharged into flue or elutriation gas causing catalyst losses, opacity problems, maintenance problems with downstream equipment. Particularly a problem where catalyst is continuously re-generated.	Fewer catalyst losses; less maintenance on downstream equipment; compliance with environmental standards.
		Liquid/Liquid Coalescers	
11.	Final Product Filtration	Refinery final products contain particles and water.	Improved product quality; fewer re-processing and contamination costs; lower maintenance and disposal costs.
12.	Catalyst Protection	Solid particulates plug and deactivate catalyst bed (5-10 μm); water can deactivate some catalysts.	Improved conversion efficiency and profitability; fewer catalyst change outs; lower maintenan-ce, labor, and catalyst costs.
13.	Treating	Caustic or amine carries over into product stream causing off-specification product; carried-over caustic can form a precipitate downstream resulting in equipment fouling.	Improved profitability; lower reprocessing costs; reduced maintenance and labor costs.

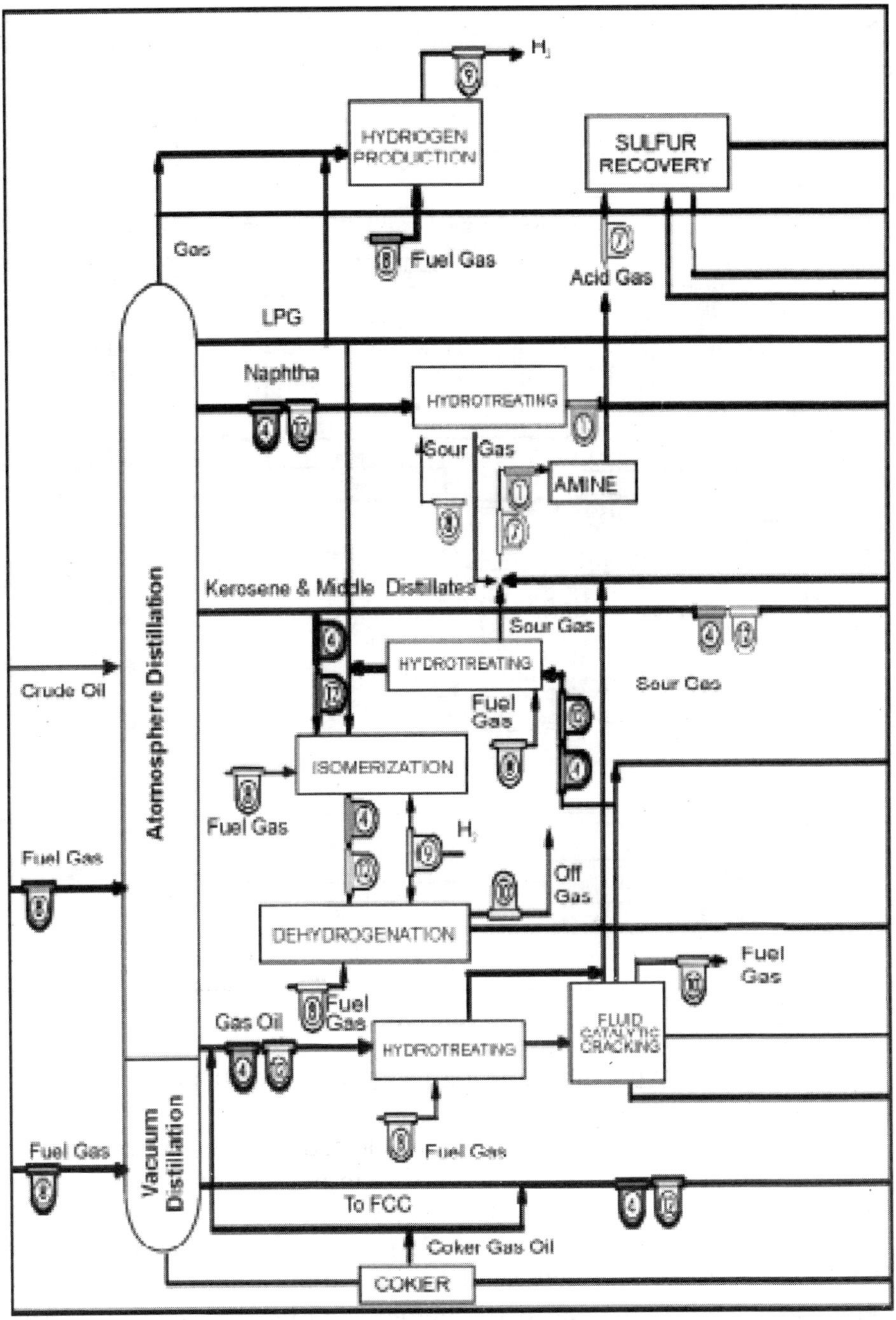
H2
HYDROGEN PRODUCTION
SULFUR RECOVERY
Gas
Fuel Gas
Acid Gas
LPG
Naphtha
HYDROTREATING
Sour Gas
AMINE
Kerosene & Middle Distillates
Sour Gas
HYDROTREATING
Sour Gas
Crude Oil
Atomosphere Distillation
Fuel Gas
ISOMERIZATION
Fuel Gas
H2
Fuel Gas
Off Gas
DEHYDROGENATION
Fuel Gas
Fuel Gas
Gas Oil
HYDROTREATING
FLUID CATALYTIC CRACKING
Fuel Gas
Fuel Gas
Vacuum Distillation
Fuel Gas
To FCC
Coker Gas Oil
COKER

(Contd.)

(*Contd.*)

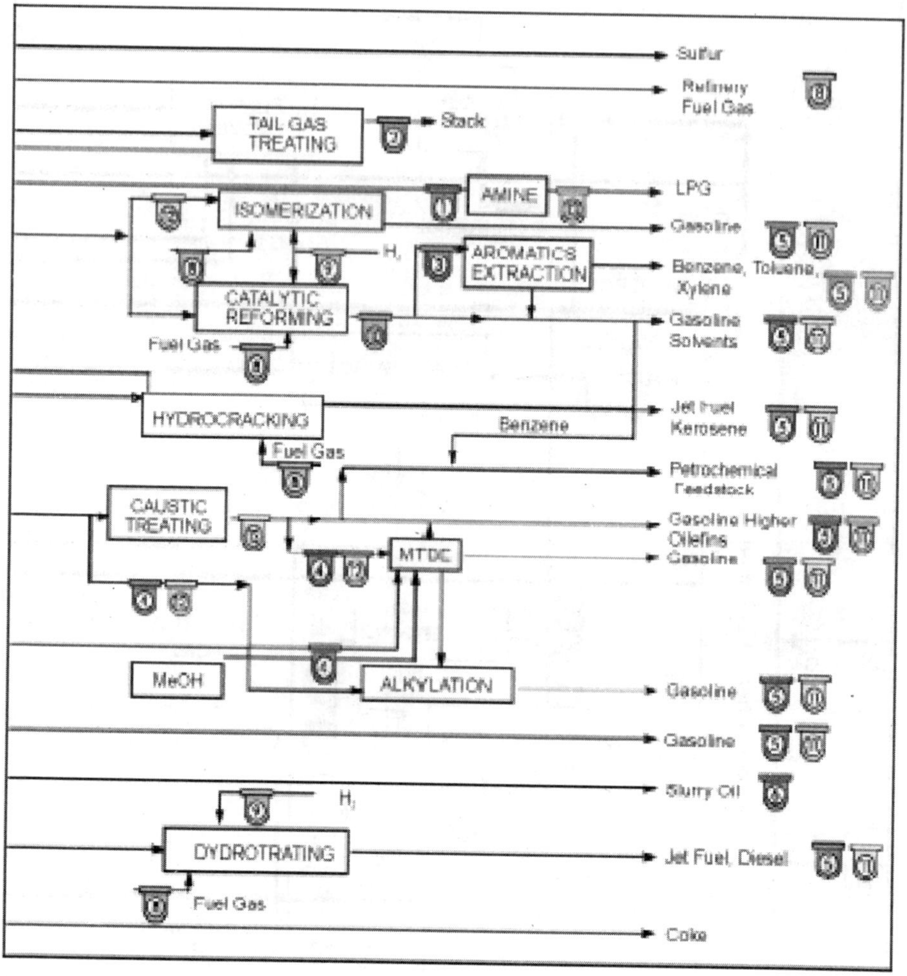

Figure 6.1:

SEPARATION OF SOLIDS FROM LIQUIDS

A variety of filtration methods can be selected for specific solid/liquid separations applications in refineries, due to the differing concentration of solids in liquid streams. For high solids loading, a backwash system may be the optimum filtration solution, while disposable filter media would be used for fairly low solidsloading applications. In some cases, removal of solids may require a combination of backwash followed by a disposable filter for polishing. Whichever filtration product or combination provides the best overall performance and economy, has the capability of providing you with the solution. originally developed backwash technology in the 1960's in response to the needs of hydrogenated chemical producers. This technology started to gain acceptance in refineries in the late 1980's when refiners needed to upgrade the value of slurry oil in Fluid Catalytic Cracking units by removing catalyst fines. This continually evolving backwash technology is now used in other refining processes for the purpose of extending the life of catalyst and improving conversion in hydrotreaters and hydrocrackers. Backwash Systems offer significant benefits over conventional wedgewire systems that are used to protect catalyst beds.

Backwash Systems

A backwash system is designed to remove and/or collect suspended solids from a liquid process stream while periodically regenerating itself. A porous filter medium with suitable pore size will efficiently collect solids on its surface, where they form a permeable cake. During backwash, a reverse flow will be initiated at a predetermined filter pressure drop and/or time interval, discharging the collected solids to recovery. The filter will then be returned to full forward flow. Its pressure drop just after backwash will remain essentially constant through backwash cycles. While vessel design, tubesheet assembly, piping, instrumentation, and valve selection all play an important role in their smooth functioning, backwash capability is the central feature of Backwash Systems. With a variety of porous filter media, both metallic and nonmetallic, and backwash techniques, Backwash Systems are designed and optimized for specific refinery applications. For example, many hydrotreaters use backwash filters to remove solids from the feed and protect the catalyst bed. In the past, wedgewire media were used in these systems. These media only provide about three to five per cent void volume in this application. Replacing the wedgewire elements in the filter vessel with Rigimesh elements provides approximately seven times higher void volume and twice the filter area. An upgrade should be based on a need for finer filtration and higher throughputs prior to backwash cycles.

Gas-Assist Backwash Method

The gas-assist method is used when process flow rates are high, or, continuous, uninterrupted flow is required. For backwash, one vessel is isolated and the downstream side of the vessel is pressurized with a controlled quantity of filtered air or other suitable gas. The vessel drain port is rapidly opened, resulting in a hydraulic pulse that "bumps" the collected solids from the filter surface. Forward flow is restored to this vessel, and the remaining filter vessels are backwashed sequentially. During gas-assist backwash, the expanding gas bubble forces the liquid through the elements in the reverse direction (inside-out) at a velocity as high as seven times the normal forward flow velocity. This effectively dislodges the accumulated cake from the elements, while significantly reducing the volume of liquid required. Thus, the concentration of the solids discharge is high. Combining the gas-assist backwash with the high dirt-holding capacity of backwash media results in long cycle times between backwashing.

Metallic Backwash Filter Selection

Porous Metal PSS Filters

Manufactured of sintered stainless steel powder, the PSS "S" medium offers exceptionally uniform permeability, and absolute removal efficiencies. These elements are seamless in construction with very high void volume (up to 60 per cent in some grades). This provides very high dirt-holding capacity and low pressure loss, especially in very fine grades, which permit design at high flux to help reduce capital costs. In addition to the standard product made from 316L stainless steel, we supply PSS medium in Inconel, nickel, nickel molybdenum, and aluminide alloys.

PMF Metal Fiber Filters

PMF filters are manufactured of finediameter 316L stainless steel fibers that are sintered at their points of contact to produce a uniform, strong, tapered-pore medium with exceptionally high dirtholding capacity while one or more inner layers provide absolute-rated filtration. Rigimesh Stainless Steel Woven Wire.

Rigimesh Stainless Steel Woven Wire Mesh Filters

Patented process permits the use of finer-diameter wires in the manufacture of the Rigimesh stainless steel medium. This results in low pressure drop, more pores per unit area, and better dirt-holding capacity than that of any other woven metal filter. The medium is sintered for superior tensile, yield, shear, and fatigue strength. Rigimesh maintains a uniform pore size and exhibits no media migration, even under high temperature and pressure conditions.

Nonmetallic Backwash Filter Selection

ProSep Filters

ProSep filters are manufactured of polyolefin or polyaramid fiber, both of which provide outstanding temperature and chemical compatibility. These filters have an outer (upstream) section in which fiber diameter and pore size decrease gradually to the element's innermost layer where the pore diameter is constant. Designed for use with or without a precoat, beta-rated ProSep filters are the appropriate choice in applications in which the process fluids have a high solids content.

Septra Filters

Septra filters incorporate a pleated, high-area absolute-rated polymeric medium in a rugged element construction designed for operation (without use of a precoat) in automated, clean-in-place fluid/solid separation systems. The higharea filter medium is a nonwoven construction that is available with polyolefin, polyaramid, or fiberglass fibers to suit operation up to 500°F (260°C).

Disposable Filters

When the solids concentration is relatively low (< 5 to 10 ppm), disposable filters will efficiently and economically remove solids from liquid streams. filters are used to remove solids from finished refinery products and from processing fluids such as amines, glycol, and sulfolane. filters are characterized by high dirt-holding capacity, and lot-to-lot uniformity. They provide long service life and improved protection for equipment and personnel. filter elements are designed to prevent both unloading of trapped particles and media migration. They are self-aligning in filter vessels, with positive sealing to eliminate fluid bypass. Disposable filters are designed and manufactured for long, effective, trouble-free service life. Their advantages include:

- Pore Size Uniformity for Full Utilization of the Filter Surface,
- Strength and Durability to Withstand the Changing Process Operations,
- Chemical and Thermal Compatibility with Process Conditions to Ensure Long Life,
- Absolute Rating for Reliable, Replicable Performance, Backed by Documented Performance Data.

Disposable Filter Selection

Profile II Filters

Profile II is a polymeric depth-type medium produced by a patented process that allows an upstream continuously graded section and a downstream absolute-rated section in a single element.

Ultipleat Profile Filters

The unique crescent-shaped construction of the Ultipleat Profile filter element provides longer life than many pleated polypropylene media. Optimized for the removal of gels and other viscous fluids, the Ultipleat Profile filter provides excellent chemical compatibility with low extractables. Ultipleat Profile filter elements are the appropriate choice for a wide range of applications in the refining industry.

Epocel Filters

Epocel cartridges have a pleated, high-area construction which provides long service life and consistent production. Constructed of epoxy-resinimpregnated cellulose, this fixed-pore construction eliminates unloading and media migration. Epocel cartridges provide a broad range of chemical compatibility and are recommended for the clarification of a wide range of fluids and gases.

Ultipor GF Plus Filters

The Ultipor GF Plus medium consists of resin-bonded glass fibers supported by upstream and downstream polymeric substrates. This unique construction provides a strongly bonded, migration-free, highdirt-capacity medium.

HDC II Filters

The HDC II medium consists of multiple layers of continuous fiber, nonwoven polypropylene pleated together, with no binders or extraneous material. HDC II cartridges have an extremely high dirt capacity, due to their taperedpore construction, and a wide range of chemical compatibility.

Profile Bag Filters

High-efficiency, beta-rated Profile bag filters are suitable for a wide variety of applications. They contain an outer (downstream) section that has a constant pore diameter to provide consistent, betarated filtration. The inner (upstream) section continuously increases in pore diameter to provide effective prefiltration.

The tapered-pore structure and depth of the medium combine to provide long service life.

SEPARATION OF SOLIDS FROM GASES

Dealing with aggressive environments, extreme temperatures, and high contaminant concentrations, has continually developed state-of-the-art technology for solid/gas separation. To efficiently separate solid particulates from gas streams has worked with both refiners and process licensors to develop blowback systems. A filter medium with sufficiently small pores is selected for this application. Solids form a permeable cake on the filter's surface that is dislodged at a predetermined pressure drop (a function of cake thickness and compressibility) by initiating a reverse pulse (blowback). The dislodged solids are purged from the filter system, where they may be returned directly to the process for reuse, or removed from the process stream and sent to a storage or collection unit. The filter is then returned to full forward flow and to an initial pressure drop that remains essentially constant through repeated blowback cycles.

Blowback Filter Selection

Porous Metal PSS Filters

Manufactured of sintered stainless steel powder, the PSS "S" medium offers exceptionally uniform permeability, and absolute removal efficiencies. These elements are seamless in construction with very high void volume (up to 60 per cent in some grades). This provides very high dirt-holding capacity and low pressure loss, especially in very fine grades, which permit design at high flux to help reduce capital costs. In addition to the standard product made from 316L stainless steel, we supply PSS medium in Inconel, nickel, nickelmolybdenum, and aluminide alloys.

PMF Metal Fiber Filters

PMF filters are manufactured of finediameter 316L stainless steel fibers that are sintered at their points of contact to produce a uniform, strong, tapered-pore medium with exceptionally high dirtholding capacity while one or more inner layers provide absolute-rated filtration.

Rigimesh Stainless Steel Woven Wire Mesh Filters

Patented process permits the use of finer-diameter wires in the manufacture of the Rigimesh stainless steel medium. This results in low pressure drop, more pores per unit area, and better dirt-holding capacity than that of any other woven metal filter. The medium is sintered for superior tensile, yield, shear, and fatigue strength. Rigimesh maintains a uniform pore size and exhibits no media migration, even under high temperature and pressure conditions.

Vitrosep Blowback Filters

The rugged construction of the Vitrosep clean-in-place blowback filter allows jet pulse cleaning and operation at high flux. This reduces the system footprint and lowers installed cost of the equipment. Equipped with a one micron rated Teflon coated fiberglass medium, Vitrosep is constructed of a stainless steel core, outer mesh wrap, and end caps.

Vitropore Ceramic Filter Candles

This rigid, highly efficient silicon carbide medium possesses high temperature and excellent corrosion resistance for gas clarification and solids recovery. The ability of every Vitropore candle to withstand corrosion and fracture at high temperatures (to 1000°C) proves it has the necessary characteristics to clean up hot gases. At temperatures above 650ºC, Vitropore candles are more resistant to corrosion than most metal filters.

Gas Particle Disposable Filter Selection

HDC II Filters

The HDC II medium consists of multiple layers of continuous fiber, nonwoven polypropylene pleated together, with no binders or extraneous material. HDC II cartridges have an extremely high dirt capacity, due to their taperedpore construction, and a wide range of chemical compatibility.

Epocel Filters

Epocel cartridges have a pleated, high-area construction which provides long service life and consistent production. Constructed of epoxy-resinimpregnated cellulose, this fixed-pore construction eliminates unloading and media migration. Epocel cartridges provide a broad range of chemical compatibility and are recommended for the clarification of a wide range of fluids and gases.

Ultipor GF Plus Filters

The Ultipor GF Plus medium consists of resin-bonded glass fibers supported by upstream and downstream polymeric substrates. This unique construction provides a strongly bonded, migration-free, highdirt-capacity medium.

Ultipleat Profile Filters

The unique crescent-shaped construction of the Ultipleat Profile filter element provides longer life than many pleated polypropylene media. Ultipleat Profile filter elements are the appropriate choice for a wide range of applications in the refining industry.

Profile II Filters

Profile II is a polymeric depth-type medium produced by a patented process that allows an upstream continuously graded section and a downstream absolute-rated section in a single element.

Jet Pulse Blowback Method

The jet pulse blowback method is cost effective because it eliminates the need for large vessel isolation valves. Full forward flow is maintained at all times. Groups of elements are blown back sequentially by directing a high-pressure pulse of gas into the throat of each element. The pulses last between 0.1 and 1.0 second, at two to three times the process pressure. During this period, the flow to the elements being cleaned is reversed momentarily by a high-pressure jet pulse. The shock wave set up by the reverse pulse, enhanced by the venturi in the element throat, effectively removes the accumulated cake from the elements.
There are numerous applications for Blowback Systems, including:

Catalytic Reforming

Blowback Systems are used in catalytic reforming units where catalyst is continuously regenerated to protect against erosive wear and fouling of the recycle compressor. These systems filter the elutriation gas that is recycled back to the regenerator. For example, PSS porous stainless steel elements (rated one micron in gaseous service) with jet pulse (blowback) in-situ cleaning are used. Ultipor GF Plus Filters

The Ultipor GF Plus medium consists of resin-bonded glass fibers supported by upstream and downstream polymeric substrates. This unique construction provides a strongly bonded, migration-free, highdirt-capacity medium.

Catalytic Cracking

Blowback Systems used to remove catalyst fines from FCC flue gas allow the refiner to:

- Meet emission guidelines, and
- Fully protect downstream process equipment.

In the U.S., the permissible particulate emission for a fluidized-bed catalytic cracker (FCC) was set at 0.1 gm/nm^3 in 1990. However, emission standards in states such as California are expected to be reduced to 0.006 gm/nm^3 by 1995, with similar emission targets proposed in Europe and Japan. Depending on the economics at a specific refinery, there are several blowback options:

(a) Where turbo expanders are in service, a hot gas filter is necessary upstream of the expander to maintain maximum power recovery and extend turbine blade life. In hot gas applications (up to 850ºC, 1560ºF) as related to coal gasifiers and fluid bed coal combustors with an even more corrosive and hotter environment than FCC, Vitropore ceramic candle filters have proven to be efficient and reliable. As is not the case with cyclones, the removal efficiency remains constant during variations in gas flow rate.

(b) If turbo expanders are not used, Vitrosep clean-in-place blowback filters are recommended to meet emission guidelines consistently. Unlike that of electrostatic precipitators, which are often used for this service, the removal efficiency of Vitrosep elements remains constant during flow surges (upset conditions) and is not affected by changes in the electrostatic charge of the particulate matter.

(c) If third- and fourth-stage cyclones are in place, replacing the fourth stage with Vitropore filters will improve solids capture, which may, of itself, be sufficient to meet emission guidelines at minimum capital investment.

Dehydrogenation Processes

PSS seamless porous metal elements process the catalyst regenerator off-gas at dehydrogenation plants operating fluid bed reactors. The Blowback System reduces particulate emissions, in a single step, to levels at or below regulatory guidelines. Older plants had used two stages of cyclones followed by flue gas scrubbers to reduce such particulate emissions.

SEPARATION OF LIQUIDS AND SOLIDS FROM GASES

LG coalescers have been used in refineries and gas processing plants for more than twenty years. The LG coalescer provides maximum liquid and solid removal at low saturated pressure drop to reduce maintenance and operating costs associated with contaminated gas. The element's high effective filtration area and resin-bonded pleated cartridge construction is surrounded by nonwoven polymeric support and drainage layers. The coalescer is rated 0.3 μm (99.97 % removal efficiency) for solid particles, and produces downstream liquid aerosol concentration as low as 0.003 ppmw. These specifications ensure long service life, minimized operating costs with minimal labor requirements, as well as greatly improved equipment reliability.

LG coalescers merge, or coalesce, small droplets of liquid into larger drops. A gas is forced to flow through several layers of filter media, each layer having a progressively larger mean pore opening. As droplets compete for the open pores, they coalesce, and the process continues until the larger drops continually collect and drain into a collecting sump. LG coalescers remove virtually all liquids in gas streams (down to 0.003 ppmw liquid in the effluent gas). In addition, LG coalescers have a patented oleophobic/hydrophobic treatment that ensures efficient removal and quick recovery from process upsets that send slugs of liquid downstream. There are numerous applications, including:

Hydrogen Compressor Protection

LG coalescers are installed in front of hydrogen compressors to protect blading or cylinders and prevent costly unscheduled downtime. Refiners have been able to extend regular maintenance schedules by installing LG coalescers upstream of compressors. Because refinery hydrogen streams are particularly prone to changes in purity and liquid concentration, oleophobic/hydrophobic treatment is extremely beneficial in this application.

Burner Protection

Liquid and solid contamination in fuel gas streams is a universal problem in refineries. By removing such contaminants, LG coalescers significantly reduce the maintenance associated with plugged burner tips in process furnaces. Low NOx burners are even more susceptible to plugging. Because the concentration of liquids in fuel gas can change instantaneously, oleophobic/hydrophobic treatment again is particularly beneficial.

Lube Oil Recovery

Most compressors have an oil lubricating system. The lube oil often is discharged into the process gas in aerosol form. An LG coalescer placed at the discharge of a compressor will recover the lube oil and prevent contamination of downstream catalyst, desiccant, and activated carbon beds.

Gas Treating Processes/Sulfur Recovery

In amine treating units, foaming can be initiated by liquid hydrocarbon contaminants in the feed gas. Installation of an LG coalescer upstream of the amine contactor can reduce the costs associated with foaming such as reduced processing capacity and increased amine losses. LG coalescers can also recover any amine that is carried over into the acid gas or the treated gas.

Liquid/Gas (LG) Coalescer

LG coalescers perform with maximum removal efficiency and economy that reduces equipment downtime and labor and maintenance costs. The unique filter medium provides for single-stage coalescing; separating liquid aerosols smaller than 0.1 micron in size. Patented coalescer treatment improves the drainage of liquids through the coalescer, allowing for smaller assemblies, and minimizing up-front capital costs. The treatment also lowers operating costs by operating at a lower liquid saturated pressure drop and by recovering quickly from process upsets.

Separation of Liquids from Liquids

When refinery customers alerted us to recurring problems with hazy fuel and caustic carryover, worked with them to quickly develop the AquaSep and PhaseSep liquid/liquid coalescers. We consulted with our customers during every step of product development and design to ensure that the AquaSep and PhaseSep coalescers met their every need.

Today, AquaSep and PhaseSep coalescers are installed in refineries throughout the world for use in several applications, including:

- Removal of Water from Gasoline, Diesel, Kerosene, and Jet Fuel,
- Protection of Catalysts from Water Contamination,
- Removal of Carried-Over Caustic from Caustic Treating Processes,
- Removal of Carried-Over Amine from LPG.

The AquaSep and PhaseSep coalescers are multiple-stage systems. They first remove particulate matter, then coalesce and separate the water or liquid contaminant from a hydrocarbon stream. An AquaSep or PhaseSep coalescer will remove entrained water to a level at or below 15 ppmw with influent conditions of:

- Inlet water concentration as high as 10 per cent water by weight.
- Interfacial tensions 2 dyne/cm and above PhaseSep liquid/liquid coalescer has been demonstrated to reduce the sodium concentration downstream of a caustic treating unit to below 0.5 ppmw of sodium.

In addition, liquid/liquid coalescers do not disarm in the presence of surface active agents in the fuel.

Unique Stack Design

AquaSep and PhaseSep coalescer element is stacked on top of a separator element. This optimizes the flow distribution from the coalescer to the separator, ensuring that each separator has equal flow. In conventional two-stage systems, the separators are located at different distances from the coalescer, causing unequal distribution of flow to the separator. These conventional two-stage systems require several coalescer elements for each separator. Stack design results in overall smaller assembly size and a longer coalescer/separator life.

Low Interfacial Tension (IFT)

The ability to remove water improves as the IFT between the two phases increases. The IFT effectively measures the stability of an emulsion or dispersion. The IFT is a critical factor when considering liquid/liquid coalescence because the largest possible stable droplet size that will form by the coalescence process will be dictated by IFT. A system with a high IFT (i.e., > 20 dyne/cm) can sustain a large stable coalesced droplet size. Systems with low IFT (i.e., water in fuels with additives: < 20 dyne/cm) form smaller stable coalesced droplets and require a high-efficiency coalescer/separator.

Disarming

Surfactants in fuels have a tendency to form fuel/water hazes and can degrade the performance or disarm conventional glass fiber coalescers. Liquid/liquid coalescers do not contain glass fiber, but are constructed of polymeric material using a unique patented process. This results in long, reliable, low-maintenance service life when compared to conventional liquid/liquid coalescers. They are immune to disarming caused by surface active components like naphthenate or sulfonate carryover, or the addition of corrosion inhibitors, dispersants, and static dissipators.

AquaSep and PhaseSep Liquid/Liquid Coalescers

Both of these coalescers were developed to efficiently separate very stable liquid/liquid dispersions to provide high fluid quality and value. They are compatible with highly acidic and basic fluids. The high-performance stack design allows an even flow distribution which permits a high flow rate in a smaller assembly. Also, the long life of the cartridge results in fewer changeouts, which reduces maintenance and disposal costs.

Chapter 7

Technology Development

Global refinery throughput is forecast to increase from 3.45 GTOE in 2002 to 4.5 in 2010 and more than 5.0 in 2020, with particularly high regional growth rates in South-east Asia and Latin America. But the structure of demand is changing and the refining industry is dealing with a range of other challenges.

In terms of demand structure, the share of heavy fuels is declining, while that of gasoline, diesel and jet fuel is rising. Demand for the latter two is expected to increase more rapidly than for gasoline. On the supply side there will be a need to handle more sour crudes. Then there are new fuel specifications, emissions controls and taxation policies, which are part of the drive towards sustainable development.

New fuel specifications started with a lead ban, which is not yet fully in place worldwide, and moved on to sulphur reduction and aromatics limitations. Regional standards vary, but changes in the European Union illustrate the impact on the refining industry. For example, the 2005 EU specification for gasoline fixes a sulphur limit of 50 parts per million (ppm) compared with 150 ppm in 2000, and an aromatics volume of 35 per cent compared to 42 per cent. For diesel oil sulphur content must fall from 350 ppm to 50 ppm. In the longer term the sulphur content of both gasoline and diesel will certainly be reduced to 10 ppm and this could even be mandatory by 2007/8. In the US by 2006 the Federal gasoline limit will be 30 ppm and the highway diesel limit 15 ppm. Some states such as California are going even further with a ban on MTBE, the original replacement for lead.

Besides changing controls on the content of the end products, waste emissions of sulphur dioxide, nitrous oxides, particulates and CO_2 will also be subject to new limits. Advances in catalysis and catalytic technologies, fully integrated with reactor and process development, are ensuring that the industry can adapt to meet these challenges. They are described below, ranging from heavy to light products. Resid conversion and treating and heavy crude treating Sincor, a Venezuelan venture of PDVSA, Statoil and Total, was set to up exploit the heavy crude of the country's Orinoco Belt and opened its treatment plant in March 2002. Sincor has a production capacity of 200,000 b/d, which is converted into 180,000 b/d of sweet Zuata crude along with 860 tonnes of sulphur and 6,000 tonnes of coke.

There are other ongoing projects to exploit sour crudes and treatment of bitumen oil sands is continuously being improved. Resid hydrocracking/ hydrotreating has benefited from improvements in catalysts. In most cases these have allowed conversion and desulphurisation of the different cuts to be improved; 0.3 per cent wt sulphur can be achieved for the residual oil. Coupling with integrated gasification combined cycle (IGCC) plants is also going to be more interesting.

Besides these improvements of the existing fixed, moving or ebullating bed processes, some more important modifications have been claimed by ENI for slurry processes (ENI Slurry Technology—EST); the modifications mainly concern the possibility of obtaining high conversions while maintaining the stability/processability of the residues.

Aqua conversion to convert heavy crudes into transportable crudes has been demonstrated by Intevep and UOP. R&D work by several teams is underway with the use of supercritical water to reach high heavy crudes conversion.

Hydrocracking

Hydrocracking appears to be the tool of choice to produce more and cleaner diesel oil. The recent trends have been to improve the selectivities of catalysts toward middle distillates as well as cycle lengths. A new generation of catalysts has allowed middle distillate yields to be improved by several points and further improvement is expected in the future.

Hydrodenitrogenation (HDN) performance of the pretreating catalyst is the key issue. The improvements obtained in recent years have been significant and even bigger ones are expected to be brought onto the market by Akzo Nobel. In the second stage of the hydrocracking unit the denitrogenated feedstock is cracked over dedicated catalysts. These can be tailored as to activity and selectivity with in general a trade-off between middle distillate selectivity and activity.

Zeolyst has introduced several new catalysts that offer higher middle distillate selectivities and lower gasmakes than those of the previous generation, with clear economic benefits to the refiner. Mild hydrocracking was developed in the past mainly to prepare feeds for fluid catalytic cracking (FCC). Recently, some refiners have considered it to produce an increased amount of diesel oil and so to get more flexibility in their gasoline/middle distillate ratio and reduce the sulphur content of the overall products. Until now, mild hydrocrackers have had fixed bed reactors and only authorise rather low conver-sion (10-30 per cent). The use of an ebullating bed reactor will allow much higher conversions.

Deep Gas Oil Desulphurisation

In the last few years there has been a tremendous amount of R&D work devoted to the deep desulphurisation of gas oil. Hydrotreating has been the preferred—but not the only—approach. The insight into the chemistry of deep desulphurisation has been developed. The identification of individual alkylated benzothiophenes as well as substituted dibenzothiophenes has become well known. Kinetic studies have shown that the rate of desulphurisation is linked to the structure of the compounds: rates decrease in the following way sulphides > thiophenes > benzothiphene > substituted benzothiophenes.

The difficulty of deep desulphurisation is linked mainly to the presence of such a structure having substituents close to the sulphur atom (sterical hindrance). The conclusion has been that a much more active catalyst would be required and that any process conditions increasing reaction rates should be used.

A combination of catalyst improvement as published by CC&T and advanced reactor internals as developed by Shell Global Solutions International has been able to bridge most of the increased demand in workload in hydroprocessing. However, further improvements in overall performance are required if one wishes to avoid extra reactors.

Higher hydrogen pressure units have been implemented, purified hydrogen and inter stage H2S removal have been introduced as well as an optimised fluid distribution. Based on a deep analysis of pathways for dibenzothiophene desulphurisation different new catalysts have been developed by most of the catalyst manufacturers. Strategies for their uses may be dependent on the pressure of the reactors, but the clear goal is to get higher volumic activities. Specific arrangements are claimed to give enhanced desulphurisation activities. Even if the constraints on aromatics have in most places been rather low there has been some demand for low aromatics city diesel. Synsat has been an elegant process answer in the past and has developed with new catalysts more recently.

Bimetallic noble-metal catalysts are claimed, which increase the activity by factors higher than five compared to platinum ones. Saturation of polynuclear aromatics only increases the cetane numbers slightly. In order to improve the cetane number of light cycle oil (LCO) or naphthtemic crudes, it is also necessary to open the rings. Classical hydrocracking leads to low yield. Synshift largely improves them and there is a large ongoing effort to develop selective ring-opening catalysts.

Deep hydrodesulphurisation (HDS) has an average cost of between 5 and 50 cents per gallon, a part of it is linked to hydrogen consumption.

Several alternate routes have been explored and some have reached industrial development. Biodesulphurisation has made huge progress in the last 10 years. It seems technically viable but the economics are not yet at the required level. Another interesting route is the conversion of sulphur compounds into sulphones which can be extracted by water; the hydrocarbon retained sulphones being adsorbed on alumina or other adsorbents. The oxyding agent may be hydrogen peroxide, organic peroxides (UniPure) or even ultrasonically activated water as developed by SulphCo. Recently the use of ConocoPhillips' S Zorb, which was initially developed for gasoline, has been extended to the finishing treatment of hydrotreated products or to rather low sulphur containing streams. Results presented show that a treatment of a 3,000 ppm sulphur feed to less than 10 ppm can be done with no chemical hydrogen consumption and no significant change of the physical properties.

Other work is going on to develop the use of adsorbents at least for "polishing" operations from a few hundred ppm to less than 10. SK Corporation has developed a process to remove nitrogen compounds from diesel oil in order to get a feed more easy to desulphurise; required reactor temperatures can be reduced by 20°C.

FCC

The steady development of FCC technologies has seen improvements in the mixing zone, catalyst separation and reactor stripping. The most recent developments are oriented toward short contact times (less than a few seconds) in order to improve gasoline yields and to make more olefins. Short contact time FCC and dropper technologies have been claimed. Their use has not yet been largely developed.

Fortum has come up with a new technology called Neste catalytic cracking (NExCC) for which increased product qualities (more olefins), reduced investment and limited catalyst attrition and equipment erosion are claimed. To cope with emissions regulations new technologies have been implemented like improved flue gas scrubbing and better particulate removal devices.

But the biggest effort in recent years has been devoted to improving the quality of the products: LCO (sulphur, heavy polyaromatics, cetane) and gasoline (sulphur, research octane number and motor octane number).

FCC Gasoline Desulphurisation

FCC gasoline represents about 35 per cent of the gasoline pool although it contributes to more than 95 per cent of its sulphur content. A 50 ppm sulphur in the gasoline pool can be achieved with a severe pretreatment of the FCC feed but it does not seem economic to reach 10 ppm through pretreatment alone.

Olefins are mainly in the light cut and sulphur in the heavy one. Except for mercaptans removal refiners are used to eliminate sulphur compounds through hydrodesulphurisation. Saturation of olefins will decrease the octane numbers of the gasoline, while H_2S can recombine with olefins to produce mercaptans.

Taking this into account, process licensors and catalyst manufacturers together with refiners have come up with different solutions based on hydrotreating of the full gasoline or of specific cuts only.

The first answers on the market were based on deep HDS of the gasoline and a recovery of the octane rating loss through reactions like isomerisation, cracking and alkylation.

ExxonMobil's OCTGAIN and ISAL (which was jointly developed by INTEVEP, SA and UOP) are the main processes developed in that line. Selective hydrotreating has been developed in different versions. ExxonMobil's Selective Cat Naphtha hydrofining (SCANfining) is based mainly on a highly selective catalyst (jointly developed by ExxonMobil and Akzo Nobel) and takes advantage of carefully selected operating conditions. Octane rating losses of between 2.1 and 2.8 octane numbers have been obtained in low sulphur operations (15-20 ppm). Optimal improvements like splitting the light cracked naphtha (LCN) or removal of heavy mercaptans with the Exomer process, which ExxonMobil developed jointly with Merichem, allow octane losses to be reduced noticeably. Axens has developed Prime-G+, which is based on a combination of a selective hydrogenation, removing diolefins and light mercaptans, a splitter and a selective hydrodesulphurisation of the medium cracked naphtha (MCN) and heavy cracked naphtha (HCN) cut through a dual catalytic system. With 10 ppm sulphur in the gasoline octane losses are in the range of 1.0 to 1.5.

CDTech has configured a process scheme that uses two stages of catalytic distillations; the first, CD Hydro column fractionates the full range naphtha into LCN, which is treated in the top of the column on a nickel catalyst, the MCN/HCN is fed to the CD HDS column loaded with HDS catalysts. Average octane losses of about 1.0 have been obtained with a 1,000 ppm feed at 85 per cent sulphur conversion.

Olefinic alkylation of thiophenic sulphur (OATS) technology developed by BP uses a catalytic system to convert the light sulphur compounds into heavier ones via alkylation of thiophenic compounds with olefins. At the mean time some olefins are oligomerised leading to a lower Reid vapour pressure (RVP) of the full range gasoline. The naphtha is splitted and the heavy end hydrotreated. An alternate solution to separate the reactive olefins from the sulphur has been developed by GTC Technology. GT-DeSulf is based on solvent extraction of the sulphur and the aromatics from the light and part of the medium naphtha cuts, leaving a sulphur/aromatics free light/medium cut, mixing the extract with the rest of the naphtha and sending it to a hydrotreater.

Other processes based on adsorption on solids have been developed. Black & Veatch's IRVAD process consists of a multistage fluidised adsorber. The adsorbent is circulated to a continuous reactivator where it is brought into cross-current contact with reactivating gas at elevated temperature. A reduction of sulphur from a few hundred to less than 10 ppm has been demonstrated in pilot plants. The most successful adsorption technology is ConocoPhillips' S Zorb. This uses a proprietary adsorbent that adsorbs sulphur molecules and removes the sulphur. The unique chemistry requires only low hydrogen consumption.

Octane losses lower than 1.0 have been obtained at 10 ppm sulphur starting from 675. The feed mixed with hydrogen is put in contact with the reduced adsorbent in a fluidised bed, sulphided adsorbent is transferred to an oxidising zone and finally the oxidised solid is reduced in a third zone.

There will be efforts in the future to reduce the cost of existing technologies and to make them well adapted to the production of gasoline with a sulphur content of less than 10 ppm. Associations of technologies, SCANfining and Exomer, Prime-G+ and OATS may be some of the solutions. CO_2 taxes could, depending on their level, disturb the competition in favour of the low hydrogen consuming and low energy intensive technologies.

MTBE alternatives/solid alkylation Concerns about the contamination of groundwater from gasoline containing MTBE have led some areas such as California to ban MTBE and several alternative technologies have been announced. UOP's Indirect Alkylation (InAlk) process uses commercial, solid catalysts for reacting iso-butane with light olefins to produce a high octane, paraffinic gasoline component similar to traditional alkylate. Fortum has combined its NExOCTANE process for converting existing MTBE-producing units to environmentally benign iso-octane with NExSAT hydrogenation to obtain up to 90 per cent isooctane starting from iso-butene. Selectivity is obtained by the use of tertiary butyl alcohol generated by water addition. Solid alkylation has not yet found its way for an industrial utilisation. Haldor Topsøe and UOP re ready to license but the economics do not show a clear advantage today. Nevertheless work is continuing and ABB, Akzo Nobel and Fortum are operating a pilot plant.

Paraffin isomerisation/catalytic reforming The use of paraffins isomerisation is growing continuously. Chlorided alumina catalysts have been improved and their utilisation is increasing, while a new generation of catalyst, more environmentally friendly but less active has been introduced. The need for new catalysts is important. The required properties are mainly higher activity and the possibility to handle heptanes or higher products.

Associations of isomerisation and separation techniques (distillation, adsorption) in order to recycle the low octane linear paraffins are growing. This is a line where further progress is expected. Catalysts for catalytic reforming have been steadily improved in order to get continuously the hydrogen needed to satisfy the increasing hydrotreating demand. If in the future the total aromatics content in gasoline is reduced to less than about 35 per cent catalytic reforming will be hurt and certainly other hydrogen sources will have to be implemented.

Chapter 8

Pollution Prevention

Waste Characteristics

Boilers, process heaters, and other process equipment are responsible for the emission of particulates, carbon monoxide, nitrogen oxides (NO_x), sulfur oxides (SO_x), and carbon dioxide. Catalyst changeovers and cokers release particulates. Volatile organic compounds (VOCs) such as benzene, toluene, and xylene are released from storage, product loading and handling facilities, and oil-water separation systems and as fugitive emissions from flanges, valves, seals, and drains. For each ton of crude processed, emissions from refineries may be approximately as follows:

- Particulate matter: 0.8 kilograms (kg), ranging from less than 0.1 to 3 kg.
- Sulfur oxides: 1.3 kg, ranging 0.2–06 kg; 0.1 kg with the Claus sulfur recovery process.
- Nitrogen oxides: 0.3 kg, ranging 0.06–0.5 kg.
- Benzene, toluene, and xylene (BTX): 2.5 grams (g), ranging 0.75 to 6 g; 1 g with the Claus sulfur recovery process. Of this, about 0.14 g benzene, 0.55 g toluene, and 1.8 g xylene may be released per ton of crude processed.
- VOC emissions depend on the production techniques, emissions control techniques, equipment maintenance, and climate conditions and may be 1 kg per ton of crude processed (ranging from 0.5 to 6 kg/t of crude).

Petroleum refineries use relatively large volumes of water, especially for cooling systems. Surface water runoff and sanitary wastewaters are also generated. The quantity of wastewaters generated and their characteristics depend on the process configuration. As a general guide, approximately 3.5–5 cubic meters (m^3) of wastewater per ton of crude are generated when cooling water is recycled. Refineries generate polluted wastewaters, containing biochemical oxygen demand (BOD) and chemical oxygen demand (COD) levels of approximately 150–250 milligrams per liter (mg/l) and 300–600 mg/l, respectively; phenol levels of 20–200 mg/l; oil levels of 100–300 mg/l in desalter water and up to 5,000 mg/l in tank bottoms; benzene levels of 1–100 mg/l; benzo (a) pyrene levels of less than 1 to 100 mg/l; heavy metals levels of 0.1–100 mg/l for chrome and 0.2–10 mg/l for lead; and other pollutants. Refineries also generate solid wastes and sludges (ranging from 3 to 5 kg per ton of crude processed), 80 per cent of which may be considered hazardous because of the presence of toxic organics and heavy metals.

Accidental discharges of large quantities of pollutants can occur as a result of abnormal operation in a refinery and potentially pose a major local environmental hazard.

Pollution Prevention and Control

Petroleum refineries are complex plants, and the combination and sequence of processes is usually very specific to the characteristics of the raw materials (crude oil) and the products. Specific pollution prevention or source reduction measures can often be determined only by the technical staff. However, there are a number of broad areas where improvements are often possible, and site-specific waste reduction measures in these areas should be designed into the plant and targeted by management of operating plants. Areas where efforts should be concentrated are discussed here.

Reduction of Air Emissions

- Minimize losses from storage tanks and product transfer areas by methods such as vapor recovery systems and double seals.
- Minimize SO_x emissions either through desulfurization of fuels, to the extent feasible, or by directing the use of high-sulfur fuels to units equipped with SO_x emissions controls.
- Recover sulfur from tail gases in high-efficiency sulfur recovery units.
- Recover non-silica-based (i.e., metallic) catalysts and reduce particulate emissions.
- Use low-NO_x burners to reduce nitrogen oxide emissions.
- Avoid and limit fugitive emissions by proper process design and maintenance.

- Keep fuel usage to a minimum.

Elimination or Reduction of Pollutants

- Consider reformate and other octane boosters instead of tetraethyl lead and other organic lead compounds for octane boosting.
- Use non-chrome-based inhibitors in cooling water, where inhibitors are needed.
- Use long-life catalysts and regenerate to extend the catalysts' life cycle.

Recycling and Reuse

- Recycle cooling water and, where cost-effective, treated wastewater.
- Maximize recovery of oil from oily wastewaters and sludges. Minimize losses of oil to the effluent system.
- Recover and reuse phenols, caustics, and solvents from their spent solutions.
- Return oily sludges to coking units or crude distillation units.

Operating Procedures

- Segregate oily wastewaters from stormwater systems.
- Reduce oil losses during tank drainage carried out to remove water before product dispatch.
- Optimize frequency of tank and equipment cleaning to avoid accumulating residue at the bottom of the tanks.
- Prevent solids and oily wastes from entering the drainage system.
- Institute dry sweeping instead of washdown to reduce wastewater volumes.
- Establish and maintain an emergency preparedness and response plan and carry out frequent training.
- Practice corrosion monitoring, prevention, and control in underground piping and tank bottoms.
- Establish leak detection and repair programs.

Target Pollution Loads

Implementation of pollution prevention measures can yield both economic and environmental benefits. However, a balance on energy usage and environmental impacts may have to be struck. The production-related targets described below can be achieved by measures such as those detailed in the previous section. The values relate to the production processes before the addition of pollution control measures.

New refineries should be designed to maximize energy conservation and reduce hydrocarbon losses. A good practice target for simple refineries (i.e., refineries with distillation, catalytic reforming, hydrotreating, and offsite facilities) is that the total quantity of oil consumed as fuel and lost in production operations should not exceed 3.5 per cent of the throughput. For refineries with secondary conversion units (i.e., hydrocrackers or lubricating oil units), the target should be 5–6 per cent (and, in some cases, up to 10%) of the throughput.

Fugitive VOC emissions from the process units can be reduced to 0.05 per cent of the throughput, with total VOC emissions of less than 1 kg per ton of crude (or 0.1% of throughput). Methods of estimating these figures include emissions monitoring, mass balance, and inventories of emissions sources. Design assumptions should be recorded to allow for subsequent computation and reduction of losses.

Vapor recovery systems to control losses of VOCs from storage tanks and loading areas should achieve 90–100 per cent recovery. Plant operators should aim at using fuel with less than 0.5 per cent sulfur (or an emissions level corresponding to 0.5% sulfur in fuel). High-sulfur fuels should be directed to units equipped with SOx controls. Fuel blending is another option. A sulfur recovery system that achieves at least 97 per cent (but preferably over 99%) sulfur recovery should be used when the hydrogen sulfide concentration in tail gases exceeds 230 mg/Nm^3. The total release of sulfur dioxide should be below 0.5 kg per ton for a hydroskimming refinery and below 1 kg per ton for a conversion refinery.

A wastewater generation rate of 0.4 m^3/t of crude processed is achievable with good design and operation, and new refineries should achieve this target as a minimum. The generation rate of solid wastes and sludges should be less than 0.5 per cent of the crude processed, with a target of 0.3 per cent.

TREATMENT TECHNOLOGIES

Air Emissions Control of air emissions normally includes the capture and recycling or combustion of emissions from vents, product transfer points, storage tanks, and other handling equipment. Boilers, heaters, other combustion devices, cokers, and catalytic units may require particulate matter controls. Use of a carbon monoxide boiler is normally a standard practice in the fluidized catalytic cracking units. Catalytic cracking units should be provided with particular removal devices.

Steam injection in flaring stacks can reduce particulate matter emissions. Liquid Effluents Refinery wastewaters often require a combination of treatment methods to remove oil and contaminants before discharge. Separation of different streams, such as stormwater, cooling water, process water, sanitary, sewage, etc., is essential for minimizing treatment requirements. A typical system may include sour water stripper, gravity separation of oil and water, dissolved air flotation, biological treatment, and clarification.

A final polishing step using filtration, activated carbon, or chemical treatment may also be required. Achievable pollutant loads per ton of crude processed include BOD, 6 g; COD, 50 g; suspended solids, 10 g; and oil and grease, 2 g. Solid and Hazardous Wastes Sludge treatment is usually performed using land application (bioremediation) or solvent extraction followed by combustion of the residue or by use for asphalt, where feasible. In some cases, the residue may require stabilization prior to disposal to reduce the leachability of toxic metals.

Oil is recovered from slops using separation techniques such as gravity separators and centrifuges. Emissions Guidelines Emissions levels for the design and operation of each project must be established through the environmental assessment (EA) process on the basis of country legislation and the Pollution Prevention and Abatement Handbook, as applied to local conditions. The emissions levels selected must be justified in the EA and acceptable to the World Bank Group.

The guidelines given below present emissions levels normally acceptable to the World Bank Group in making decisions regarding provision of World Bank Group assistance. Any deviations from these levels must be described in the World Bank Group project documentation. The emissions levels given here can be consistently achieved by well-designed, well-operated, and well-maintained pollution control systems. The guidelines are expressed as concentrations to facilitate monitoring. Dilution of air emissions or effluents to achieve these guidelines is unacceptable. All of the maximum levels should be achieved for at least 95 per cent of the time that the plant or unit is operating, to be calculated as a proportion of annual operating hours.

Air Emissions

The emissions levels presented in Table 8.1 should be achieved.

Table 8.1: Emissions from the Petroleum Industry (milligrams per normal cubic meter)

Parameter	*Maximum value*
PM	50
Nitrogen oxides*	460
Sulfur oxides	150 for sulfur recovery units; 500 for other units
Nickel and vanadium (combined)	2
Hydrogen sulfide	152

* Excludes NO_x emissions from catalytic units.

Liquid Effluents

The emissions levels presented in Table 8.2 should be achieved.

Effluent requirements are for direct discharge to surface waters. Discharge to an offsite wastewater treatment plant should meet applicable pretreatment requirements.

Table 8.2: Effluents from the Petroleum Industry (milligrams per liter)

Parameter	*Maximum value*
pH	6–9
BOD	30
COD	150
TSS	30
Oil and grease	10
Chromium Hexavalent	0.1
Total	0.5
Lead	0.1
Phenol	0.5
Benzene	0.05
Benzo (a) pyrene	0.05
Sulfide	1
Nitrogen (total)[a]	10
Temperature increase	£ 3°C[b]

a. The maximum effluent concentration of nitrogen (total) may be up to 40 mg/l in processes that include hydrogenation.

b. The effluent should result in a temperature increase of no more than 3° C at the edge of the zone where initial mixing and dilution take place. Where the zone is not defined, use 100 meters from the point of discharge, provided there are no sensitive ecosystems within this range.

Solid Wastes and Sludges

Wherever possible, generation of sludges should be minimized to 0.3 kg per ton of crude processed, with a maximum of 0.5 kg per ton of crude processed. Sludges must be treated and stabilized to reduce concentrations of toxics (such as benzene and lead) in leachate to acceptable levels, for example, below 0.05 milligram per kg.

Ambient Noise

Noise abatement measures should achieve either the levels given below or a maximum increase in background levels of 3 decibels (measured on the A scale) [dB(A)]. Measurements are to be taken at noise receptors located outside the project property boundary.

	Maximum allowable log equivalent (hourly measurements), in dB(A)	
	Day	*Night*
Receptor	(07:00–22:00)	(22:00–07:00)
Residential, institutional, educational	55	45
Industrial, commercial	70	70

Monitoring and Reporting

Frequent sampling may be required during startup and upset conditions. Once a record of consistent performance has been established, sampling for the parameters listed in this document should be as described below.

Air emissions from stacks should be monitored once every shift, if not continuously, for opacity (maximum level, 10%). Air emissions of hydrogen sulfide from a sulfur recovery unit should be monitored on a continuous basis. Annual emissions monitoring of combustion sources should be carried out for sulfur oxides (sulfur content of the fuel monitored on a supply-tank basis) and for nitrogen oxides.

Liquid effluents should be monitored daily for all the parameters listed above, except that metals should be monitored at least monthly. Monitoring data should be analyzed and reviewed at regular intervals and compared with the operating standards so that any necessary corrective actions can be taken. Records of monitoring results should be kept in an acceptable format. The results should be reported to the responsible authorities and relevant parties, as required.

Chapter 9

FAQ's (Frequently Asked Questions)

Q.1. What is the gasoline octane number?

Ans. The performance of a spark-ignited gasoline-fueled engine (i.e., an Otto cycle engine) is improved when the gasoline burns and advances quietly with a smooth flame front as it drives the engine's cylinders' pistons rather than knocking or pinging, which indicates that the fuel is detonating instead of burning smoothly. The octane number is a measure of the anti-knock quality of gasoline.. in other words, the ability of the gasoline to resist knocking when it is burned in a spark-ignited Otto cycle engine. Pure iso-octane, which has an excellent anti-knock quality, has been designated as having an octane number of 100. Samples of gasoline are tested by running them in a single-cylinder engine mounted in a laboratory. For example, if a gasoline's anti-knock performance during the test is 85 per cent as good as the performance of pure iso-octane in the same test, then that gasoline is said to have an octane number of 85. As another example, if a gasoline's anti-knock performance during the test is 10 per cent better than that of pure iso-octane, then that gasoline is said to have and octane number of 110. There are two different sets of engine operating conditions used during the tests to measure the octane number of a gasoline. One set of conditions is used to determine the so-called Research octane number (RON), and another set of conditions is used to determine the so-called Motor octane number (MON). The average of the RON and MON is abbreviated to (R + M)/2 and that is the octane rating which is usually posted at service stations in the US.

Q.2. How does the petroleum industry define a "barrel"?

Ans. The "barrel" is a volumetric unit commonly used in the petroleum industry and one barrel is equivalent to 42 U.S. gallons.. or 34.97 Imperial gallons... or 158.99 liters... or 5.615 Cubic feet. Please note from this that an Imperial gallon is 1.20 times as large as a U.S. gallon.

Q.3. How long does it take to produce gasoline in a refinery?

Ans. Many of the various process units in a refinery have feedstock and/or intermediate product storage tanks. If we neglect the time spent in those storage tanks, the actual time spent in refining the crude oil to produce automotive gasoline is about an hour or perhaps even less. If we do include the time spent in those storage tanks, the total time might be 1 to 4 weeks

Q.4. What are the consequences of allowing water to come in contact with gasoline that has been blended with ethanol?

Ans. The water causes the ethanol to become insoluble in the hydrocarbon so two layers form. They are gasoline and ethanol/water.

Q.5. Why is ethanol added to gasoline?

Ans. Ethanol is blended into automotive gasoline for two reasons: (1) To provide a blend component with a high octane number, which improves the performance of the gasoline. (2) To provide an oxygenated blend component, which reduces certain environmentally undesirable emissions from the automotive engines

Q.6. What is the Lo-cat Process?

Ans. The Lo-Cat process is licensed by ARI Technologies Incorporated. Basically, it is a process for removing H_2S from gas streams and converting the H_2S into elemental sulfur. There is an "aerobic" version for removing H_2S from air streams. There is also an "anaerobic" version for removing H_2S from raw natural gas or from process gas streams in refineries, petrochemical plants, coke oven gas, etc.

Q.7. What is the typical frequency for a flare inspection in a refinery?

Ans. Flare inspection programs are more often dictated by practical rather than theoretical considerations. For example, if your refinery is served by a single flare, you have no option but to schedule the inspection of the flare along with the maintenance and inspection turnaround schedule for the rest of the process units. In such a scenario, you should inspect the flare whenever you are allowed to do so. Some refineries have a spare flare stack. You can, then, offer a stack for inspection after a year of operation. A good flare stack maintenance program, must, of course, include an inspection of the flare burner tip, pilot burner tip and the molecular seal. It is also a good idea to carry out a thickness survey of the refinery flare header and unit flare headers whenever you can. Flare headers can sometimes get choked with debris from upsets/ continous operation. You can perhaps break a flange to see the available flow area in the header. Experience with flare headers indicate that it is a good idea to keep a slight nitrogen purge in the header during a maintenance and inspection shutdown to prevent air ingress. The headers always contain pyrophoric iron which can ignite upon exposure to oxygen. You should carry out routine checks (monthly) on the pilot burner ignitor circuit. Finally, during an inspection shutdown, the pilot and fuel gas header lines should be flushed, preferably with nitrogen. If you have a nitrogen crisis, you can use LP steam, but make sure you drain out the condensate.

Q.8. What is the best way to measure the flow in flare stack?

Ans. An orifice plate is generally not suitable for this service. The reason is that the the flare line is always sized for the maximum possible flare flow, in an upset condition. The flow to the flare under normal conditions will be much smaller, by a factor of 100 to 1000, or more. The pressure drop across the orifice under normal conditions will not be high enough to permit accurate flow measurement. (Don't forget that pressure drop varies with the square of the flow). Over the years, there have been many attempts to find a suitable flowmeter for this service. I believe that the only successful type has been the utrasonic flowmeter. It also has the advantage of providing no obstruction in the line. The best-known supplier is Panametrics. They have an excellent reference list, on refineries and chemical plants. The only disadvantage is the cost - in the order of $30,000 per installation.

Q.9. What is a good source of information about catalyst cooler operation, reliability and design for fluidized catalytic crackers.

Ans. UOP, Kellogg Brown and Root (KBR) and Stone and Webster all have catalyst cooler designs that are in commercial operation (the Stone and Webster Design was actually developed by the Bejing Design Institute in China). Each can probably supply you with contacts at refineries using their designs. All three cooler designs use essentially the same basic principles, but the hardware differs from licensor to licensor. From this standpoint, each design has both positive and negative aspects, and these differences must be weighed against your particular unit and needs. All three designs remove heat from a dense bed of catalyst located in the cooler vessel. Heat removal is through the generation of steam in tubes submerged in the dense bed. Hot catalyst enters the cooler from the regenerator bed. In most designs, cooled catalyst is returned to the regenerator using a standpipe and possibly a lift line. UOP also offers a "backmixed" design that does not require catalyst return lines (catalyst circulation to and from the cooler is controlled by fluidization air). Compared to flow through coolers, back mixed coolers have lower heat transfer rates. These dense bed coolers are considerably more reliable than older dilute phase coolers where the catalyst is blown through the tubes using air.

Q.10. What parameters and tools do refiners use to measure two stage desalter performance?

Ans. Performance of desalting process can be quantified on the basis of dehydration and thereby desalting efficiency levels acheived. For evaluating the factors affecting these efficiencies we shall divide the parameters into two classes: 1) Controllable parameters 2) Non controllable parameters Controllable parameters include 1) Wash Water rate 2) Mix valve energy or presure drop 3) Wash water quality that affects emulsion stability 4) Interface level 5) Injection points of wash water 6) Slop oil addition 7) Demulsifier /Desalting chemical dosage and injection points, if being used. Non Controllable parameters include: 1) Pressure 2) Temperature 3) Crude charge rate that directly affects residence time 4) Crude type that includes its salt level, sediment and other impurities like metal content and corrosion products 5) Electrical potential 6) Design of desalter, which may effect residence time

Q.11. How can the performance of a LPG caustic treatment plant be improved?

Ans. The most important factor is the air quantity which is used for regeneration of the spent caustic. It is important to maintain oxygen content of 10 to 12 per cent in vent gases at the disulfide separator in the LPG Merox Unit. It may require 2~3 times the theoretical required air quantity.

Q.12. How can I reduce the solids content of FCC decanter oil?

Ans. There are several technologies that can be used to remove solids from the decanted oil. Slurry settlers were very popular in the past. These were standard equipment on many older FCC's. Most have been taken out of service so a used settler may be easy to find. Both Pall and Mott sell filter systems to remove catalyst from decanted oil. The Gulftronic Separator uses beds of electrically charged quartz beads to separate catalyst. Hydroclones (liquid cyclones) are also popular for reducing the catalyst level in the decanted oil. All of these methods concentrate the catalyst into an underflow stream that is returned to the FCC. One side effect of this catalyst return is the possibility a slight increase in the catalyst losses from the regenerator.

Q.13. What are some methods of improving filtration in amine plants?

Ans. First, a good amine solution should be transparent, almost colorless. To keep the amine clean, be sure that hydrocarbon contamination is minimized upstream of the amine unit. As far as filtration is concerned:

1. A lot of refiners deem it good sense to route at least 10 per cent of amine flow through the filter section.
2. Change over cartridge type filters at least once every three months

even if you do not read a high pressure drop. Remember, damaged and torn filters do not restrict flow or catch solids.

3. If you have a backflow wash circuit on the charcoal filter, run it at least once every two weeks.
4. Install pressure drop cells across each filter if not already installed.
5. If you load a fresh stock of charcoal in the activated charcoal filter, wash the bed with DM water to wash out any fines. Bring the filter online only after wash effluent is clear. For the first 48 hours after commissioning a new charcoal bed, change over cartridge filters every four hours and get them cleaned.

Q.14. How can the LPG levels in dry gas be controlled in FCC plants?

Ans. LPG (C_3's and C_4's) in the dry gas are normally controlled by the operation of the primary absorber. Operationally, increasing the lean oil rate to the absorber should decrease the LPG in the dry gas. If the unit has recycle lean oil from the debutanizer then this can be increased. If there is no recycle lean oil, then decreasing the main fracticnator overhead accumulator temperature or increasing its pressure would increase the amount of liquid condensed and thus the lean oil rate to the primary absorber. Decreasing the stripping rate in the stripper (deethanizer) will also decrease the C_3's and C_4's lost to dry gas. The limit to this option is normally the build up of non-condensables in the debutanizer overhead drum. Modifications that would increase the recovery of C_3 and C_4 material (and thus decrease the loss) would include the addtion of intercoolers on the primary absorber or chilled water systems to cool the lean oil to the absorber. Other modifications may also be possible depending on the configuration of the current gas plant.

Q.15. How many different types of crude are there in the world?

Ans. There are hundreds of different crude oils, perhaps even thousands. Some crude oils have lighter specific gravities than others, some have a lower total sulfur content than others, some can be refined to produce more gasoline than others, some have lower pour points than others. The physical properties of any specific crude oil are obtained by a series of laboratory tests which result in what is called a "crude oil assay". Some people classify crude oils as: paraffin base; naphthene base; aromatic base; asphalt base; or mixed base. The general elementary composition of crude oil falls within these approximate weight per cent ranges: Carbon = 80 to 87, Hydrogen = 11 to 14, Nitrogen = 0.2, and Sulfur = 0 to 3.

Q.16.What is the pour point of API 10 crude oil?

Ans. Crude oils with the same API gravity can have different pour points, if they are from different oil fields worldwide. In other words, a 10 deg API crude from California in the USA can have a different pour point than a 10 deg API crude from the Orinoco region of Venezuela. With that in mind, a 10.3 deg API crude from California has a pour point of 0°C.

Q.17. If a hydrogen line is laid beside a fuel oil line which is heat traced would it pose any hazard?

Ans. If it is steam tracing, there is probably no hazard from the tracing. If the tracing is electrical and since hydrogen has such a wide range of flammability (i.e., from 4% to 75% by volume in air), then even a very slight possiblity of electrical sparking from the tracing would pose considerable hazard in the event of a hydrogen leak caused by an earthquake or any other reason. It is quite common for hydrogen (and for that matter other flammable gases) lines to be laid next to high temperature pipes in refinery settings. Safety precautions that should be used include: killed carbon steel piping, welded fittings (hydrogen leaks very easily), piping inspection and a mechanical integrity program for vessels, and any electrical equipment should be of appropriate for the area classification

Q.18. What would be the effect of cracking about 10 per cent atmospheric gas oils blended with vacuum gas oils in the Fluidized Catalytic Cracking Unit (FCCU)?

Ans. If by 'atmospheric gas oil' you are referencing material boiling below 344 0C, then addition of this material to the FCC will generally cause an increase in catalyst circulation, a decrease in regenerator temperature and an increase in over all conversion. Much of this effect is due to heat balance considerations. The lighter gas oils generally do not produce sufficient coke to supply the heat necessary to vaporize them and raise them to reactor temperature. Thus, the unit heat balance forces higher conversions to produce the necessary coke from the heavier gas oils. The atmospheric gas oil does crack, but not to the extent of the heavier materials. It also tends to produce a lower octane cat naphtha. It is not uncommon to find this material in FCC feed, though it is more often the result of poor separation in the crude unit rather than a conscious operating decision. Since many of the effects are heat balance related, they are difficult to predict without a well tuned FCC model. If a model is used, it is essential that it correctly predicts the effect of conversion and cat/oil on coke yield.

Q.19. What is petroleum pitch?

Ans. Solid petroleum pitch is an asphalt concentrate from crude oil, where enough oil has been removed from the asphalt to produce a coal-like solid of high melting point. Petroleum pitch differs from coal and petroleum coke in that it has a melting or softening point. This softening point is high enough, however (about 350°F), to allow pitch to be stored in a 40 foot piles at 170°F without the agglomeration of particles. Petroleum pitch is a concentrate of the asphaltic material in crude. By removing or eliminating oil which is included with normal vacuum—reduced crudes, a pitch can be produced whose asphaltene concentration is sufficiently high to impart hardness properties, which permit satisfactory handling, storing, grinding, and burning as a solid fuel. From certain crudes which have very hard asphalts, it possible to produce solid petroleum pitch by simply increasing the severity of normal refinery vacuum distillation by 50 to 100° F (corrected to 760 mm of Hg). By varying the severity of distillation, pitches of varying softening point can be obtained; however, pitches of higher softening point are more difficult and costly to produce and the resulting gas oils have higher carbon residues and metals content. It is believed that pitches of 350° F minimum softening point can be handled and burned in most burners designed for either coal or heavy fuel oil with but minor modifications, and it is possible that lower softening point (275 to 325° F) pitches can be handled and burned in specially designed equipment.

Q.20. Can kerosene be catalytically cracked?

Ans. Kerosene is not normally used as a feed to catalytic cracking units. Due to its relatively low molecular weight, the reaction rate for cracking reactions would be low. Thus, there would be little conversion in a typical catalytic cracking unit. Kerosene can be cracked, but the unit would need to be designed for high catalyst/oil ratios, high contact times and higher than normal reaction temperatues. Essentially, you would need a custom built or specially modified unit to maximize conversion of kerosene. Since this is not a typical FCC feed, there is little published data.

Q.21. What are the most common methods of cracking crude streams via a catalyst?

Ans. There are two main cracking processes in petroleum refining that use catalyst - catalytic cracking and hydrocracking. Catalytic cracking, uses catalyst and heat to break large hydrocarbon molecules into smaller molecules. The reaction proceeds through carbenium ion formation followed by beta scission of the ion. This forms an olefin and a second carbenium ion. The primary use of catalytic cracking is to produce gasoline (aka petrol) blend stocks from vacuum gas oils and atmospheric resids. Paraffins, Olefins and Naphthene molecules can be cracked, but aromatic rings cannot. Thus, aromatics tend to concentrate in the heavier products.

For this reason, the diesel range material produced in the cat cracker is of poor quality. A major by product of catalytic cracking reactions is a solid carbonacious material - coke. This is deposited on the catalyst and burned off in the regenerator. The heat from coke combustion is used to raise the feed oil to reaction temperature and to provide the endothermic heat of cracking. Catalytic cracking is a carbon rejection process. That is, it improves the quality of the feed oils by removing carbon (coke). Most cat crackers use a powdered catalyst that is kept in a fluidized state. These units are refered to as fluid catalytic cracking units or FCCU's. There are still a few moving bed units in operation. These are refered to as TCC's or Houdry units. Hydrocracking uses catalyst and heat to crack hydrocarbon molecules in the presence of hydrogen. The cracking reaction mechanism is similar to that found in fluid catalytic cracking, but both the feed and the products are also subject to hydrogenation reactions. These reactions saturate both the olefins produced by cracking and many of the aromatic molecules found in the feed. Hydrocrackers can produce a wide range of products from LPG to diesel. Because the products are hydrogenated, they contain no olefins and few aromatics. Naphtha from hydrocrackers must be treated further to raise its octane value before it can be used as a gasoline blend stock. The kerosene and diesel range materials, however, are of superior quality. Hydrocrackers are widely used to produce jet fuel and low sulfur diesel. Most hydrocrackers are fixed bed units and the feed is limited to materials boiling below 510°C. Ebulating bed hydrocrackers (H-Oil, LC-Fining) can process heavier feeds but generally produce somewhat lower quality products.

Q.22. How can steam be utilized to reduce coking in crude furnaces?

Ans. The steam should be injected into each individual pass of the heater. Injecting steam in the common line will cause problems with the individual pass control valves since they are not suited to controlling two phase flow. The hard part is determining where, exactly, to put the steam into the passes. Injection too far upstream may lead to a high pressure drop across the heater that you may not be able to tolerate. Injection too far downstream may be too late to prevent coking as oil may have already started cracking. Steam should be injected just upstream of where oil vaporization starts to suppress high film temperatures. There is not a rule on how much steam you need; that would be determined specifically at the same time that the injection location would be determined. Finally, the steam pressure is dictated by the pressure at the injection point. You just need to have a high enough steam source to get it into the heater while allowing for pressure drop across a control valve.

Q.23. How do I remove the odor from kerosene?

Ans. The odor in your kerosene is probably due to mercaptans being present in the kerosene. The mercaptans can be removed by scrubbing the kerosene with a 10-15 Baume aqueous solution of caustic (i.e., sodium hydroxide). A fixed-bed Merox sweetening unit could also be used.

Q.24. Which is more economical, in-situ or ex-situ catalyst regeneration for a naphtha reformer?

Ans. This depends on who you ask. Most people agree that ex-situ regeneration is only economical during a planned shutdown. Some people claim that ex-situ regeneration results in a better regeneration of the catalyst while others don't believe that this is the case. Most companies probably tend to use in-situ regeneration. It's easier to control the oxygen level and there is no additional cost that is usually associated with dumping the catalyst (not to mention the safety concerns that must be address during catalyst dumping

Q.25. Can molten sulfur pose a safety hazard in a sealed tank?

Ans. It depends on the type of tank in which the molten sulfur is stored. Molten sulfur is probably not stored in pressure vessels. It is more likely stored in cone roof, flat bottom tanks and such tanks can only withstand a vapor space pressure of 0.5 psig or less whether they are storing molten sulfur, or water, or fuel oil, or anything else. Sulfur per se does not generate explosive gases at normal conditions, I would like to make this clear. But the system (in reality) could emit this type of gases as molten sulfur (from refining industries) usually contains dissolved quantities of such a lot of gases (such as H_2S). Your liquid is subject to stabilization (gas evacuation) while in the tank, so you must provide for removing the generated free gases (H_2S?) as they are formed and pose a potential over-pressurization hazard. To design for this problem you can: a. Use a convenient and available vacuum system to vent the accumulated gases as they are released and route them through the vacuum system to a flare header or through your scrubber. If you are in a refinery area, you should have access to this type of convenient answer. b. If you don't have a vacuum flare header available, you can supply your own blower system and evacuate the generated gases on demand as the pressure increases in the storage tank. For this, you need a dead band for the instruments to be able to function reasonably. In other words, you should have a design tank pressure ample enough to activate the blower operation and be able to shut the blower off at a lower pressure. This lower pressure should be safe-guarded from a vacuum condition which you cannot tolerate. Normally, you must be prepared to break a potential vacuum with atmospheric air. If your gases cannot be brought into contact with atm. air, you are forced to use an inert gas (Nitrogen) to break the vacuum or maintain as an inert blanket. Either way, you are going to have to confront this need to identify the compatability of your gases (whatever they may be) with air. Otherwise, you cannot design any further. This will be the first and number one issue when you meet for your Process Hazard Analysis. If you apply Nitrogen, and evacuate at a rate to keep the vapor space purged continuously, the pressure could be a variety of values only subject to the tank's MAWP (Max. Allowable Working Pressure) which is set by you.

Q.26. What is a good method of cleaning a crude tower?

Ans. Here is a typical progression: 1. Steaming: The steaming is done after all liquid hydrocarbons have been drained from the column and associated piping. The objective of steaming is to make the column and associated piping free of residual hydrocarbons. The column vent and pump strainers in the side draw piping are de-blinded and steaming is started from utility connections at the bottom of the column. Generally, steaming is continued for about 20 to 24 hrs, ensuring the column top temperature remains more than 100 °C throughout the operation. 2. Hot Water Washing: When clear steam is observed exiting the column vents, water washing of the column should be started. With steam still in commission, water is sent to the column, usually via reflux lines, and it is drained from the column bottom, associated pump strainers, etc. The water flow rate should be adjusted so that steam still comes from the vent (i.e. water should not result in condensing of all steam before it reaches the column top). Water flow should be stopped for 2-3 hrs and then resumed. This cycle of steaming and washing should be repeated several times for a total of about 15 to 20 hours. Injection of a turpene-based detergent into the steam can also be considered. The condensate-soap solution can be collected and circulated through the various side cuts. 3. Blinding: When clear water is observed at side draw pump strainers, etc., associated piping should be isolated by installing blinds wherever isolation is possible. 4. Cold Water Washing: The hot water wash should be followed by a cold water wash (i.e. steam should be fully closed). The cold water washing is done for about 20-24 hrs. 5. Chemical Injection for Removal and Neutralization of PIS Deposits: During the cold-water wash or after washing is over, chemical injection for removal of pyrophoric sulfides should be considered. The various options for chemical treatment are discussed below: Acid cleaning - This procedure involves pumping in an acid with some corrosion inhibitor. The acid dissolves sulfide scale and releases hydrogen sulfide gas. It is effective and inexpensive, however, disposal of hydrogen sulfide gas can be a problem, as can corrosion (when the system contains more than one alloy). Dilute hydrochloric acid solutions may also be used. The resulting iron chloride turns bright yellow, acting as an indicator for removal of the iron sulfide. Acid plus hydrogen sulfide suppressant - Additional chemicals can be added to the acid solution to convert or scrub the hydrogen sulfide gas. Chelating solutions - Specially formulated, high pH, chelating solutions are quite effective in dissolving the sulfide deposits without emitting hydrogen sulfide, but this is an expensive application. Oxidizing chemicals - Oxidizing chemicals convert sulfide to oxide. Potassium permanganate ($KMnO_4$) has been used commonly in the past to oxidize pyrophoric sulfide. Generally the potassium permanganate is added to the tower during the cold water washing as a 1 per cent solution. At various intervals, samples are taken and checked for color. The colors of the fresh $KMnO_4$ and the spent MnO_2 are purple and brown respectively. If the color of the solution becomes brown, additional $KMnO_4$ is needed. The reaction is judged complete when the solution color remains purple. It takes approximately 12 hours to complete the job. In recent times, the use of potassium permanganate as an oxidizer has raised serious safety and health concerns

Q.27. What technologies are commonly employed to purify hydrogen from a reforming unit?

Ans. There are three (3) commonly used technologies to purify hydrogen from reformers. The determination of which one to use depends on a number of factors. If the hydrogen must be 99 per cent by volume pure, then Pressure Swing Adsorption (PSA) is the most common choice. The other two technologies are rarely capable of this purity. Instead, they're generally able to reach 95-98 per cent by volume purity. If the purity required is in the range of 95-98 per cent by volume and the feed gas is at a pressure of 2-5 times that required for the product hydrogen stream, then membrane separation can be employed. Obviously the pressure loss in such a system is substantial. If the feed gas stream contains substantial byproducts that are to be recovered and a hydrogen purity of 95-98 per cent by volume is acceptable, then a cryogenic unit is a good choice. Byproducts can be recovered from PSA unit off-gases but the low pressure at which this stream is available often make byproduct recover uneconomical. Another consideration is the capacity of the product hydrogen stream. PSA units are built for 5-100 MMscfd of hydrogen. PSA units and cryogenic units have excellent economies of scale for large capacities. Membrane units are generally a better option for smaller capacities

Q.28. What are some common causes for pressure build up in the vapor space of a LPG storage vessel?

Ans. The following are four (4) common causes for pressure build up in such a tank: a. The tank undergoes a heat up or temperature change b.

Non-condensibles are introduced (Ex/ N2 for inerting) c. Liquid product is pumped into the tank faster the the vapor can achieve equilibrium with its liquid content. d. A significantly different composition of LPG is introduced into the tank (a lighter component) Vapor balance lines are often employed on such tanks so that the pumping rate to the tank is NOT controlled by equilibrium. Remember that if 100 per cent LPG gas is being stored, the pressure in the vapor space will be the vapor pressure of the liquid and this pressure will vary only the tank temperature and the composition of any other incoming liquid

Q.29. What usually happens to benzene that is removed from gasoline during the refining process?

Ans. Refiners typically have several options. Three (3) of the more popular options include: A. Use an extraction unit to purify the benzene in a benzene-rich stream and sell as petrochemical-grade benzene. B. If a propylene feedstock is available, alkylation of the benzene to produce cumene is an attractive option. The cumene is then used to produce phenol and acetone. Note: An earlier use of cumene was as an aviation fuel because of its high octane value. C. A third option is to use an isomerization unit to saturate the benzene with hydrogen to produce cyclohexane. The UOP Bensat process is one such example.

Q.30. What is the significance of lubricity in diesel fuel?

Ans. The lubricating properties of diesel fuel are important for rotary and distributor type fuel injection pumps where moving parts are lubricated by the fuel itself as it moves through the pump (not by the engine oil). The fuel lubricates the walls between the reciprocating pieces (plunger or needle) and its container. Lubricity is an indication of the amount of wear that two pieces of metal will experience as they contact one another while covered with the fuel. Lubricity is sometimes mistakenly connected to the viscosity of a fuel. In fact, there are polar compounds in the fuel which prevent wear on contacting metal surfaces. Since 1993, specification changes and changes to how diesel fuel is manufactured have cause a decline in the lubricity of the available fuel. The Fuel Injection Equipment (FIE) manufacturers have adopted the use of a test known as the High Frequency Reciprocating Rig (HFRR, ISO 12156-2:1998) for benchmarking diesel lubricity. They recommend that all diesel fuels meet a limit of 460 microns maximum Wear Scar Diameter (WSD)

Q.31. How can the higher heating values (HHV) for petroleum products be estimated from the specific gravity?

Ans. The following three general expression are used for estimating the higher heating values for petroleum products. These expressions are for the heating values in Btu/lb. Multiple the result by 2.3 to get KJ/kg.

For fuel oil: HHV = 13050 + (5600/SG)
For gasoline: HHV = 12720 + (5600/SG)
For kerosene: HHV = 12840 + (5600/SG)

Q.32. What is crude oil?

Ans. The naturally occurring product that is extracted from a reservoir through an oil well is called Crude Oil. This is often black, but can be a wide range of colors, including almost clear or yellow, depending upon the mixture of hydrocarbons in it.

Crude Oil is a complex mixture of organic chemicals, mainly hydrocarbons. Over 500 compounds have been found in crude oil.

Average crude oil is made mostly of the following elements:

Carbon	85 - 90 per cent
Hydrogen	10 - 15 per cent
Oxygen	0 - 1.5 per cent
Sulfur	0.1 - 3.0 per cent
Nitrogen	0.1 – 0.5 per cent

Some crude oils have much higher levels of sulfur, nitrogen and other elements. These crude oils need more complex refining processes to safely produce petroleum products such as petrol for cars (gasoline).

Some components are very light in weight and will evaporate at low temperatures. Others are very heavy, like bitumen (tar).

The density of crude oil varies, depending upon the amount of heavy and light components. Typical values, in Kg per cubic meter are:

Mexican crude oil	973
Texas crude oil	873
Gasoline	737
Kerosene	817
Diesel	840
Automobile oil	880 - 940
Pure water (at 4°C)	1000

People have used crude oil for thousands of years. The Babylonians called it "Naphtha".

Q.33. Can the cuttings from drilling be washed or otherwise disposed of, particularly when using oil based mud?

Ans. It all depends on local environmental regulations, which vary from country to country and have generally become stricter over time. It also depends on the type of oil, particularly how toxic it is and how biodegradable. Diesel is very different from refined mineral oil, which itself is very different from fully synthetic fluid.

As an example, the maximum amount of oil on cuttings that can legally be discharged during drilling in the North Sea is about 1 per cent by mass. Cuttings with higher concentrations of oil are treated, either onshore, or increasingly offshore, to remove excess hydrocarbons. In some cases, some cuttings can be re-injected into the well

Q.34. Why is oil found in anticline traps?

Ans. Oil in porous rock layers (such as sandstone) is less dense than the water and solid material. Therefore, it migrates upward. The roof of an anticline trap is made up of nonporous material (such as clay), so the oil can rise no further.

This would be similar to an upside down bucket trapping bubbles of air as they rise in a pond of water

Q.35. What are cuttings in drilling term?

Ans. The cuttings are the pieces of rock that are cut from the formation during drilling.

They are removed from the borehole by the circulation of the drilling fluid.

Q.36.What is the origin of natural oil and gas?

Ans. Most scientists agree that hydrocarbons (oil and gas) are biogenic - formed from the remains of plants and animals that lived hundreds of millions of years ago. This theory is supported by the fact that many underground oil deposits contain chemicals that are only known in plants that live on the Earths surface.

Organisms, mostly algae and tiny animals called plankton, died and sank to the bottom of the seas in which they lived. They were covered with mud and sand then, over millions of years, buried deep under other sediments. Temperature in the Earth's crust increases with depth. This has warmed the plant and animal remains, causing them to produce oil and gas.

Oil forms at temperatures between about 50°C (120°F) and 175°C (350°F). At higher temperatures, gas is formed and any oil that has already been produced starts to turn into lighter oils and eventually into Methane gas, the lightest and simplest hydrocarbon. At temperatures above about 260°C (500°F), plant and animal remains turn completely to carbon and no more oil or gas are produced.

Q.37. What is the economic importance of oil?

Ans. Oil and gas are burned in huge volumes all over the world. Countries that have no oil or gas must pay to import it. In 2000, the USA paid about US$100 billion to import oil while Saudi Arabia gained about US$80 billion in oil exports.

Most governments use a license system to encourage private companies to explore for oil and gas in their countries. Land and sea areas that could contain oil are divided into "Blocks". These are roughly square areas typically between 5 and 20 kilometers (3—12 miles) wide. Companies offer money to the government for the right to lease blocks for a period (e.g. 10 years), during which they are obliged to explore for oil using seismic imaging and drilling.

In some cases, oil companies must agree to a "production sharing contract", in which case the government will get a per centage of the value of any oil produced. In all cases, the government will benefit by taxes raised from oil companies. In addition, oil and gas production generates jobs and reduces the need for imports.

Q.38. What is the measuring unit for oil?

Ans. Oil is measured in "Barrels", shortened to "bbl".

One oil barrel is:

- 42 US Gallons
- 34.97 Imperial Gallons
- 159 Litres
- 0.159 cubic meters
- 0.136 metric tonnes weight (approximate—depends of the density of the oil)

The energy in one barrel of oil depends on the type of oil. It is approximately equivalent to:

- 5,659 cubic feet of dry natural gas
- 0.26 tonnes of coal
- 1,700 Kwh of electricity—sufficient to burn a 100 watt bulb for 17,000 hours

Gas is often measured in "Barrels of oil equivalent" shortened to "boe" This is a measure of energy approximately equal to 6.12 gigajoules (GJ), 5.8 million British Thermal Units (Btu). and is equivalent to 1,700 Kwh of electricity (the same as a barrel of oil).

Boe is often used to measure gas reserves. The conversion factor between volumes of oil and volumes of gas varies. 1 barrel of oil equivalent (boe) is approximately equal to:

- 160 cubic meters of gas
- 5659 cubic feet gas

1000 cubic metres of gas is approximately equal to 1 cubic meter of oil equivalent.

The conversion factor from volumes of gas to boe depends upon the type of gas. Many countries and oil companies report their underground gas reserves in boe based on a conversion factor of six thousand cubic feet of gas per barrel of oil equivalent

Q.39. How is remote sensing used for oil exploration?

Ans. Many types of remote sensing are used to identify areas of the Earth that are mostly likely to contain oil and gas.

Many oil and gas reservoirs naturally leak tiny amounts of hydrocarbons at the Earths surface. These are mostly light gases such as Methane and Ethane. Common biological processes can make Methane, but Ethane is only made in underground oil and gas reservoirs. A very sensitive infrared laser sensor has been developed that can detect tiny amounts of Ethane. This has been used to locate gas leaking from underground, which can indicate commercial oil and/or gas reservoirs.

Most oil or gas is in sedimentary rocks. These are less dense and contain less magnetic minerals than igneous rocks. Unpredictable variations (anomalies) in gravity and magnetism measured from boats or airplanes can indicate the depth of sedimentary rock layers that might be worth further investigation.

Seismic imaging from the Earth's surface provides detailed pictures of underground rock structures and can sometimes also indicate whether rocks are likely to contain oil or gas. Seismic surveys are usually performed before drilling into possible oil or gas reserves

Q.40. How is seismic used to find oil and gas underground?

Ans. The reflection seismic method is used at the Earths surface to image rock structures and estimate rock types that might form underground oil and gas reservoirs.

The reflection seismic method is based on measuring the time that it takes for sound waves to travel from a seismic source (for example a man-made explosion) through the earth to a rock layer that reflects the sound back to a recording device.

It is similar, in principle, to medical imaging methods that image the inside of a human body, such as NMR and Ultrasound scanning.

Seismic data does not directly identify the age of or type of rock. For this, it is necessary to match the seismic reflections with data from drilled wells and knowledge of rocks that can be seen at the surface. The process of labeling seismic reflections with specific rocks is "Interpretation". The process was once performed using colored pencils to draw lines on paper plots of the seismic. Today it is usually done on a computer screen or 3D visualization system.

Q.41. What is an oil well?

Ans. An oil well is a hole that is drilled deep into the ground to reach rocks containing oil or gas that can be taken out.

Seismic surveys provide images of rock layers many kilometers underground. Geologists analyze these images to find areas that might contain oil or gas. A rig is used to drill boreholes into the locations underground that are most likely to contain oil, thus making an oil well. Different types of rig are used on land, in shallow water and in deep water.

Boreholes are drilled many kilometers underground through solid rock. A drill bit is the tool used to cut through rock. Most bits work by scraping or crushing the rock, or both, usually with rotational motion. The drill bit is at the end of the drill stem. The drill stem is made from lengths of steel pipe screwed together. Pressure is applied from the rig to the top of the drill stem to push it into the ground. When the top of the drill stem gets near to the rig floor, another length of drill pipe is added at the top. The process continues until the drill bit reaches the rock layers thought to contain oil.

Mud is used to cool and lubricate the drill bit and remove rock chips. It is mostly a mixture of naturally occurring minerals and water or oil. The mud must be disposed-of without damaging the environment. This can involve special cleaning processes and burial of waste onshore. Water-based mud is used whenever possible, because it can be disposed-of in the sea without damaging marine life.

Electronic tools between the drill bit and the drill stem analyze the rocks being drilled. Other tools behind the drill bit can be used to gradually change drilling direction. There are boreholes over 10 kilometers (6miles) deep and some have been steered over 10 kilometers (6 miles) sideways, deep underground.

The borehole is lined with lengths of steel tube that join together to make a continuous empty hollow tube. Special cement is pumped to the bottom of the tube. The cement is forced under pressure up the outside of the steel tube to fill any gaps between the tube and the rock. Holes are made through the steel tube at the depth where rocks are expected to contain oil and gas. This is called perforating.

Valves and electronic monitoring devices are put deep in the well. Information about the amount of oil, gas and water flowing up the well can be sent electronically via satellite to experts in an office far away. Using this information, valves in the oil wells can be opened and closed by remote control, so as to maximize oil and gas production over the whole oilfield

Q.42. How do we know what rocks are in a borehole?

Ans. While drilling a borehole, small pieces of rock come out at the surface, where geologists study them. Electronic devices make measurements both during and after drilling. These measurements identify the depth of rock layers and which, if any, contain oil or gas.

The popular image of oil squirting uncontrolled from a newly drilled borehole is not accurate. Such as "blow-out" would damage the environment and be very dangerous for people on or near a drill rig. Heavy mud prevents a blowout while drilling and, after drilling is complete, valves and pipes are put in place before oil is allowed to flow.

We cannot see what the rocks look like inside a deep borehole. The drilling mud brings rock chips to the surface. These are analyzed to determine their age and whether they contain oil or gas, but it is not possible to tell exactly where in the borehole they came from.

Some measurements of the physical properties of rocks deep underground can be made in real-time while an oil well is being drilled. Tools housed at the end of the drill stem, just a few meters behind the drill bit, measure the electrical resistivity of fluids within the rock, reaction of the rock to gamma ray bombardment, production of gamma rays from fluids in the rock due to neutron bombardment; and the natural gamma radiation of the rocks. These measurements while drilling (MWD) can indicate rock type, porosity and the presence of oil or gas.

More accurate measurements, including samples of fluids in the rocks, can be made after drilling by lowering tools into a borehole on a long steel wire called a wireline.

Q.43. Who discovered oil, and when?

Ans. Natural oil seeps have been present since before the days of dinosaurs about 200 million years ago. Lighter oils evaporate in air leaving behind the heavier oils in "tar pits". People have used this naturally occurring oil since the beginning of recorded human history. Oil can also be made from animal fat, and it is not always clear whether mention of oil in ancient records refers to oil from the ground or from animals.

Ancient Greek texts describe how they would pour oil onto the sea to set fire to their enemies fleets. The Bible refers to a thick form of oil called "Pitch" which was used to waterproof Noah's ark and the baby Moses' basket. The American Indians also used pitch to waterproof canoes and to make warpaint and medicines. These examples are probably uses of oil from the ground.

The word "Petroleum" come from the Greek word for rock and the Latin for oil or fat. It literally means "oil that comes from rock".

petra / petros (Greek) = rock
oleum (Latin) = oil / fat

Crude oil was pumped from the ground in Sichuan, China, 2500 years ago, but the history of oil wells as we know them today is much younger. A brief history starting in 347 A.D:

347: Oil wells are drilled in China up to 800 feet deep using bits attached to bamboo poles.
1264: Mining of natural oil seeps in medieval Persia is witnessed by Marco Polo on his travels through Baku.
1500's: Seep oil collected in the Carpathian Mountains of Poland is used to light street lamps.
1594: Oil wells are hand dug at Baku, Persia up to 35 meters (115 feet) deep.
1735: Oil sands are mined and the oil extracted at Pechelbronn field in Alsace, France.
1815: Oil is produced in United States as an undesirable by-product from brine wells in Pennsylvania.
1848: First modern oil well is drilled in Asia, on the Aspheron Peninsula north-east of Baku, by Russian engineer F.N. Semyenov.
1854: First oil wells in Europe are drilled 30- to 50-meters deep at Bóbrka, Poland by Ignacy Lukasiewicz.

Oil historians in the USA give credit for the first modern commercial oil well to Colonel Edwin L. Drake. His well reached a depth of 22m (72-ft). It was drilled in "Oil Creek" near the town of Titusville, slightly east of Pittsburgh, Pennsylvania, USA and started producing oil on August 28,1859. There were no automobiles in those days; the main market for petroleum was for medicine. It was called Rock Oil and sold for about $40 a barrel, which is about the same as a barrel of oil costs today, so it would have been worth a lot of money in 1859.

There are several other claims for "the first oil well", including a well drilled in 1858 in Wietze, Germany. This area, locally nicknamed "Little Texas" now houses a petroleum museum (the Deutsches Erdölmuseum).

The first offshore oil well was in the bayous (swamps) of Louisiana, USA during the 1950s. The first drilling in open sea was done in 1955 for Shell Oil in the Gulf of Mexico just south of New Orleans. This used a barge with a drill rig attached named "Mr. Charlie", which continued to drill in the Gulf of Mexico for 32 years.

Q.44. Why is Petrophysics important in the study of oil fields?

Ans. Petrophysics is the study of the physical and chemical properties that describe the occurrence and behavior of rocks, soils and fluids. To accurately characterize an oil or gas reservoir, measurements such as resistivity, neutron and density are made, from which effective porosity, saturations, and permeability can be quantified

Porosity is the porous volume by unit of volume of a formation. The porosity of a formation can vary considerably. The dense carbonates (limestone) and evaporites can have zero porosity for all practical effects. On the other hand, consolidated sandstones can have from 10 per cent to 15 per cent of porosity, whereas non - consolidated sandstones can have 30 per cent or more porosity. Finally, shales and clays can have a high porosity of 40 per cent, but the individual pores are generally so small that the rock is impermeable to fluid flow. Effective porosity is the important measurement - the interconnected pore volume in a rock that contributes to fluid flow in a reservoir.

Water saturation can help to determine the probability of there being hydrocarbons in a rock formation and the existing volume of hydrocarbons in a determined reservoir, that is, calculation of reserves.

Permeability is a measurement of a rock's ability to transmit fluids. To be permeable, the rock must have interconnected pores or fractures, therefore, there is a general relation between porosity and permeability.

To study an oil field, you need to know these and other important properties of the reservoir... and for this we have Petrophysics

Q.45. What alternative theories are there for the origin of natural oil and gas?

Ans. Most scientists agree that hydrocarbons (oil and gas) are biogenic —formed from the remains of plants and animals that lived hundreds of millions of years ago.

Some other people think that hydrocarbons are non-biogenic.

For example, Tom Gold (born 1920, died 2004) of Cornell University, USA, suggested that oil and gas, like water, have existed since the formation of the earth. This is based on the discovery of hydrocarbons on other planets and in some volcanic rocks on earth. He believes that some hydrocarbons on earth have been contaminated by bacteria living deep underground; a combination which, in his opinion, has incorrectly led scientists to assume that the oil and gas come from organic matter. Tom Gold also believed that there is hundreds of times more oil underground than has been predicted by oil producing companies and organizations such as OPEC.

Few scientists believe with the non-biogenic theory. Most think that the scientific evidence strongly supports the biogenic theory; that oil and gas come from ancient plants and animals. Most scientists also agree that oil and gas reserves are running out fast.

Q.46. What happens to the space underground after oil and gas have been removed?

Ans. Oil and gas come from small spaces inside solid rock, so there is no big empty space left underground after oil and gas have been removed.

Oil and gas in underground reservoirs is mostly inside small spaces (pores) between grains of sand in sedimentary rocks. These sand grains are packed tightly together and are usually stuck together by minerals such as calcium carbonate, making solid rock. The pores of most sedimentary rocks are filled with salt water. Only very small areas of sedimentary rocks have pores filled with oil and gas, and these areas are nearly always usually above and near much larger areas filled with water. When oil and gas are removed, water flows in to fill the space.

Water is less compressible than oil and gas, so less is needed to fill the space. The only likely effect of oil and gas production is a reduction in pressure. This can be small or large, temporary or permanent. Change in pressure can lead to tiny movements in the rocks, causing microseismic events (small noises). These are sometimes analyzed to see where pressure is changing inside a reservoir, which can help to ahow from where in the underground oil and gas are flowing.

Hundreds of cubic kilometers of oil and gas have been removed from underground reservoirs. Although large, this is very small compared to the Earth's volume, even when only the crust is considered. This is one reason why finding oil is difficult.

In a very few cases, a drop in underground pressure due to oil extraction has caused the Earth's surface to sink measurably. An example is Ekofisk in the Norwegian North Sea, which has sunk 8 meters since oil production started in 1971.

Q.47. How can we get more oil from a reservoir?

Ans. There are several ways to get more oil out of a reservoir after it has stopped coming out under its own pressure.

Underground oil reservoirs are usually under pressure due to the weight of the rocks above them. After an oil well has been drilled and completed, oil usually flows out naturally under its own pressure. As oil is taken out, pressure will decrease and the flow of oil will slow down and eventually stop. Pumps deep inside or at the top of the well can help to get more oil out.

Another way to get more oil from a reservoir is to drill boreholes (injection wells) around the edge of the oil producing area. Water is pumped into these holes to increase pressure underground, forcing out more oil. The water pumped into injections wells is often the same water that is produced with the oil. Putting it back in the ground is an environmentally friendly way to manage oilfield water, which is often too full of salts and minerals to discharge safely into the sea.

If the oil is very thick, or needs more help to flow through the reservoir rock, there are several methods of "Enhanced Oil Recovery" (EOR) including:

1. Heating the oil, usually by pumping steam into the injection wells (steam injection). This increases pressure and makes the oil less viscous, the same way that butter melts in a hot pan. The increased pressure at the injection wells pushes oil towards the producing wells.
2. Adding a solvent such as liquefied petroleum gas (LPG), nitrogen, alcohol or carbon dioxide (CO_2). Carbon dioxide dissolves easily in crude oil and makes it swell, thus increasing pressure, and thinner, so more able to flow. Pumping CO_2 into the ground has the extra benefit of removing greenhouse gas from the atmosphere.
3. Adding surfactants, which act like detergent. In the same way that a little soap helps oil to be washed from your hands, low concentrations of surfactants in water can help to wash oil from the reservoir.
4. Some oil reservoirs already contain a lot of bacteria. In other cases bacteria can be injected into a reservoir for Enhanced Oil Recovery. Nutrient, such as molasses (from sugar cane), ammonium nitrate, nitrogen, oxygen and phosphorus are put into the reservoir. These feed the bacteria, which grow, producing gases that increase pressure. The bacteria also create chemicals that make crude oil thinner.

Q.48. What products are made from oil?

Ans. Crude oil is mostly a mixture of many chemical compounds of carbon and hydrogen (hydrocarbons). The different hydrocarbons are separated in an oil refinery. Hydrocarbons are a major source of energy. They are burned for domestic and industrial heating. They fuel most land vehicles, ship and airplanes. Hydrocarbons have many other uses, and are a raw material for many products including:

- ➢ Rubber, Plastic, Paint, Synthetic cloth, such as nylon,
- ➢ Bitumen (asphalt or tar) for road surfaces and building,
- ➢ Lubricating oil for engines and other machines,
- ➢ Drugs and creams (e.g. Vaseline).

Q.49. What alternatives to oil and gas are there as sources of energy?

Ans. There are many alternative sources of energy, but few are as easy, versatile reliable and relatively cheap as oil and gas.

Approximate costs of electricity in US Cents per kilowatt hour are:

Natural gas	—	5
Coal	—	5
Nuclear	—	4
Onshore wind farm	—	6
Offshore wind farm	—	10
Wave and tide	—	12
Chicken waste	—	12

Fossil fuels (oil, gas and coal) took millions of years to create and are not renewable. We are likely to use most of the oil and gas within the hundred years from 1930 to 2030. It is currently quite cheap, but the cost will soon rise. Fossil fuels also generate carbon dioxide, which is considered to be the main cause of global warming. There are several projects around the world in which carbon dioxide is being pumped into underground oil reservoirs. In the future, this might enable us to produce clean energy (hydrogen) from coal without releasing carbon dioxide into the atmosphere.

Nuclear fission is a widely used source of energy, but it creates large volumes of waste that will remain very toxic for thousands of years. Experiments are being done to try to get energy from nuclear fusion. This is the process that takes place in the sun and stars, where at temperatures of 10-15 million degrees Celsius, Hydrogen is converted to Helium, a safe by-product.

Some countries with mountains and high rainfall get a lot of energy from hydroelectric power. Others are using crops such as sugar cane to make ethanol for vehicle fuel.

Oil and gas power stations can start and stop on demand. A major challenge with sun, wind and wave power is that they are not predictable and cannot provide energy exactly where and when it is needed. It is likely that in the future we will use a variety of renewable energy sources combined with oil and gas power to fill in the gaps between supply and demand.

Q.50. What is seismic acquisition?

Ans. Seismic acquisition is the artificial generation and recording of seismic data.

A seismic source, such as a dynamite explosion, compressed air gun or vibrator unit, generates energy that travels into the Earth as vibrations passing through underground rock layers. Different types of rock filter the seismic waves and some energy returns to the surface due to refraction or reflections from at surfaces between different rock layers.

The returning seismic energy is measured by receivers, which record the seismic signals as electronic waveforms. Geophones are one type of receiver, used on land, on the sea floor or inside a wellbore (as in a vertical seismic profile) to record the seismic signal. Geophones use a moving magnet or coil, as used in some microphones, to measure small vibrations. Hydrophones are another type of receiver, used to measure changes in pressure as sound travels through water. Hydrophones are usually towed a few meters underwater inside "streamers" that are typically 3 – 6 km (abut 2 – 4 miles) long. Hydrophones can also be suspended vertically or laid on the sea floor.

Different types and configurations of receivers are used depending on local environmental conditions and the underground geological features that are to be imaged. The layout of the receivers is also designed to minimize the effect of noise that can otherwise mask the seismic signal. So long as appropriate receiver configurations are used, special computer processing techniques can be applied to the recorded data to remove noise and enhance the seismic signal.

Q.51. What can seismic velocities be used for?

Ans. Seismic velocity is the speed at which seismic waves travel through a medium. P-waves travel at the "speed of sound", which varies dependent mainly upon the stiffness and density of the material, and is also affected by temperature and pressure. Typical values are:

- Air 330 m/s (740 mph)
- Sea water 1,500 m/s (3,360 mph)
- Sandstone 2,500 m/s (5,600 mph)
- Granite 5,500 m/s (12,300 mph)

Seismic velocities are analyzed and used during computer processing of seismic data to correct for moveout, which results from the varying distances between seismic sources and receivers.

High-quality seismic velocities can be used to predict underground rock stresses and pore pressure, which can reduce the cost and risks encountered during drilling. These velocities are obtained using seismic tomography, which is like a medical X-ray that is shone through the body to build up a picture of its internal structure. Similarly, seismic reflection tomography allows us to make maps of the Earth's interior to visualize the velocity field, which in turn reveals variations and distribution of pore pressure.

Measuring changes in seismic velocity over time can indicate the movement of reservoir fluids due to production. This is of particular interest to a reservoir engineer, who only has production data at the wells. The availability of seismic data between wells can enhance the reservoir engineer's understanding of the whole reservoir.

Q.52. Is crude oil a mineral?

Ans. A rock is any natural deposit forming a part of the earth's crust, whether consolidated or not when in natural beds, so soft layers of sand, earth and clay are rocks. Rocks are usually solid but can be liquid or gas.

A mineral is a naturally occurring, homogeneous inorganic solid substance having a definite chemical composition and characteristic crystalline structure, color, and hardness.

Naturally occurring oil (Crude Oil) found underground is a mixture of substances with a very variable chemical composition, which makes it a rock, not a mineral.

Q.53. What sources of errors are there in Petrophysics?

Ans. In the search for oil and gas, Petrophysics is used to estimate properties of rocks (e.g. fluid content, porosity, pressure, permeability) from measurements of their physical properties (e.g. resistivity, natural radiation and sonics).

There are many sources of errors, including tolerance of the measuring devices, quality of measurements (e.g. due to variable borehole conditions) and the reliability of models (equations, algorithms and other mathematical processes) that are used to predict rock properties from the measurements.

Measurements of actual production from a borehole (oil, gas & water) can be compared with predictions from Petrophysics. These measurements, plus those from laboratory experiments, are used to continuously improve oilfield technology and modeling algorithms, leading to better management of errors and improved reliability of predictions.

Q.54. What does "well completion" mean?

Ans. "Well Completion" includes the activities performed in a well after it has been drilled to prepare it for safely producing oil or gas.

After drilling through rock to where oil or gas is expected, a well is lined with lengths of steel tube (casing) that join together to make a continuous empty hollow tube. Cement is pumped to the bottom of the casing and forced under pressure up the outside of the steel tube to fill the gap between the tube and the rock.

In many wells, holes are made through the casing at the depth where oil and gas are expected. This is done with a perforating gun, which contains explosive charges and pointed bullets that blast through the steel casing to make holes through which oil can flow.

For wells in soft sand, steel screens are used in place of solid casing. These screens act like a sieve, letting oil and gas through while stopping the sand from collapsing into the well.

Some wells do not need any special filtering systems. These "openhole" (or "barefoot") completions are performed when the geology allows that oil and gas can flow safely and efficiently from the bare rock below the cased part of the hole.

A device called a "Christmas tree" is fitted to the top of wells. This contains valves that control the flow of oil and gas and can quickly stop the flow in the event of an emergency. Some wells also have measuring devices and valves deep inside the well that can be remotely controlled to adjust the flow from one or more rock layers so as to maximize oil and gas production over a whole oilfield. This is called an "intelligent completion".

Q.55. How much CO_2 is produced by burning 1 gallon of petrol?

Ans. The exact value will vary depending upon the type of petrol and the per centage of carbon in the petrol that is converted to CO_2.

The approximate answer is:

- 10.7 Kg per imperial gallon—about 5.4 m^3 - roughly equal to the volume of the passenger compartment of a medium sized car.
- 8.9 Kg per US gallon—about 4.5 m^3 (159 ft^3)
- 2.4 Kg per liter - 1.2 m^3 (1,200 liters)

We can consider the calculation at a molecular level: Carbon has a molecular weight of 12. Petrol is a mostly a mixture of C_5 - C_{10} hydrocarbons. If we consider C_8 hydrocarbon as the average: C_8 is 8 x C and 18 x H giving a molecular weight of (8*12) + 18 = 114 with carbon 84.2 per cent of the mass. Petrol has a density of about 0.75 gm/cc so one imperial gallon (4.55 liters) weighs 3.4 Kg of which 2.9 Kg is carbon. Oxygen has a molecular weight of 16, so CO_2 has a molecular weight of 44. The imperial gallon of petrol will therefore produce 2.9 * 44/12 = 10.6 Kg CO_2.

Q.56. What is an oilfield?

Ans. An oilfield is an area of rock under the ground that is full of oil or gas or both. Oilfields are typically 10-20 km (6-12 miles) wide.

An oilfield needs a special combination of geological features:

Oil is created in a source rock, along with water and gas. Over millions of years, the oil and gas float upwards above the water along a migration path. Oil and gas often rises all the way to the surface of the earth. In other cases, it collects in a reservoir; a rock that has spaces where oil can collect. Sandstone is a good example, in which oil collects in the spaces (pores) between the sand grains. There must also be a trap - a rock structure that forms a seal above and around the reservoir rock. A seal is impermeable - that is, fluids such as oil cannot flow through it. Clay and shale are impermeable and make good seals.

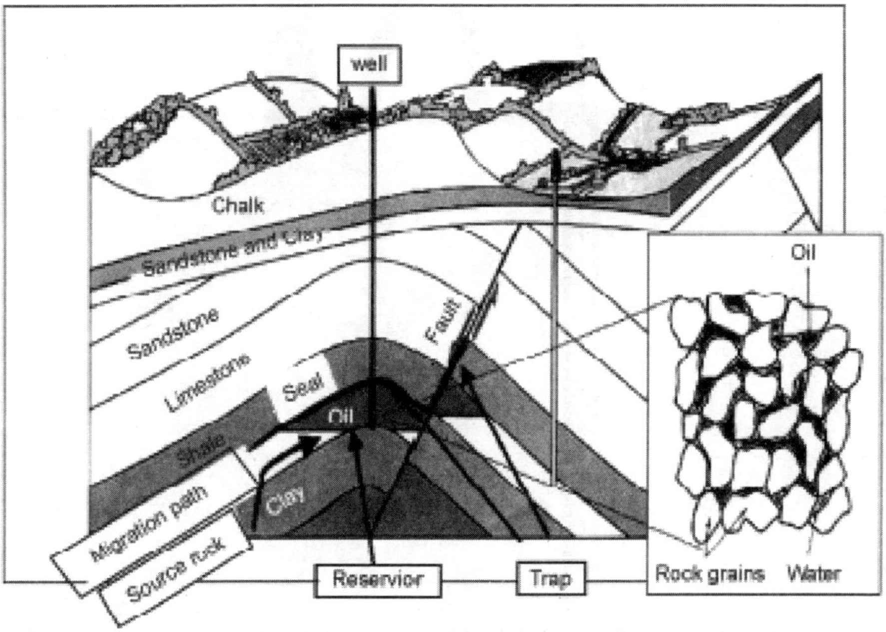

Fig 9.1: Cross section through an oilfield

Some oil is in structural traps; for example where rocks have been folded into a dome shape, in which oil will collect at the top of the dome. Other structural traps are sealed by faults, where tilted blocks of rock have slid up or down and become sealed by layers of clay.

Some oil is in stratigraphic traps, where areas of sand have been laid down within areas of clay - for example in ancient river channels.

Q.57. How does a nodding donkey lift oil from underground?

Ans. A Nodding Donkey (also known as a "horsehead pump" or "pump jack") provides the up/down motion (stroke) that powers a submersible pump in a borehole. The submersible pumps oil in a similar way to which a bicycle pump pumps air into a tyre.

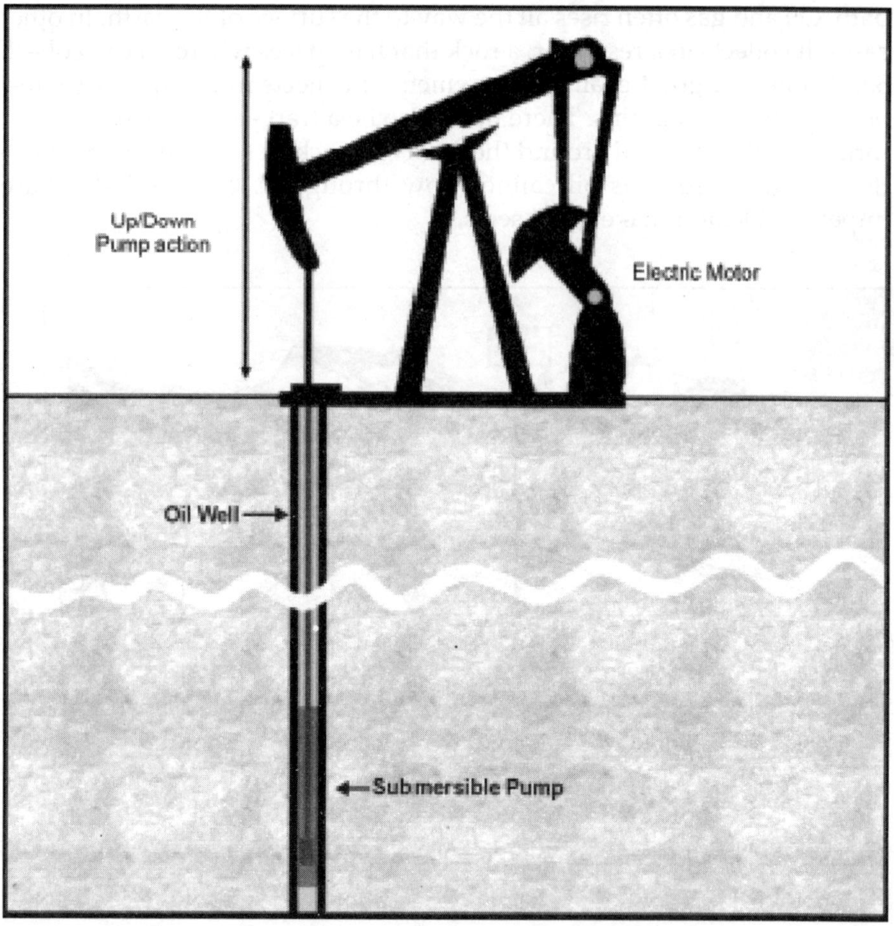

A motor, usually electric, turns a pair of cranks, which by their action raise and lower one end of a metal beam. The other end of the beam, which often has a curved end, is attached to a metal rod that moves up and down. The rod, which can be thousands of feet long, attaches to a pump deep in an oil well.

Q.58. Where in the world is oil and gas found?

Ans. Oil and natural gas can be found in many areas of the world but there are only a few areas where it can be produced for an economical cost. There are few oil reserves in Africa, South America and Asia. Europe and North America have already used most of their oil reserves. The Middle East has over 63 per cent of the worlds proven remaining oil reserves, with 23 per cent in Saudi Arabia. North America has also used most of its natural gas supplies. There are still large reserves in Russia and the Middle East.

Remaining Oil Reserves (Thousand million barrels)

Saudi Arabia	263	23%
Iran	131	11%
Iraq	115	10%
United Arab Emirates	98	9%
Kuwait	97	8%
Venezuela	78	7%
Russian Federation	69	6%
Libya	36	3%
Nigeria	34	3%
USA	31	3%
Rest of the world	197	17%
TOTAL WORLD	1148	—

Remaining Gas Reserves (Trillion cubic metres)

Russian Federation	47	27%
Iran	27	15%
Qatar	26	15%
Saudi Arabia	7	4%
United Arab Emirates	6	3%
USA	5	3%
Nigeria	5	3%
Algeria	5	3%
Venezuela	4	2%
Canada	2	1%
Rest of the world	43	24%
TOTAL WORLD	176	—

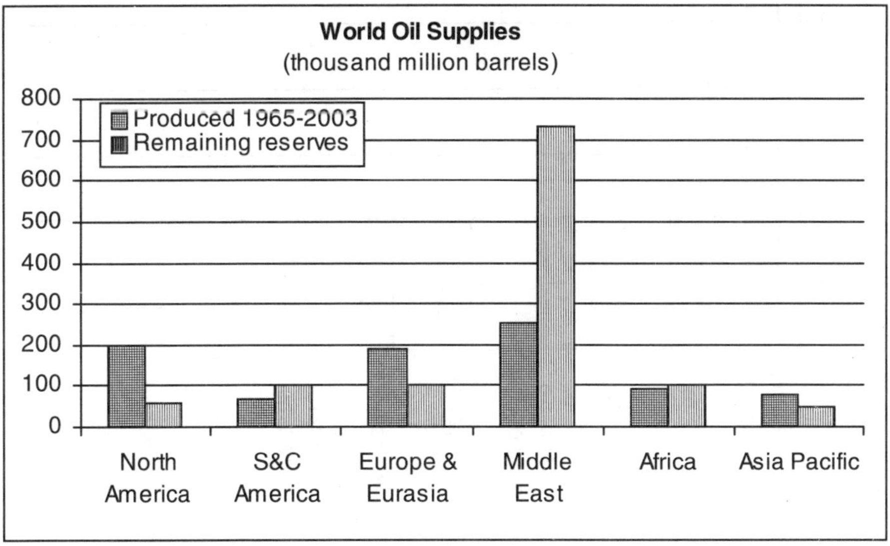

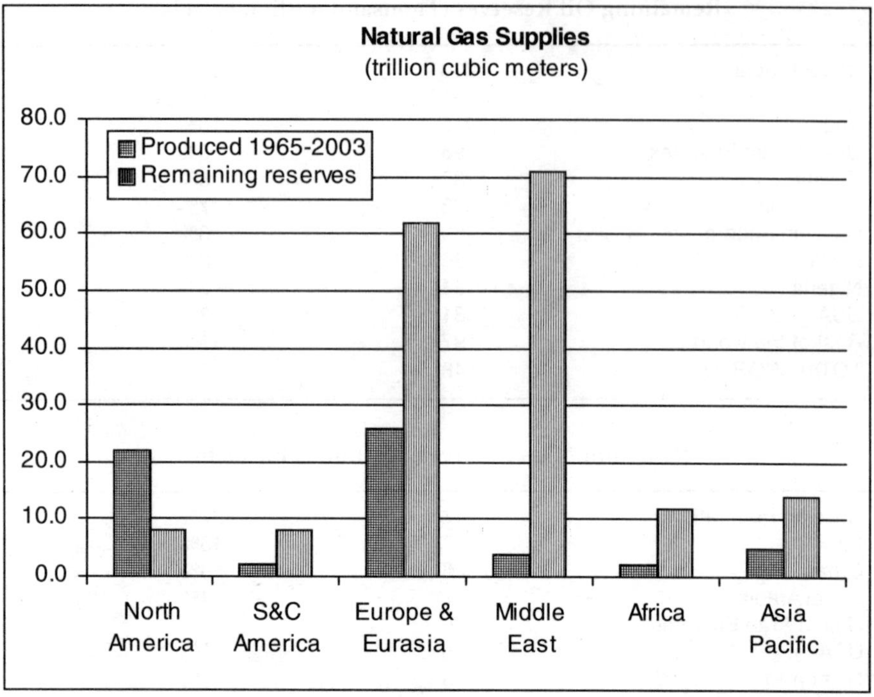

Natural Gas Supplies
(trillion cubic meters)
80.0
70.0
60.0
50.0
40.0
30.0
20.0
10.0
0.0
Produced 1965-2003
Remaining reserves
North America
S&C America
Europe & Eurasia
Middle East
Africa
Asia Pacific

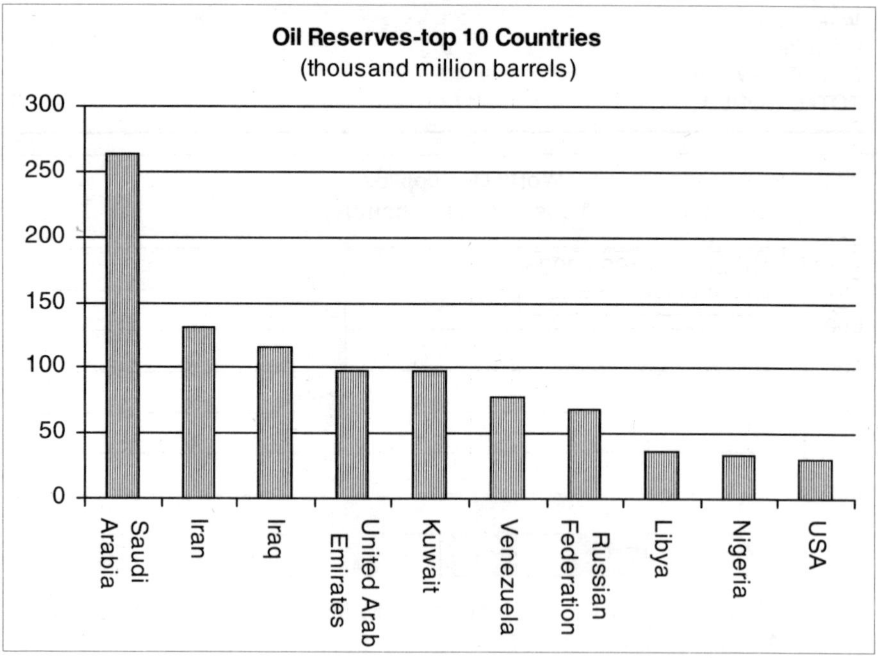

Oil Reserves-top 10 Countries
(thousand million barrels)
300
250
200
150
100
50
0
Saudi Arabia
Iran
Iraq
United Arab Emirates
Kuwait
Venezuela
Russian Federation
Libya
Nigeria
USA

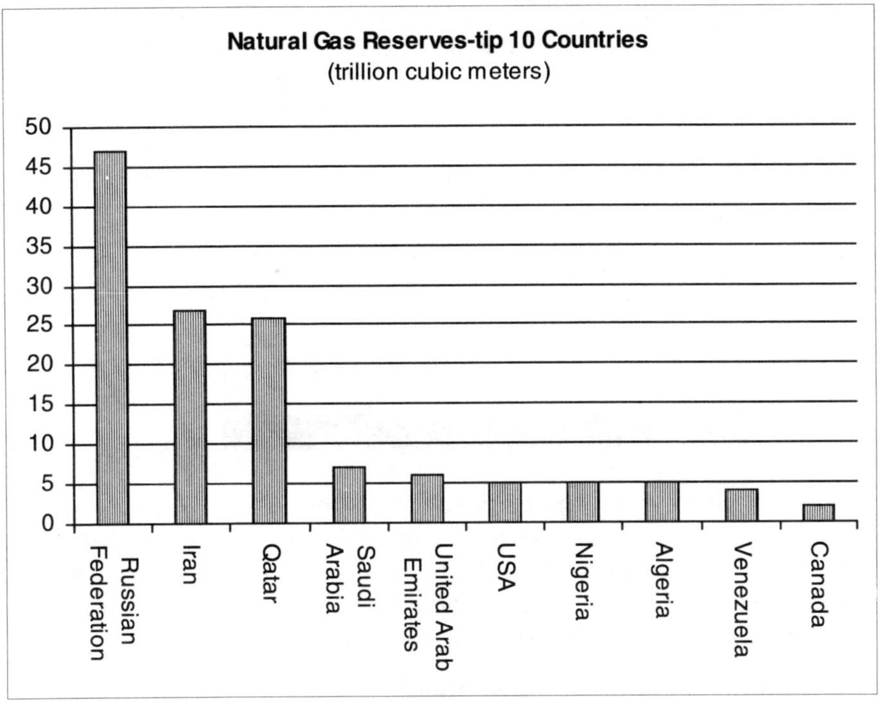

Q.59. How do we get oil and gas out of the ground?

Ans. Oil and gas can get trapped in pockets underground such as where the rocks are folded into an umbrella shape.

Oil and gas can move through the porous rocks (rocks with gaps between the grains). The oil and gas move upwards from the source rock where they were formed. When they met a layer of cap rock (a rock with no spaces between the grains) the oil and gas are trapped.

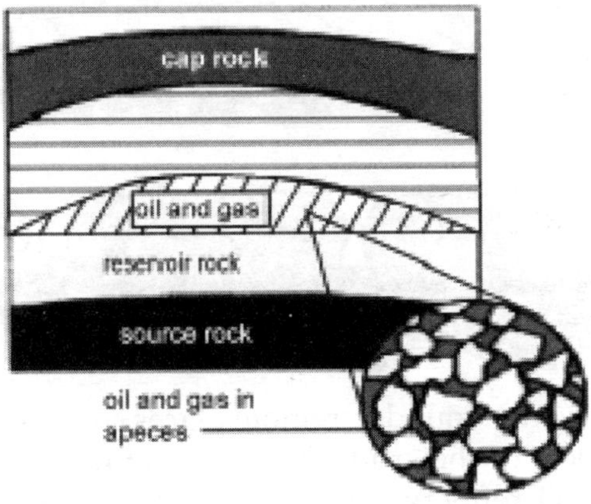

A well is drilled so that the crude oil and other liquids travel up the bore hole. When it comes to the surface the crude oil has to be moved closer to where it is needed.

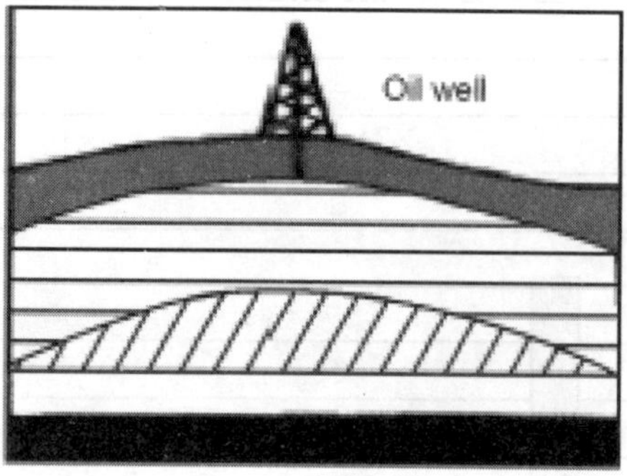

Finding oil and gas trapped deep underground and drilling a well are very complicated and expensive. It costs millions of pound to drill a well and only a few are successful. The liquids found underground can be a complicated mixture of water, crude oil and gas. The crude oil and gas need to be separated before they can be transported safely.

Q.60. How are oil and gas transported?

Ans. Oil and gas are often found far away or under the sea. They have to be transported to an oil refinery. This is often through a pipeline or in a tanker.

The crude oil is often found in remote places such as deserts, jungle or the Arctic. Transport of the crude oil to the refinery is sometimes very complicated. The oil refinery turns crude oil into useful products and materials. These are transported all over Britain or abroad. The products can travel through pipelines, by road, rail or by boats around the coast or along rivers and canals.

Q.61. How is crude oil turned into finished products?

Ans. Crude oil is often a dark, sticky liquid that cannot be used without changing it. The first part of refining crude oil is to heat it until it boils. The boiling liquid is separated into different liquids and gases in a distillation column. These liquids are used to make petrol, paraffin, diesel fuel etc.

Crude oil is a mixture of different chemical called hydrocarbons. The boiling oil turns into a mixture of gases in the column. The gases flow up the column which is hottest at the bottom and cooler at the top. The gases cool down as they go up the column until they condense (turn back into liquid again). The separated liquids and gases, after cleaning and further processing, are used to make many products.

Liquids from refining oil still have to be changed to make them more useful. Sometimes it's to make them clean enough to be used. Sometimes it's to turn some of the unwanted liquids into things people want to buy.

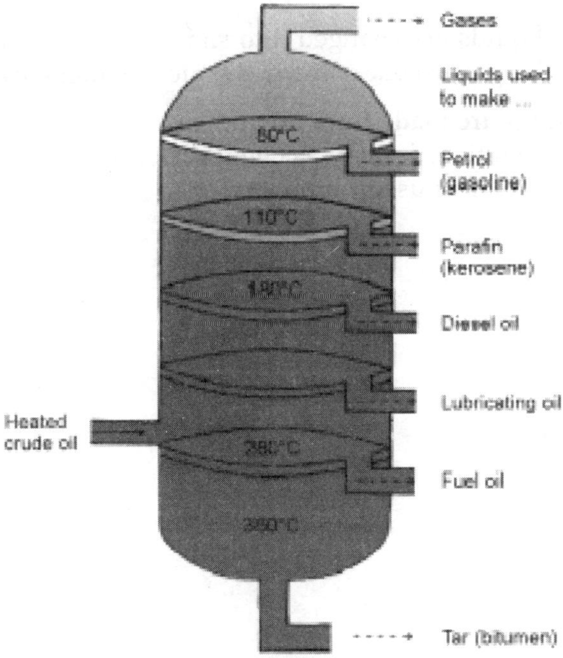

The heavier liquids are in less demand from customers so are turned into lighter products that are in demand. One of the processes is called catalytic cracking. It breaks down some of the heavy liquids from the distillation column.

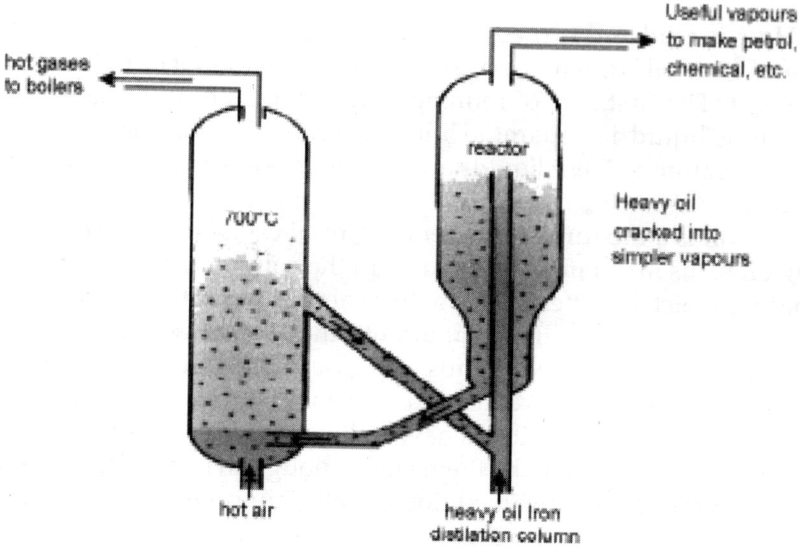

The heavy liquids are changed into simple and more useful liquids and gases. Cracking is just one of many chemical changes in an oil refinery.

Q.62. What things are made from oil and gas?

Ans. In an oil refinery and in a chemical works, crude oil and gas are turned into hundreds of useful products.

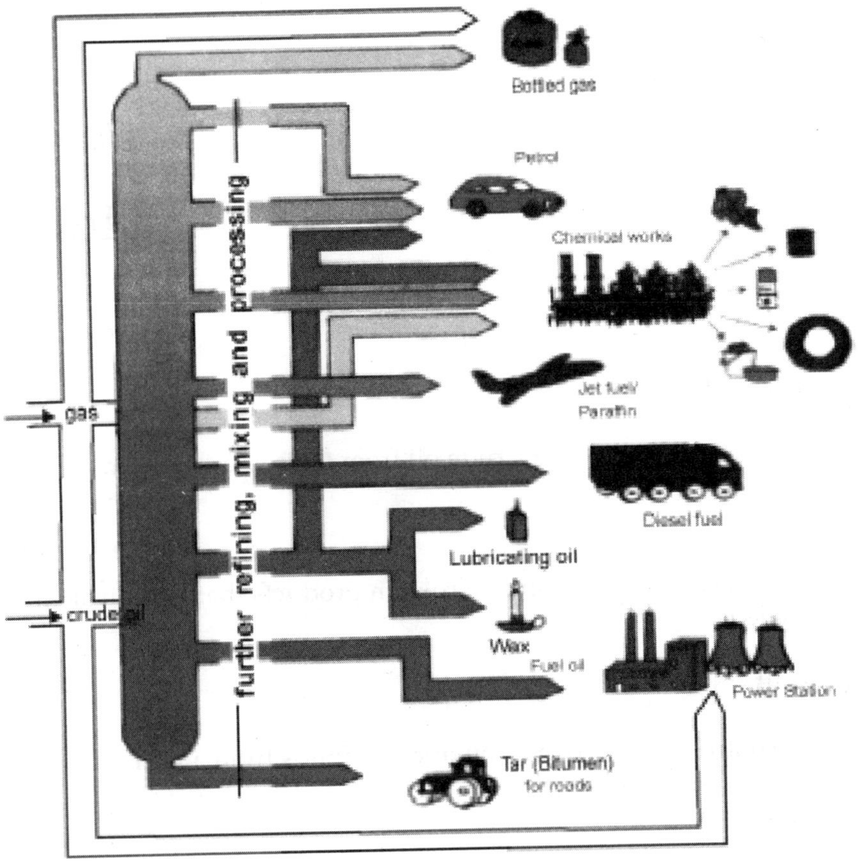
Bottled gas
Petrol
Chemical works
Jet fuel/
Paraffin
Diesel fuel
Lubricating oil
Wax
Fuel oil
Power Station
Tar (Bitumen)
for roads
gas
crude oil
further refining, mixing and processing

Chapter 10

Glossary

Associated Gas

Natural gas produced with crude oil from the same reservoir.

Barrel

A unit of measure for oil and petroleum products that is equivalent to 42 U.S. gallons.

Blowout

The uncontrolled flow of gas, oil or other fluids from a well.

Blowout Preventer (BOP)

The equipment installed at the wellhead to control pressures in the annular space between the casing and drill pipe or tubing during drilling, completion, and workover operations. See also Christmas tree.

British Thermal Unit (BTU)

A measure of the heating value of a fuel

Casing

Metal pipe inserted into a wellbore and cemented in place to protect both subsurface formations (such as groundwater) and the wellbore. A surface casing is set first to protect groundwater. The production casing is the last one set. The production tubing (through which hydrocarbons flow to the surface) will be suspended inside the production casing.

Christmas tree

The assembly of valves, pipes, and fittings used to control the flow of oil and gas from a well.

Coiled Tubing

A long, small diameter pipe flexible enough to be stored on and deployed from a large, truck-mounted roll. Used to replace jointed pipe in certain types of drilling, completion, and workover operations.

Compressor

An engine used to increase the pressure of natural gas so that it will flow more easily through a pipeline

Darcy

A unit of permability. That value which will allow flow of 1 milliliter/sec of 1 centipoise viscosity through 1 cm^2 under pressure of 1 atmosphere per cm.

Deconvolution

The process of undoing the effect of a filter.

Deepwater

Generally defined as operations in water depths of 1000 ft. or greater.

Deepwater Upper Completions

Offshore wells in water depths greater than 1000 feet are considered "deepwater". These wells are characterized by high drilling and intervention costs. Both "dry tree" and "wet tree" completion methods are common.

Depth Migration

Special seismic data processing used to focus subsurface events to their proper location in depth. Prestack depth migration, a computing-intensive process, has been highly successful for complex imaging, including subsalt formations.

Desander

A centrifugal device for removing sand from drilling fluid to prevent pump abrasion.

Development Well

A well drilled within the proved area of an oil or gas reservoir to the depth of a stratigraphic horizon known to be productive; a well drilled in a proven field for the purpose of completing the desired spacing pattern of production.

Deviation

The angle between the wellbore axis (in the direction of the end, or bottom, of the well) and the downward vertical. Deviation values are always positive. Also called Inclination.

Dip

The angle that a refractor or reflector makes with the horizontal. Also, the angle of inclination of a geologic layer or sedimentary bed.

Directional Drilling

The method of guiding a well along a predetermined path to a specific target. A directional drilling company provides technology and rig site supervision to efficiently meet directional drilling objectives.

DMO– Dip Moveout Offset

The difference in arrival times at different seismic detectors due to reflector dip.

Downhole

A term used to describe tools, equipment, and instruments used in the wellbore, or conditions or techniques applying to the wellbore.

Downhole Factory

A Baker Hughes concept for the future in which many tasks associated with hydrocarbon production are performed in the wellbore instead of on the surface. Downhole oil/water separation and automated well maintenance are examples.

Downhole Motor

A tool directly above the drill bit in a drill string that converts the hydraulic energy of the circulating drilling fluid into mechanical energy to turn the bit independently of drill string rotation. May include a bent section to perform directional drilling.

Downstream

When referring to the oil and gas industry, this term indicates the refining and marketing sectors of the industry. More generically, the term can be used to refer to any step further along in the process.

Drill Collar

Heavy-walled sections of pipe included at the bottom of the drill string to apply weight to the drill bit during drilling.

Drill Cuttings

The small pieces of rock created as a drill bit moves through underground formations while drilling.

Drill Ship

A vessel designed for drilling in deep water without legs or anchors holding it to the sea floor and using dynamic positioning to hold it over the subsea wellhead.

Drill Stem

All components in a rotary drilling assembly from the swivel to the bit.

Drill Stem Testing

A method of determining the potential productivity of a subsurface formation in either open or cased hole.

Drill String

The total string of drill pipe with attached tools and bit.

Drill-in Fluid

A specialty drilling fluid designed to minimize formation damage in the reservoir.

Drilling Fluid

Fluid used in the wellbore to lubricate and cool the bit, control bottom-hole pressures, and remove cuttings.

Dry Gas

The volume of gas remaining after all water and natural gas liquids have been removed.

Dry Hole

Any exploratory or development well that does not find commercial quantities of hydrocarbons.

DZO

Demigration to zero offset—in seismic processing an improvement over DMO in situations where velocity varies significantly as a function of depth.

E&P

Exploration and production. The "upstream" sector of the oil and gas industry.

Elastomer

An elastic synthetic rubber or plastic material—often the main component of packing material in downhole packers.

Electrical Submersible Pump (ESP)

A system comprised of a downhole pump, a downhole electric motor, cabling, and surface controller to lift larger quantities of fluids from wells that do not flow under their own pressure.

Emulsifier

A material that causes water and oil to form an emulsion, i.e.; fine oil droplets suspended in the water.

Enhanced Oil Recovery (EOR)

Refers to a variety of processes to increase the amount of oil removed from a reservoir, typically by injecting a liquid (e.g., water, surfactant) or gas (e.g., nitrogen, carbon dioxide).

Expandables

Expandables refers to any of a series of systems and products based on expandable solid tubular, which relies on forcing an expansion device (i.e., mandrel or pig) through standard oilfield tubulars to permanently expand the diameter of the pipe downhole.

Exploration well

A well drilled in search of an undiscovered reservoir or to greatly extend the limits of a known reservoir.

Exploratory well

A hole drilled: a) to find and produce oil or gas in an area previously considered unproductive area; b) to find a new reservoir in a known field, i.e., one previously producing oil and gas from another reservoir, or c) to extend the limit of a known oil or gas reservoir.

Field

An area consisting of a single reservoir or multiple reservoirs all grouped on, or related to, the same individual geological structural feature or stratigraphic condition. The field name refers to the surface area, although it may refer to both the surface and the underground productive formations.

Fishing

The process of recovering equipment lost or stuck in the wellbore. Tools and services that perform specialty and repair work downhole. Fishing activities include retrieving lost tools and repairing wellbore damage.

Float equipment

Check valves that permit fluid to pass downward, but not upward through the casing. A float collar is a special coupling device inserted one or two joints above the bottom of the casing string. A float shoe is positioned at the end of the casing.

Flocculation

The coagulation of solids in a fluid, produced by special additives or contaminants.

Flow Control Equipment

Are mechanical devices that use conventional wireline techniques for the express purpose of directing, managing and controlling the flow of produced or injected fluids in the production tubing.

Fluid

A substance which deforms continuously under the action of a shear force, however small. Wellbore fluids include oil and water (with or without gas in solution) and free gas.

Foaming agent

A chemical used in gas wells to lighten the water column to promote gas production. Also, a chemical used while drilling wells with air or gas as the drilling fluid, to force water with the air and cuttings.

Fold

In seismic processing, the number of traces with different source-to-receiver separations summed into a single trace.

Formation damage

The reduction in permeability in reservoir rock due to the infiltration of drilling or treating fluids into the area adjacent to the wellbore.

FORMation Junction

Aa Baker Oil Tools system using formed metal technology to create a multilateral junction of two 7-in. casing strings, the overall outside diameter (OD) of which is no more than 12.0 in. The technology allows the junction to pass through 13 3/8-in. casing or a 12 1/4-in. open hole

Fracturing (FRAC)

A method of stimulation production by opening new flow channels in the rock surrounding a production well by pumping proppant and fluid into the well at high pressure and volume.

Full Bore Isolation Valve (FBIV)

Is a normally closed disc valve, designed to provide isolation within the tubing string, holding pressure from both above and below.

Gamma-ray log

Well log that records natural radioactivity of formations around the wellbore.

Gas-to-Liquids (GTL)

The conversion of natural gas to a liquid form so that it can be transported easily. Typically, the liquid is converted back to natural gas prior to consumption.

Gather

A display of the input data to a stacking process arranged to show all the seismic traces corresponding to some criterion are displayed side by side.

Gauge

The diameter of a bit or the hole drilled by the bit.

Geophone

A device to transform seismic energy (movement) to an electrical voltage—voltage that is proportional to the velocity of the seismic wave motion.

Geosteering

A subset of horizontal drilling in which measurements of formation properties are used to place the wellbore in specific geologic targets.

GeoThermal

High temperature energy from the Earth located where the earth's major Tectonic Plates are joined.

Gravel Pack

A completion technique used to control production of sand from loosely consolidated formations.

Gravity

A standard adopted by the American Petroleum Institute for measuring the density of a liquid. Gravity is expressed in degrees with lower numbers indicating heavier liquids and higher numbers indicating lighter liquids.

Gyroscopic Survey

Survey conducted inside a well's casing to measure is position and trajectory. The steel's casing prevents use of magnetic readings to compute the directional and coordinate information.

Highly Deviated Wells

A class of nonvertical wells where the deviation angle is approximately 30-80 degrees.

Holdup

The fraction of the total volume of the borehole sampled by the measurement device that contains a particular phase; more specifically, oil holdup, water holdup, and gas holdup

Horizontal Drilling

A subset of directional drilling in which the angle of deviation of the wellbore reaches at least 80 degrees from the vertical, maximizing the length of wellbore exposed to the formation.

Horizontal Well

A class of nonvertical wells where the wellbore axis is near horizontal (within approximately ten degrees of the horizontal), or undulating (fluctuating above and below 90 degrees deviation

Horner Plot

Formation parameters are calculated by analyzing the shut-in build-up using the techniques developed by Horner in his classic pressure build-up paper. The Horner Plot is used to extrapolate a build-up that has not yet stabilized to enable us to predict the virgin reservoir pressure. The slope of the extrapolated line (designated 'm') is also used in the calculations because it reflects the nature of the reservoir in terms of the rate of the steady state increase of pressure. That is to say, the rate of pressure response is due to the nature of the reservoir rock and the fluids flowing in the system.

Hydrocyclone

A cone-shaped device for separating mixed liquids (e.g., oil and water).

Hydrophone

A pressure-sensitive sensor to transform changes in (water) pressure to an electrical voltage—voltage which is proportional to the velocity of the seismic wave

Inclination

The measurement of a well's deviation from vertical, in degrees. Also the angle between the direction of the produced fluid flow and the horizontal. Inclination values are positive for fluid flowing upward and negative for fluid flowing downward.

Infill Drilling / Well

Wells placed between known producing wells to further exploit a reservoir.

Inflatable Systems

Inflatable tools are designed with a nitrile based bladder that is expanded by means of hydraulic pressure much in the same manner as blowing up a 'balloon'.

Integrated

When applied to an oil company, it indicates a firm that operates in both the upstream and downstream sectors (from exploration through refining and marketing)

Intelligent Completions

A completion technology in which formation property measurements are made and the completion dynamically adjusts itself to maximize production from the well.

Intelligent Well Technology

Technology that combines sensors and remote control to detect production problems and solve them without costly well intervention.

Jackup Drill Rig

A mobile, bottom-supported offshore drilling structure. Legs or columns rest on the seafloor and the platform is raised or adjusted by moving up or down on the legs.

Kickoff Point

The depth in a vertical wellbore at which some type of directional drilling operations (e.g., deviated or slant hole) begin.

Kriging

The geostatistical method of applying known values in one sample to produce an unbiased estimate of values in another

Lateral Bore

Normally referred to as the deviated or horizontal extension in the drilling of a horizontal well or multilateral well.

Lease

A legal document conveying the right to drill for oil and gas, or the tract of land on which a lease has been obtained where the producing wells and production equipment are located.

Lifting Costs

The cost of producing oil from a well or lease.

Liner

A string of pipe used to case open hole below existing casing, overlapping inside the upper string and held in place by a liner hanger packer.

Lithology

1. the study of rocks, usually macroscopic. 2. the individual character of a rock in terms of mineral composition, structure, and so forth.

Log

To conduct a survey inside a borehole to gather information about the subsurface formations; the results of such a survey. Logs typically consist of several curves on a long grid that describe properties within the wellbore or surrounding formations that can be interpreted to provide information about the location of oil, gas, and water. Also called well logs, borehole logs, wireline logs.

Logging-While-Drilling (LWD)

A variation of measurement-while-drilling in which the LWD tool gathers information (i.e.; resistivity, density, porosity, gamma ray) about the formation while the well is being drilled.

Main bore

The main casing string from which subsequent directional drilling operations or openhole operations are initiated. Lateral bores extend from the main bore to the desired target depth.

Measurement-While-Drilling (MWD)

Measuring directional information (azimuth, inclination, and tool orientation) downhole to adjust the drilling process and guide the wellbore to a specific target.

Midstream

A term sometimes used to refer to those industry activities that fall between exploration and production (upstream) and refining and marketing (downstream). The term is most often applied to pipeline transportation of crude oil and natural gas.

Migration

A computational process applied to seismic data that returns reflection events to their origin in the subsurface.

Multilateral

The construction of two or more wellbores into one or more reservoirs for the purpose of managing and optimizing fluid movement within the reservoir(s). These lateral wellbores are connected back to a common main bore that extends to surface. Various levels of completion systems can be installed in a multilateral well to enable it to produce from several zones simultaneously.

Naptha

Petroleum distillate with a boiling range of approximately 20 to 200°C; also an aromatic solvent obtained from coal tar.

Natural gas Liquids (NGL)

The portions of gas from a reservoir that are liquified at the surface in separators, field facilities, or gas processing plants. NGL from gas processing plants is also called liquified petroleum gas (LPG).

NMO

Normal moveout offset—the difference in arrival times of reflected seismic data at different detectors due to different source and detector separations.

Non-associated Gas

Natural gas produced from a reservoir that does not contain significant quantities of crude oil.

North Sea Brent

Crude oil produce from the Brent field in the British sector of the North Sea. Often quoted as a benchmark for world oil prices

Naphthenes (alicyclic hydrocarbons or cycloalkanes)

Alkylcyclohexanes and alkylcyclopentanes found in crude petroleum.

OCS

Outer Continental Shelf, a term used primarily in the U.S. for the offshore areas under federal jurisdiction.

Oilfield chemicals

Chemicals used to treat produced fluids and control corrosion and deposition in producing wells.

Olefins (alkenes)

Unsaturated hydrocarbons with one double bond, having the molecular formula CnH2n. They may be thought of as derivatives of ethylene.

Oligomer

Low-molecular-weight polymer made up of two to four monomer units.

Open Hole

Any wellbore in which casing has not been set.

Oxygenates

Substances which, when added to gasoline, increase the amount of oxygen in that gasoline blend. Ethanol, methyl tertiary butyl ether (MTBE), ethyl tertiary butyl ether (ETBE), and methanol are common oxygenates.

Motor Octane Number (MON)

Measure of uniformity of burning (resistance to knocking) of gasoline under laboratory conditions which simulate highway driving conditions.

P&A (plugged and abandoned)

A depleted well or dry hole that has been (typically) filled with cement and marked, with all surface equipment removed.

paraffins (alkanes)

Saturated aliphatic hydrocarbons with the molecular formula C_nH_{2n+2}.

Packer

Open and cased-hole devices used to create seals to control fluid flow.

PDC Drill Bits

Use fixed position polycrystalline diamond compact cutters that shear the formation instead of grinding it. In many applications, PDC bits offer higher penetration rates and longer life than Tricone bits.

Perforate

To open holes through casing walls and cement into a formation so that fluids can flow into the borehole, or vice-versa.

Perforating Gun

A device fitted with shaped charges or bullets that is lowered to a desired depth in a well and fired to create penetrating holes in casing, cement, and formation.

Permeability

A measure of the ease with which a fluid can pass through the pore spaces of a formatio

Permeability

A measure of the ability of a rock to transmit fluid through pore spaces.

Porosity

A ratio between the volume of the pore space in reservoir rock and the total bulk volume of the rock. The pore space determines the amount of space available for storage of fluids.

Produced water

The water extracted from the subsurface with oil and gas. It may include water from the reservoir, water that has been injected into the formation, and any chemicals added during the production/treatment process. Produced water is also called "brine" (and may contain high mineral or salt content) or "formation water." Some produced water is quite fresh and may be used for livestock watering or irrigation (where allowed by law).

Raffinate

Portion of a treated stream that is not removed.

Reforming Conversion

Reforming Conversion of naphtha to more volatile products of higher octane via simultaneous combination of polymerization, cracking, dehydrogenation, and isomerization.

Reformulated Gasoline (RFG)

The Clean Air Act requires gasoline to be "reformulated" in order to burn cleaner [i.e., to reduce emissions of ozone-forming (smog) and toxic air pollutants]. Gasoline is made up of different hydrocarbon compounds, including aromatics, olefins, and benzene, all of which contribute to ozone and toxic air pollution. The levels at which these components are present in RFG will be lower, resulting in fewer harmful emissions. RFG will contain "chemical oxygen" (oxygenates).

Proppant

A granular substance that is carried into the formation by the fracturing fluid and helps keep the cracks open after a fracture treatment.

PS-wave (or converted wave)

Seismic energy which has traveled partly as a P wave and then as an S-wave; having been converted upon reflection at an interface.

P-wave

A wave in which particle motion is in the direction of source propagation. Also called compressional wave, primary wave, pressure wave, and longitudinal wave.

Rate of Penetration (ROP)

The speed (rate) with which the bit drills the formation, measured in feet or meters per hour.

Raypath

A line everywhere perpendicular to wavefronts (in isotropic media).

Resistivity

A measurement of a formation's resistance to electrical current used to determine whether the formation holds hydrocarbons or water.

Reynolds Number

The dimensionless ratio of inertial forces to viscous forces in flowing fluids. It may be viewed as a ratio of the shear stress due to turbulence to the shear stress due to viscosity. Flow with a Reynolds number less than 2000-4000 is laminar flow; that with a Reynolds number greater than 2000-4000 is turbulent flow.

Rotary Drilling

The method of drilling oil and gas wells in which the entire drill string is rotated from the surface to turn the drill bit, and cuttings are removed from the hole by a circulating fluid.

Rotary Steerable System

A drilling system that can perform directional drilling operations without interrupting drill string rotation. Rotary steerable systems include downhole guidance systems that continually adjust the drill bit's course.

Royalty

A per centage interest in the value of production from a lease that is retained and paid to the mineral rights owner.

Run in

To go into the hole with tubing, drill pipe, tools, or other devices.

Semi-submersible rig

A mobile offshore drilling unit that floats on the water's surface above the subsea wellhead and is anchored in place. The semi-submersible rig gets its name from pontoons at its base which are empty while being towed to the drilling location and are partially filled with water to steady the rig over the well.

Separation

The process of separating liquid and gas hydrocarbons and water. This is typically accomplished in a pressure vessel at the surface, but newer technologies allow separation to occur in the wellbore under certain conditions.

Shaped Charge

A small container of high explosive that is loaded into a perforating gun. The charge releases a small, high-velocity stream (jet) of particles that penetrate the casing, cement, and formation.

Shot Hole

A relatively small hole drilled into the earth to hold a high-explosive charge that creates vibrations in the earth's crust for seismic recording.

Shut in

To close valves on a well so that it stops production, or a well on which the valves have been closed.

Sonde

A well logging tool.

Sonic Log

A well log of the travel time for acoustic waves per unit of distance.

Sonobuoy

A free-floating sea buoy containing radio equipment to relay data received by one or more detectors to recording equipment typically aboard the source vessel.

Sour crude oil

Oil containing free sulfur or other sulfur compounds whose total sulfur content is in excess of 1 per cent.

Sour gas

Natural gas containing hydrogen sulfide.

Spacing

The distance between wells producing from the same reservoir. Spacing is often expressed in terms of acres, e.g., 40-acre spacing, and is often established by regulatory agencies.

Stack

A composite record made by combining traces from different seismic records. Commonly, to combine data from several detectors in a limited area.

Statics

Acorrections applied to seismic data to determine the reflection arrival times which would have been observed, were no weathering or low-velocity material present.

Steerable Motor

A downhole motor used for directional drilling which can turn the drill bit independently of drill string rotation. Placed just above the bit, a steerable motor has a bend in its housing that can be oriented to steer the well's course. During "rotary" mode the entire drill string is rotated from the surface, negating the effect of this bend and causing the bit to drill a straight course. During "sliding" mode, drill string rotation is stopped and the bit drills in the direction that it is oriented, gradually turning the well.

Stimulation

The term used for several processes to enlarge old channels, or create new ones, in the producing formation of a well designed to enhance production. Examples include acidizing and fracturing.

Stoneley wave

A type of seismic wave propagated along an interface. A surface wave in a borehole.

Straddle

The placement of completion equipment across a multilateral junction to achieve hydraulic integrity (level 5), or means to isolate the internals of the junction(s) from fluid flow and/or pressure from the reservoir.

Streamer

A marine cable containing regularly spaced hydrophones. During a marine seismic survey, several of these streamers are towed behind the survey vessel.

Surfacant

A soluble compound that concentrates on the surface boundary between two substances, such as oil and water, and reduces the surface tension between the substances. Surfacants permit mixing of substances that ordinarily remain separate.

Swage Tool

A device that passes through a Formation Junction's preformed leg, reforming the leg into a full, round shape.

S-wave

Shear wave or transverse wave. Sometimes called converted waves because the particle motion is perpendicular to the direction of propagation. Shear waves are generated by the incidences of P-waves on surfaces at other than normal incidence.

Tension-leg Platform

An offshore drilling platform attached to the seafloor with tensioned steel tubes. The buoyancy of the platform applies tension to the tubes.

Thru Tubing

Operations performed from inside the production tubing of an existing well.

Trace

A record of one seismic channel.

Transition Zone

The region between land and deep water consisting typically of surf, swamps, and bayous.

True Vertical Depth (TVD)

The vertical distance from a measured depth datum (usually the drilling rig floor) to a point in the well.

Underbalanced Drilling

Drilling under conditions where the pressure being exerted inside the wellbore (from the drilling fluids) is less than the pressure of the oil or gas in the formation.

Underground Injection

The placement of gases or fluids into an underground reservoir through a wellbore. May be used as part of enhanced oil recovery or waterflooding processes or for disposal of produced water.

Upstream

The exploration and production portions of the oil and gas industry.

Waterflooding

The injection of water into an oil reservoir to "push" additional oil out of the reservoir rock and into the wellbores of producing wells.

Well log

A record of one or more subsurface formation measurements as a function of depth in a borehole.

Well servicing

Maintenance work performed on an oil or gas well to improve or maintain the production.

Wellhead

The equipment at the surface of a well used to control the pressure; the point at which the hydrocarbons and water exit the ground

Wet Gas

Natural gas containing significant amounts of liquifiable hydrocarbons.

Whipstock

A long steel tool that uses an inclined plane to cause the bit to deflect from the original borehole at a slight angle.

Wildcat Well

A well drilled in an area where no current oil or gas production exists. Also called a "rank wildcat."

Wireline

A slender rod-like or threadlike small-diameter piece of metal used to lower tools, such as logging tools, perforating guns, valves, and fishing tools into a well. May include electrical conductors to power and control instruments and to convey data to the surface.

Workover

Operations on a producing well to restore or increase production. A workover may be performed to stimulate the well, remove sand or wax from the wellbore, to mechanically repair the well, or for other reasons.

WTI

West Texas Intermediate, a type of crude oil commonly used as a price benchmark.

Index